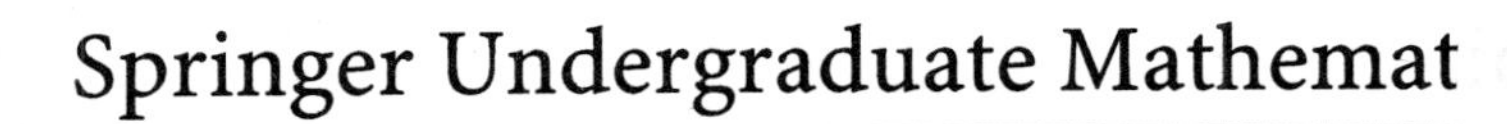

Springer
London
Berlin
Heidelberg
New York
Barcelona
Hong Kong
Milan
Paris
Singapore
Tokyo

Other books in this series

A First Course in Discrete Mathematics *I. Anderson*
Analytic Methods for Partial Differential Equations *G. Evans, J. Blackledge, P. Yardley*
Applied Geometry for Computer Graphics and CAD *D. Marsh*
Basic Linear Algebra *T.S. Blyth and E.F. Robertson*
Basic Stochastic Processes *Z. Brzeźniak and T. Zastawniak*
Elementary Differential Geometry *A. Pressley*
Elementary Number Theory *G.A. Jones and J.M. Jones*
Elements of Logic via Numbers and Sets *D.L. Johnson*
Groups, Rings and Fields *D.A.R. Wallace*
Hyperbolic Geometry *J.W. Anderson*
Information and Coding Theory *G.A. Jones and J.M. Jones*
Introduction to Laplace Transforms and Fourier Series *P.P.G. Dyke*
Introduction to Ring Theory *P.M. Cohn*
Introductory Mathematics: Algebra and Analysis *G. Smith*
Introductory Mathematics: Applications and Methods *G.S. Marshall*
Linear Functional Analysis *B.P. Rynne and M.A. Youngson*
Measure, Integral and Probability *M. Capińksi and E. Kopp*
Multivariate Calculus and Geometry *S. Dineen*
Numerical Methods for Partial Differential Equations *G. Evans, J. Blackledge, P. Yardley*
Sets, Logic and Categories *P. Cameron*
Topics in Group Theory *G. Smith and O. Tabachnikova*
Topologies and Uniformities *I.M. James*
Vector Calculus *P.C. Matthews*

Roger Fenn

Geometry

Roger Fenn, MA, PhD
School of Mathematics, University of Sussex, Falmer, Brighton BN1 9QH, UK

Cover illustration elements reproduced by kind permission of:
Aptech Systems, Inc., Publishers of the GAUSS Mathematical and Statistical System, 23804 S.E. Kent-Kangley Road, Maple Valley, WA 98038, USA. Tel: (206) 432 - 7855 Fax (206) 432 - 7832 email: info@aptech.com URL: www.aptech.com
American Statistical Association: Chance Vol 8 No 1, 1995 article by KS and KW Heiner 'Tree Rings of the Northern Shawangunks' page 32 fig 2
Springer-Verlag: Mathematica in Education and Research Vol 4 Issue 3 1995 article by Roman E Maeder, Beatrice Amrhein and Oliver Gloor 'Illustrated Mathematics: Visualization of Mathematical Objects' page 9 fig 11, originally published as a CD ROM 'Illustrated Mathematics' by TELOS: ISBN 0-387-14222-3, german edition by Birkhauser: ISBN 3-7643-5100-4.
Mathematica in Education and Research Vol 4 Issue 3 1995 article by Richard J Gaylord and Kazume Nishidate 'Traffic Engineering with Cellular Automata' page 35 fig 2. Mathematica in Education and Research Vol 5 Issue 2 1996 article by Michael Trott 'The Implicitization of a Trefoil Knot' page 14.
Mathematica in Education and Research Vol 5 Issue 2 1996 article by Lee de Cola 'Coins, Trees, Bars and Bells: Simulation of the Binomial Process page 19 fig 3. Mathematica in Education and Research Vol 5 Issue 2 1996 article by Richard Gaylord and Kazume Nishidate 'Contagious Spreading' page 33 fig 1. Mathematica in Education and Research Vol 5 Issue 2 1996 article by Joe Buhler and Stan Wagon 'Secrets of the Madelung Constant' page 50 fig 1.

Springer Undergraduate Mathematics Series ISSN 1615-2085

ISBN 1-85233-58-9 Springer-Verlag London Berlin Heidelberg

British Library Cataloguing in Publication Data
Fenn, Roger
Geometry. – (Springer undergraduate mathematics series)
1. Geometry
I. Title
516
ISBN 1852330589

Library of Congress Cataloging-in-Publication Data
Fenn, Roger, 1942-
Geometry / Roger Fenn
p. cm. – (Springer undergraduate mathematics series, ISSN 1615-2085)
Includes bibliographical references and index.
ISBN 1-85233-058-9 (alk. paper)
1. Geometry. I. Title. II. Series.
QA445 .F415 2000
516—dc21
00-044027

Typesetting: Camera ready by the author and Michael Mackey
Printed and bound at the Athenæum Press Ltd., Gateshead, Tyne & Wear
12/3830-543210 Printed on acid-free paper SPIN 10561197

Preface

I hope that this book will inspire its readers to like geometry as much as I do. Geometry is the most accessible branch of mathematics. This means that a mind fresh to mathematics can realise its beauty through geometry more quickly than through any other route. When some mathematics students come to university not knowing what an ellipse is then something is wrong and I hope that this book will go some way towards correcting this. The book is meant to be read by university undergraduates of any year but could be used by a keen sixth former or a research mathematician. The sixth former may not understand everything and the research mathematician may sniff disdainfully at some of the easier passages but I hope that both will find the book useful and perhaps revealing. It is based upon a course of geometry given at Sussex University.

It is pointless and insulting to talk down to anybody so the book contains all levels of mathematics from the easiest, which a lot of readers will have met before, to some quite tricky concepts. I have set the harder parts which can be skipped on a first or second reading in smaller type as follows.

Items in type like this can safely be left to a second or later reading.

In fact I have not followed a slavishly linear route in placing the contents of the book. If the reader wants to read a chapter out of order or feels it worthwhile to dip into another chapter then they can. Some results quoted are too difficult or inappropriate so I have left their proof out. I hope the reader will forgive me.

Some of the displayed mathematics is in bold type. This is to emphasise important results or useful formulæ.

There is no hierarchy or numbering for any results. They are referred to either by name, (e.g. the triangle inequality) or by content, (e.g. the base angles of an equilateral triangle are equal).

■ Indicates the start of a result and its end is indicated by □

In this book I have taken a very practical point of view and considered the original meaning of *geo-metry* as world-measurement. Ancient peoples needed some descriptive language to measure their fields, count the wheat in a granary, predict astronomical events and do a thousand other calculations of everyday life. They also found in their leisure time that the language of geometry was fun to speculate with and set riddles and puzzles even if the results were not always immediately applicable.

In fact speculation and studying things for their own sake quite often pay off in the future. The early Greek geometers would have been amazed that their ellipses and other conics would have governed the motion of solar system bodies. The eminent mathematician, Hardy, wrote that

"No one has yet discovered any warlike purpose to be served by the theory of numbers..."

However even as his words were first being read, in conditions of great secrecy at Bletchley the first computer and the fledgling mathematical theory of cryptography was being used to crack the German High Command enigma code. He was to die before the use of huge primes was applied to the coding of signals for the passage of information down a telephone line.

I have tried to set do-able and interesting questions together with revealing examples. Most answers are given and if anyone finds the inevitable mistake please let me know. You can send me the details via my web site

maths.sussex.ac.uk/Staff/RAF/

where mistakes will be noted.

Acknowledgements: Over the years my colleagues, both at Sussex and elsewhere, have freely shared their insights with me for which I am very grateful. My good friends Dale Rolfsen, Colin Rourke and Brian Sanderson are always ready to stimulate my imagination over a glass of something. Robert Smith has often helped my understanding of astronomy. I am particularly grateful to Peter Croyden for keeping my computer going (and Luca Giuzzi). My Ph.D tutor John Reeve was a huge source of geometric ideas. Both Andy Bartholemew and Richard Noss have pointed out typos in the text. Kate Abell and John Claisse have read through the text and questions but final responsibility lies only with me. I am acutely aware that someone else's name should probably also be mentioned and I apologise to them now.

In terms of typesetting, this book would not have been possible without Donald Knuth and his wonderful TEX program (note NOT latex). Some people can sit down and write out a book from front to back cover in one continuous stream. However I am not one of those and I need constantly to change and edit any work as it progresses. Without some typesetting program I would have found this impossible and plain TEX gives you all the flexibility you need to produce beautiful looking mathematics exactly as you want it. I should also put in a word for the program *metapost*, written by J.D. Hobby which produced the diagrams. It's free like TEX and I urge all mathematicians to try it.

Finally I would like to quote from Sir Michael Atiyah's obituary of J.A. Todd, my undergraduate geometry teacher.

> Geometry has always oscillated between the synthetic and the algebraic approach. Diagrams, pictures and mental visions are the heart of the synthetic method, where intrinsic geometric concepts and constructs are the only tools used. By contrast, ever since the time of Descartes, algebraic formulæ and manipulation have provided a mechanical alternative: technically powerful but lacking in insight. As with all fundamental dichotomies, the reality is much more complicated than antagonists allow. Geometrical argument can become lost in its own intricacies, and elegance in algebra can be cultivated. Moreover most new developments involve an appropriate combination of geometric and algebraic ideas, notation and techniques.

Roger Fenn
Sussex, England 2000

Contents

1. The Geometry of Numbers ... 1
1.1 Natural Numbers ... 2
1.2 Adding Natural Numbers ... 4
1.3 Multiplying Natural Numbers ... 4
1.4 Square and Triangular Numbers ... 5
1.5 Powers ... 9
1.6 Zero and Negative Numbers ... 9
1.7 Rational Numbers or Fractions ... 10
1.8 Powers of Rational Numbers ... 14
1.9 Rational Numbers as a Field ... 14
1.10 Real Numbers ... 15
1.11 Irrational Numbers ... 16
1.12 Four Famous Numbers: $\sqrt{2}$, π, τ, e ... 16

2. Coordinate Geometry ... 29
2.1 Coordinates ... 29
2.2 $\mathbb{R}^n$, the Space of Coordinates ... 30
2.3 The Line through Two Points ... 35
2.4 The Plane Containing Three Points ... 36
2.5 Distance and Angle ... 38
2.6 Polar Coordinates ... 44
2.7 Area ... 48
2.8 Hyperplanes ... 51
2.9 Angles between Hyperplanes and Nearest Points to Hyperplanes 55

3. The Geometry of the Euclidean Plane 63
3.1 The Life of Euclid 63
3.2 The Euclidean Axioms for the Plane 64
3.3 Angles and Lines 65
3.4 Some Basic Facts about Triangles 67
3.5 General Polygons 69
3.6 Congruences and Similarities 73
3.7 Isosceles Triangles 76
3.8 Circles 78
3.9 Triangles and their Centres 83
3.10 Metric Properties of Triangles 87
3.11 Three Surprising (and Beautiful) Theorems 89

4. The Geometry of Complex Numbers 99
4.1 What is $\sqrt{-1}$? 100
4.2 Modulus and Division 103
4.3 Unimodular Complex Numbers and the Unit Circle 104
4.4 Lines and Circles in the Complex Plane 106
4.5 Manipulating Complex Numbers 107
4.6 Infinity and the Riemann Sphere 111
4.7 Division and Inversion 115
4.8 Möbius Transformations 118
4.9 Cross Ratios 120
4.10 A Formula for the Cross Ratio 125
4.11 Roots of Unity 127
4.12 Formulæ for the nth Roots of Unity 127
4.13 Solving Cubic and Biquadratic Polynomials 130

5. Solid Geometry 137
5.1 Points and Coordinates 137
5.2 Scalar Product 141
5.3 Cross Product 142
5.4 The Scalar Triple Product 144
5.5 The Vector Triple Product 146
5.6 Planes 147
5.7 Lines in Space 149
5.8 Isometries of Space 154
5.9 Projections 162
5.10 Polyhedra 166

6. Projective Geometry 183
- 6.1 The Projective Plane 184
- 6.2 Lines in the Projective Plane 185
- 6.3 Incidence and Duality 186
- 6.4 Desargues' Theorem 188
- 6.5 Cross Ratios Again 190
- 6.6 Cross Ratios and Duality 193
- 6.7 Projectivities and Perspectivities 195
- 6.8 Quadrilaterals 199
- 6.9 Projective Transformations 200
- 6.10 Fixed Points and Eigenvectors 202
- 6.11 Pappus' Theorem 202
- 6.12 Perspective Drawing: Tricks of the Trade 204
- 6.13 The Fano Plane 207

7. Conics and Quadric Surfaces 211
- 7.1 Conic Sections 212
- 7.2 The Conic as Quadratic Curve 214
- 7.3 Focal Properties of Conics 225
- 7.4 The Motion of the Planets 231
- 7.5 Quadric Surfaces 238
- 7.6 The General Quadric Surface 244

8. Spherical Geometry 253
- 8.1 Geodesics 255
- 8.2 Geodesic Triangles 256
- 8.3 Latitude and Longitude 260
- 8.4 Compass Bearings 264
- 8.5 The Celestial Sphere 269
- 8.6 Observer's Coordinates 272
- 8.7 Time and Right Ascension 274

9. Quaternions and Octonions 287
- 9.1 Extended Complex Numbers 287
- 9.2 Multiplying Quaternions 288
- 9.3 Inverses of Quaternions 289
- 9.4 Real and Pure Parts of Quaternions 291
- 9.5 Multiplying Quaternions and Linear Transformations of $\mathbb{R}^4$ 292
- 9.6 Octonions 295
- 9.7 Vector Products in $\mathbb{R}^7$ 297
- 9.8 Octonions and Associativity 299
- 9.9 Hexadecanions? 300

Bibliography 305

Index 307

1
The Geometry of Numbers

Counting is what allowed people to take the measure of their world, to understand it better and to put some of its innumerable secrets to good use.
Georges Ifrah

Eeni, meeni, mini, mo
Ancient counting

It may seem odd that the first chapter of a book on geometry should be about numbers. But, as was mentioned in the introduction, this book will take the original meaning of *geo-metry* as world-measurement. So if we consider geometry primarily as a language of measurement then we will need to study the basic tools of measurement, *real numbers*. Everyone who reads this chapter will know or think they know what real numbers are. However this doesn't make the chapter superfluous. We will look at real numbers from a geometric viewpoint. The operations of real numbers such as addition and multiplication are all based on geometric considerations. We have all heard of square numbers but there are triangular numbers and pentagonal numbers too.

We will also look at real numbers with obvious geometric connections such as π. In later chapters we shall consider complex numbers and other generalisations which also have heavy geometric connections.

1.1 Natural Numbers

God made the integers, all the rest is the work of man.
L. Kronecker

Actually, the famous quotation above by Kronecker is false. We say that we have three beans because our minds can conceive of a bean and its exact repetition twice. In real life no two beans are exactly the same. We can look for similarities and at the same time ignore differences.

All societies have their own names for numbers and it is hard to imagine any sort of society where this is not true, as everyone needs to count. In fact languages can be related by their words for numbers since these are very stable over time.

Normally counting systems are based on 5 or 10, because we can use the fingers of one or two hands. Sometimes counting systems are binary, e.g. two pints is one quart, four quarts is one gallon, sixteen ounces is one pound. The English word *score* for twenty and the French word *quatre-vingt* for eighty suggest the remnants of a vigesimal system. The Babylonians had a system based on sixty. The consequences after 3,000 years are still with us: one hour is sixty minutes and one minute is sixty seconds, one complete revolution is $6 \times 60 = 360$ degrees, a clock face is subdivided into twelve and so on. This might seem bizarre at first (why did they not invent the decimal system?) but is based on sound geometric principles as we shall see in the sections on constructibility and elementary astronomical geometry.

The universal Arabic mathematical symbols for the natural numbers $1, 2, 3, \ldots$ were introduced into Europe via Spain about the end of the 10th century. The Arabs brought them from India about 250 years earlier. However they were not universally adopted in Europe until after the invention of printing. The set of natural numbers is indicated by

$$\mathbb{N} = \{\mathbf{1}, \mathbf{2}, \mathbf{3}, \ldots\}$$

The alert reader will have noticed the three dots in $1, 2, 3, \ldots$ and in the equation $\mathbb{N} = \{1, 2, 3, \ldots\}$. They mean *and so on*, the pattern having been established by the first few examples. This is such a useful piece of notation that neophytes should get used to it as soon as possible. The phrase "and so on" has the technical name of *induction*: some fact about the integers is verified for the integer 1 say, a linking implication from any integer n to its successor $n + 1$ is proved and then the fact is true for all integers bigger than 1.

The beauty of the Arabic notation is that not only are written calculations easier (imagine the difficulty of adding and multiplying using Roman numerals) but the notation allows the possibility of infinite repetition. So the sequence

$$1, 10, 100, 1000, \ldots$$

goes on for ever. If there were a largest number we could make a bigger one by adding a 0 on the right-hand side. Roman numerals have a symbol called the *apostrophus*, a sort of backwards C, which when placed after a number multiplies it by ten. On the whole though Roman numerals are a particularly clumsy form of notation and in mathematics, notation is all or at least nine tenths. It is difficulties like these with pre-Arabic notation which could have delayed for so long the development of a limit and the calculus.

Examples of Roman numerals in lower and upper case are given below

1	2	3	4	5	6	7	8	9	10	11	12	20	21	50	100	500	1000
i	ii	iii	iv	v	vi	vii	viii	ix	x	xi	xii	xx	xxi	l	c	d	m
I	II	III	IV	V	VI	VII	VIII	IX	X	XI	XII	XX	XXI	L	C	D	M

The signs used in Roman numerals are of great antiquity and predate any form of writing. They derive from the tally marks made on sticks by herdsmen in prehistoric times. The explanation for these symbols are as follows: I is obvious, five is the outline of a hand simplified to V, ten is twice five giving X. The Roman numerals C and M are the first letters of *Centum* and *Mille* although this just happens to be serendipitous. The origins of L and D are more obscure. An old method of representing 1000 was a bit like $\varPhi$ and the right half of that is like D. Similarly 50 was represented by $\downarrow$ which could have changed into L.

The rule for evaluating Roman numerals is roughly that a smaller or equal number placed after another constitutes addition and on a smaller number being placed before, constitutes subtraction. On inscriptions you may find the forms IIII for IV and VIIII for IX.

Exercise 1.1

What numbers are represented by DLXIII, MCCLXXXVII? Find the Roman equivalents of 796 and 1999.

In this book we retain the lower case symbols n, m, p, q, r etc. for general natural numbers in formulæ.

1.2 Adding Natural Numbers

Now of my threescore years and ten
Twenty will not come again,
And take from seventy springs a score,
It only leaves me fifty more.
A.E. Housman

If we take 5 beans and place them next to 3 beans we get 8 beans altogether: symbolically

$$5 + 3 = 8,$$

which is just mathematical notation for accumulation of objects. The geometric picture of the sum $n + m$ is shown in Fig. 1.1.

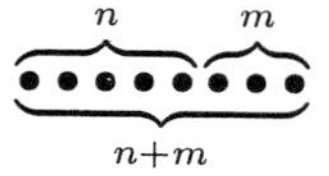

Fig. 1.1. Pictorial recipe for adding two numbers

The advantage of seeing addition geometrically is that a universal identity such as

$$n + m = m + n$$

can immediately be seen to be true by flipping over Fig. 1.1, see Fig. 1.2.

Fig. 1.2. Proof that $n + m = m + n$

The opposite of addition is subtraction, so $x = m - n$ is the solution to the equation $n + x = m$. There is no difficulty if $m > n$ but if $m \leq n$ it is necessary to introduce zero and negative integers which we will consider later.

1.3 Multiplying Natural Numbers

Multiplying natural numbers is just a numbered repetition of addition. So if 3 kings each bring you 5 beans you will have 15 beans: symbolically

$$3 \times 5 = 5 + 5 + 5 = 15.$$

The geometric picture of the mathematical notation $n \times m$, which we shall usually write in the shortened form $n \times m = nm$, is obtained by stacking the objects in m rows of n, as shown in Fig. 1.3.

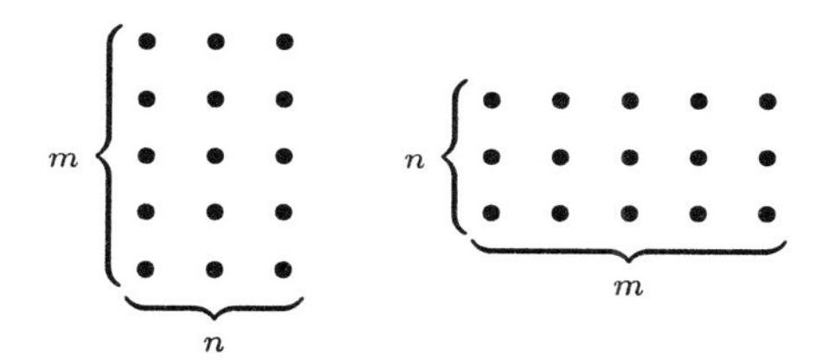

Fig. 1.3. Multiplying: $nm = mn$

Rotate the stack through 90 degrees so that the rows become the columns and vice versa. Since this is achieved without changing the total number we see immediately that multiplication is commutative: that is the order of multiplication is immaterial. So $nm = mn$.

Although the multiplication of real numbers is commutative there are many other operations whose ordering is vital. We shall meet many during the course of this book. An example from real life is the putting on of shoes and socks.

Exercise 1.2

Make pictures which convince you of the truth of

(a) $(m+n)+p = m+(n+p)$ (associativity for addition)
(b) $(mn)p = m(np)$ (associativity for multiplication)
(c) $m(n+p) = mn + mp$ (distributivity)

1.4 Square and Triangular Numbers

The opposite of multiplication is division. Any natural number which can be grouped into a solid rectangle as in Fig. 1.3 is called *compound.* Compound numbers can be divided exactly by a smaller natural number which is not 1. Otherwise a natural number is a *prime*. The primes start $2, 3, 5, 7, 11, \dots$ and go on for ever. Compound numbers of the form n^2 are called *squares* for obvious reasons.

Imagine the prime numbers ordered by size and let P_n denote the nth prime. Let

$$N = 2 \times 3 \times 5 \cdots \times P_n + 1.$$

Then N is not divisible by any prime up to the nth. But it must be divisible by some prime, maybe itself, so there is a prime bigger than P_n. In other words the set of primes is infinite.

The ancient Chinese considered the even numbers *female* and the odd numbers *male*. The number 3 was considered particularly potent because its shape echoed masculine genitalia. Three is just one of an infinite sequence of 3-sided numbers called *triangular numbers*, see Fig. 1.4.

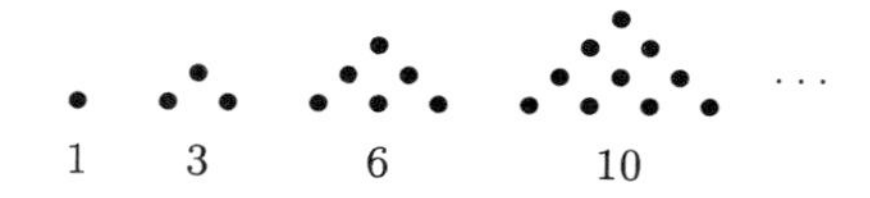

Fig. 1.4. The triangular numbers t_n

A rectangle of side n and $n+1$ is divided into two triangular numbers by a diagonal line. In symbols:

$$n(n+1) = 2t_n.$$

So solving for t_n we get

$$t_n = \frac{n(n+1)}{2}.$$

This gives the formula for the sum of the first n natural numbers,

$$\mathbf{t_n = 1 + 2 + 3 + \cdots + n = \frac{n(n+1)}{2}}.$$

This is a good time to introduce the notation $\sum_{i=1}^{n} i = 1 + 2 + 3 + \cdots + n$. So the above formula can be written

$$\sum_{i=1}^{n} i = \frac{n(n+1)}{2}.$$

Exercise 1.3

By partitioning a square, show that the first n odd numbers sum to n^2. That is

$$1 + 3 + 5 + \cdots + 2n - 1 = n^2.$$

Exercise 1.4

Show that $111111111^2 = 12345678987654321$. A calculator will not help you here. A computer might but a few moments thought will convince you that the equation is true.

If we take a cube, n^3, and remove from it the subcube, $(n-1)^3$, we get the hexagon shape shown in Fig. 1.5.

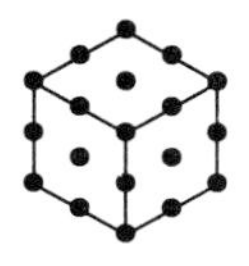

Fig 1.5.

This is made up of 3 squares of size n^2 which meet in pairs in 3 edges of length n and which have total intersection 1. Counting gives

$$n^3 - (n-1)^3 = 3n^2 - 3n + 1$$

which can be verified algebraically. Adding up this equation from $i = 1$ to $i = n$ gives

$$n^3 = 3\sum_{i=1}^{n} i^2 - 3\sum_{i=1}^{n} i + n.$$

Substituting the value of $\sum i$ obtained above we get after some manipulation,

$$\mathbf{1^2 + 2^2 + 3^2 + \cdots + n^2 = \frac{n(2n+1)(n+1)}{6}}$$

as a formula for the sum of the first n squares. Numbers of the form $n(2n+1)(n+1)/6$ are called *pyramidal numbers.*

Exercise 1.5

Imagine that you are an Egyptian pharoah and want to build a pyramid with a square base 755 feet long. (The Great Pyramid at Giza is 755 feet long.) How many stone blocks would you need if each block is a cube of side 5 feet?

Exercise 1.6

Use a similar method to the above method for squares or just plain induction to verify the following formula for the sum of the first n cubes,

$$1^3 + 2^3 + 3^3 + \cdots + n^3 = \frac{n^2(n+1)^2}{4}.$$

Exercise 1.7

It has been known for centuries that a cube cannot be the sum of *two cubes*. However verify that $6^3 = 3^3 + 4^3 + 5^3$.

The (positive) *pentagonal numbers* 1, 5, 12, ... have five sides as we see in Fig. 1.6.

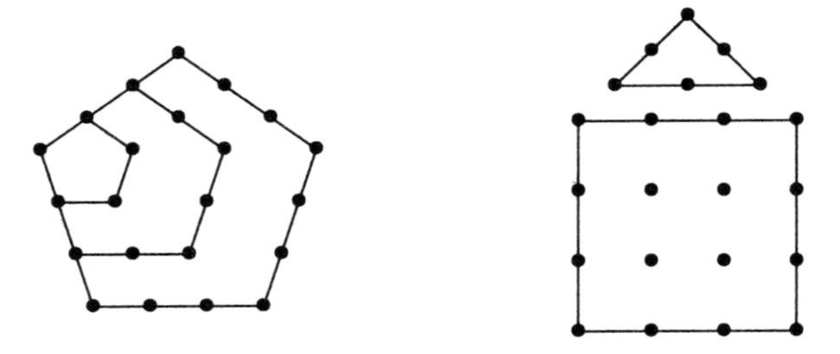

Fig. 1.6. Pentagonal numbers

The nth pentagonal number, p_n, is made up of a square and a triangular number. In symbols we have

$$p_n = n^2 + \frac{n}{2}(n-1) = \frac{n}{2}(3n-1), \qquad n = 1, 2, \dots .$$

This formula also makes sense if n is negative, giving the so-called negative pentagonal numbers

$$\frac{n}{2}(3n-1), \qquad n = 0, -1, -2, \dots$$

or

$$\frac{n}{2}(3n+1), \qquad n = 0, 1, 2, \dots .$$

Euler's pentagonal number theorem is

$$\prod_{n=1}^{\infty}(1-x^n) = \sum_{n=-\infty}^{\infty}(-1)^n x^{\frac{n}{2}(3n-1)}.$$

For example if we expand the first two factors in the infinite product on the left-hand side we get

$$(1-x)(1-x^2) = 1 - x - x^2 + x^3.$$

The first three terms of this are the same as the first three terms of the infinite sum of the right-hand side of Euler's formula. For more details of this formula see Rademacher (1973).

Exercise 1.8

* Verify that 3 is not a pentagonal number by expanding the first three factors in the infinite product on the left-hand side of Euler's formula.

This is the first example of a harder or *starred* question, so do not be put off if you cannot answer it immediately. Later on you will meet double-starred questions which are even harder.

1.5 Powers

Squares and cubes (n^2, n^3) have obvious geometric interpretations. We can inductively define higher powers by

$$n^m = n \times n^{m-1}.$$

So it is easy to believe the laws of indices,

$$n^p \times n^q = n^{(p+q)}$$

and

$$(n^p)^q = n^{pq}.$$

You can invent inductive pictures of n^m as n copies of n^{m-1} stacked on top of one another to form a higher-dimensional cube.

Many calculating programs like *Maple* or *Mathematica* use $n * m$ for $n \times m$ and $n * *m$ or $n\hat{\ }m$ for n^m.

Exercise 1.9

The expression 4^{3^2} is ambiguous. What values can be reasonably assigned to it?

1.6 Zero and Negative Numbers

> ... *the achievement of the unknown Hindu, who some time in the first centuries of our era discovered the principal of position, assumes the proportion of a world event.*
>
> Dantzig

The Sanskrit name for zero is *sunya*; the Arabic is *sifr*. Mediaeval Latin transcribed *sifr* into *cifra* and *zefirum* from which we get *zero* and *cipher*. If we call a man a mere cipher then we mean he counts for nothing, although the word cipher has other meanings.

Initially zero or nought was represented by a dot and this got expanded into 0. The importance of this innovation as an aid to calculation and as a stimulus for algebra (from the Arabic *al-jabr*) cannot be over-emphasised. With 0 as a place holder it means that the simple methods for adding and multiplying numbers such as 1207 and 340 which we learnt at school are far easier than would be the case if Roman numerals were used.

It is important in algebra that expressions are universally true if possible and are not subject to confusing special conditions or constraints. Algebra is a mindless machine which grinds out an answer to which we then have to put a meaning. For example the meaning of $n - m$ is quite clear if $n > m$. But we need to solve equations such as

$$x + m = n$$

where the possibility exists that $n \leq m$. If $n = m$ the solution is zero. The case $n < m$ leads to the idea of a *negative* number $x = -(m - n)$ as the solution of the above. For consistency we always have

$$-(-m) = m.$$

The natural numbers, zero and the negative natural numbers are called the *integers* and we denote the set of integers by

$$\mathbb{Z} = \{\ldots - 2, -1, 0, 1, 2, 3, \ldots\}$$

(from the first letter of the German word *Zahlen*). We now have the idea of *negation* which assigns to an integer n its negative version $-n$.

With the introduction of zero and negation, integers satisfy the usual operations of addition and multiplication as above together with a new operation *subtraction* or *minus*

$$n - m = n + (-m)$$

defined on *all* pairs of integers.

Exercise 1.10

Find three integers p, q, r to show that subtraction is not commutative or associative in general, i.e.

$$p - q \neq q - p \quad \text{and} \quad p - (q - r) \neq (p - q) - r.$$

What conditions must be placed on p, q, r for these laws to hold?

1.7 Rational Numbers or Fractions

I wish I had a pound for every occasion on which I have met someone new at a party, introduced myself as a mathematician and have them step back a pace or two and boldly proclaim "I never was very good at maths at school". (Usually their second remark is that they never realised there was anything new to be learnt about maths.) I am willing to bet that the defining moment which separates those who can cope from those who boast they are innumerate is the introduction of fractions and how to deal with them. Probably they are already punch drunk from dealing with -1. The notation for fractions and how to multiply, or worse, add fractions, finishes them off.

Fractions are required in order to have a notation for cutting things up into pieces, like the cake in Fig. 1.7.

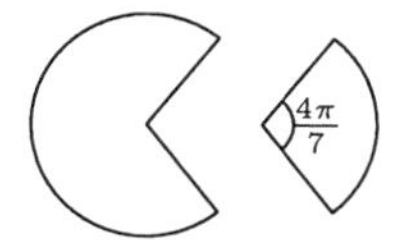

Fig. 1.7

In mathematical terms fractions are required in order to solve equations such as

$$7x = 2$$

where x is our piece of cake.

One of the difficulties of fractions or rational numbers as they are often called is that they introduce two notions of fundamental importance to mathematics which are equally important in real life and that is the notion of an *equivalence relation* and an *equivalence class.*

The notation $\frac{n}{m}$ or n/m is a representation of an equivalence class because $\frac{kn}{km}$ also represents the same number.

Equivalence relations and classes are central to our whole existence. If Peter and Jane are redheads then they are equivalent under hair colour. If I call an object in my hand a bean it is because it belongs to the equivalence class of all beans. Mathematically an equivalence relation is a generalisation of equality. If a, b are members of a set A then instead of $a = b$ we write $a \sim b$ and say that this is an equivalence relation if it satisfies the following three properties.

Reflexivity: Always $a \sim a$.

Symmetry: If $a \sim b$ then $b \sim a$.

Transitivity: If $a \sim b$ and $b \sim c$ then $a \sim c$.

The equivalence class containing a is then the set of all b equivalent to a.

So what has this got to do with fractions or rational numbers? Let A be the set of all pairs of integers (n, m) where $m \neq 0$. Write $(n, m) \sim (n', m')$ if $nm' = n'm$. You can quickly check that this is an equivalence relation. The equivalence class containing (n, m) is the rational number $\frac{n}{m}$ or n/m. The equivalence relation here is obtained by putting the two rational numbers equal,

$$\frac{n}{m} = \frac{n'}{m'}$$

and multiplying both sides by mm'. (The number $n/1$ is always equated with n.)

Exercise 1.11

State whether the following are equivalence relations on the set of "towns".
(a) Two towns both lie in the same country.
(b) Two towns are no more than 10 miles apart.
(c) Two towns are connected by a railway.
If yes, what are the equivalence classes?

The four operations on rational numbers which cause such grief at school are given by the well-known formulæ

$$\begin{aligned}\frac{n}{m}+\frac{n'}{m'}&=\frac{nm'+mn'}{mm'}\\ \frac{n}{m}\times\frac{n'}{m'}&=\frac{nn'}{mm'}\\ \frac{n}{m}-\frac{n'}{m'}&=\frac{nm'-mn'}{mm'}\\ \frac{n}{m}\div\frac{n'}{m'}&=\frac{n}{m}\times\frac{m'}{n'}=\frac{nm'}{mn'}\ .\end{aligned}$$

Note that division is possible only if $n' \neq 0$.

The important thing to note and which is easily checked is that the operations respect the equivalence relation. So that what we are operating on is the equivalence class and not just the pair of integers.

Exercise 1.12

Find all solutions in positive integers p, q of the equation

$$\frac{1}{p}+\frac{1}{q}=\frac{1}{2}$$

and the inequality

$$\frac{1}{p}+\frac{1}{q}>\frac{1}{2}.$$

Repeat for the two cases given by

$$\frac{1}{p}+\frac{1}{q}+\frac{1}{r}\geq 1.$$

There are other ways of writing rational numbers; for example

$$\frac{19}{51}=\cfrac{1}{2+\cfrac{1}{1+\cfrac{1}{2+\cfrac{1}{6}}}}.$$

In general an expression of the form

$$a_0 + \cfrac{1}{a_1 + \cfrac{1}{a_2 + \cfrac{1}{a_3 \ddots \cfrac{1}{a_n}}}}$$

is called a *continued fraction.* To evaluate a continued fraction as an ordinary fraction we note that a continued fraction c is equal to $a + \frac{1}{c'}$ where c' is a less complicated continued fraction. If we have worked out c' then it is easy to calculate c.

To write an ordinary fraction as a continued fraction we need to use the *euclidean algorithm.* This is best illustrated by an example.

$$\frac{43}{5} = 8 + \frac{3}{5} = 8 + \frac{1}{\frac{5}{3}} = 8 + \cfrac{1}{1 + \frac{2}{3}}$$
$$= 8 + \cfrac{1}{1 + \cfrac{1}{1 + \frac{1}{2}}}.$$

Exercise 1.13

Write the following continued fractions as ordinary fractions.

$$\text{(a)}\ -2 + \cfrac{1}{2 + \cfrac{1}{4 + \cfrac{1}{6 + \frac{1}{8}}}} \qquad \text{(b)}\ 1 + \cfrac{1}{2 + \cfrac{1}{3 + \cfrac{1}{4 + \frac{1}{5}}}}.$$

Exercise 1.14

Write $187/57$ and $-19/70$ as continued fractions.

Let us write $\mathbb{Q}$ for the set of all rational numbers. There are the same number of rationals as integers in the sense that there is a 1–1 correspondence between the two sets. However unlike the integers $\mathbb{Z}$, the rational numbers $\mathbb{Q}$ cannot be written down successively in their natural order because between any two, a and b, is a third, say $(a+b)/2$ and consequently an infinity of others.

1.8 Powers of Rational Numbers

Beware when taking powers involving rational numbers because there are traps for the unwary. For example 2^{-1} is defined (it is $1/2$) but $1^{\frac{1}{2}}$ has two values (± 1) and $(-1)^{\frac{1}{2}}$ is not even defined, at least as a real number. So let us treat algebra as a mindless machine and see where the laws of indices lead us for powers of rational numbers.

First of all we must have $a^0 = 1$. Secondly it then follows that $a^{-1} = \frac{1}{a}$ if $a \neq 0$. So $a^{-n} = \frac{1}{a^n}$ if n is a natural number. Generally speaking providing the powers are defined then the laws of indices hold.

If you want to know about fractional powers in general then you will have to look at an elementary book on analysis. However the symbol $\sqrt{a}$ will always mean the positive root of $a^{\frac{1}{2}}$ if $a > 0$.

1.9 Rational Numbers as a Field

The standard operations on the rational numbers have the following properties

$$
\begin{aligned}
a + (b + c) &= (a + b) + c & \qquad a(bc) &= (ab)c \\
a + b &= b + a & ab &= ba \\
a + 0 &= a & a1 &= a \\
a + (-a) &= 0 & a(b + c) &= ab + ac.
\end{aligned}
$$

In addition there is the important fact of the existence of a multiplicative inverse. So if a is a non-zero rational number then there is a rational number written a^{-1} or $1/a$ or $\frac{1}{a}$ such that $a^{-1}a = 1$.

Any set satisfying all the above is called a *field*. So the rationals $\mathbb{Q}$ are a field as are the numbers $s + t\sqrt{2}$ where s, t are rationals. On the other hand, the integers $\mathbb{Z}$ do not form a field because there is no integer solution to $2x = 1$.

The main difficulty with proving that a set of numbers is a field is usually verifying the inverse rule. For example consider the number $3 + \sqrt{2}/5$. Its inverse is

$$
\begin{aligned}
\frac{1}{3 + \frac{\sqrt{2}}{5}} &= \frac{5}{15 + \sqrt{2}} = \frac{5}{15 + \sqrt{2}} \cdot \frac{15 - \sqrt{2}}{15 - \sqrt{2}} \\
&= \frac{5(15 - \sqrt{2})}{225 - 2} = \frac{75}{223} - \frac{5}{223}\sqrt{2}
\end{aligned}
$$

which is now in the same form.

Exercise 1.15

If s, t are rational, write $1/(s + t\sqrt{2})$ in the form $s' + t'\sqrt{2}$ where s', t' are rational.

Exercise 1.16

Write $1/(2 - 3\sqrt{5})$ in the form $s + t\sqrt{5}$ where s, t are rational. Can you do something similar with the complex number $1/(2 - 3\sqrt{-5})$?

1.10 Real Numbers

The polynomial equation $x^2 = 2$ has a solution $\sqrt{2}$ which is not a rational number. Likewise the non-polynomial or transcendental equation $\cos(x) = -1$ also has a solution π which is not a rational number. Finding solutions to these and other equations leads to the *real* numbers, $\mathbb{R}$. All real numbers are a limit of rational numbers. The rational numbers can be represented by finite decimals or recurring decimals. On the other hand, the number π is the limit of a sequence which starts

$$3,\ 3.1,\ 3.14,\ 3.141,\ 3.1415.$$

Alternatively we can think of a real number as represented by a continued fraction which may go on for ever.

The number $\sqrt{2}$ can be written as the infinite continued fraction

$$\sqrt{2} = 1 + \cfrac{1}{2 + \cfrac{1}{2 + \cfrac{1}{2 + \cfrac{1}{\ddots}}}}.$$

To see this, let the right-hand side be x. Then

$$1 + x = 2 + \frac{1}{1 + x}$$

from which we get $x = \sqrt{2}$.

Exercise 1.17

Show that
$$\sqrt{3}+1 = 2 + \cfrac{1}{1+\cfrac{1}{2+\cfrac{1}{1+\cfrac{1}{\ddots}}}}.$$

Write $\sqrt{5}+2$ as an infinite continued fraction.

We depict the reals as a line, typically a horizontal x-axis, where the numbers increase from left to right (Fig. 1.8).

$-\infty$ — -2, -1, 0, 1, $\sqrt{2}$, τ, 2, e, 3, π — $+\infty$

Fig. 1.8. The real line $\mathbb{R}$ and some famous members

It is convenient to introduce the symbols $-\infty$ and $+\infty$ as the left and right limits of $\mathbb{R}$.

Real numbers are manipulated by manipulating the rational numbers which approximate to them. Of course it must be checked that this process works and that is the subject of a course on analysis.

1.11 Irrational Numbers

The reals $\mathbb{R}$ include not only the rational numbers but also the irrational numbers, say $\sqrt{2}$, which cannot be expressed as a ratio of integers. The irrational numbers include the *transcendental* numbers such as π which by definition are not roots of any polynomial equations with integer coefficients.

There are many more transcendental numbers than not. However actually verifying that any particular number is transcendental seems very hard. After a great deal of work it has been shown that π and e are transcendental. "Clearly" $e+\pi$ is transcendental but as far as I know this has never been proved.

1.12 Four Famous Numbers: $\sqrt{2}$, π, τ, e

The number $\sqrt{2}$ turns up as soon as you start to do any geometry in the plane. For example the diagonal of a square of unit side has length $\sqrt{2}$ by Pythagoras' theorem.

Alternatively if you believe the formula for the area of a square then you can see that the square on the diagonal d is made of eight quarters of the unit square into which it is divided by the diagonals (Fig. 1.9). So $d^2 = 8/4 = 2$.

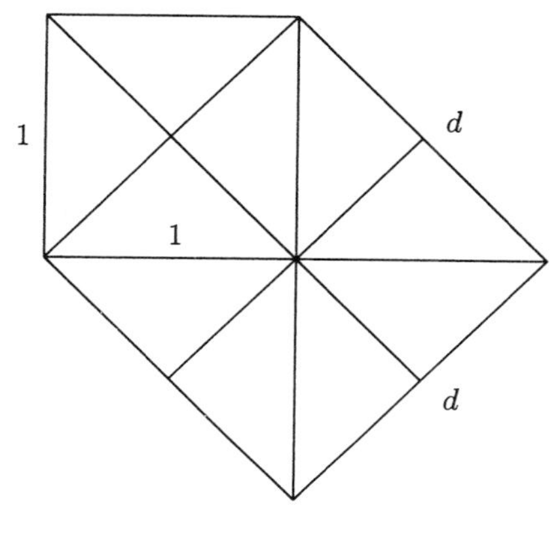

Fig. 1.9

According to legend, the fact that this number, so easily constructed, was not the ratio of two integers devastated Pythagoras and his followers. However recent scholarship suggests that this is not the case.

■ Here is Euclid's (and probably Pythagoras') ancient proof that $\sqrt{2}$ is irrational.

Proof To say that $\sqrt{2}$ is irrational is merely another way of saying that 2 cannot be expressed in the form $(\frac{n}{m})^2$. Which is the same as saying that the equation

$$n^2 = 2m^2$$

has no integer solutions.

We will assume that there is a solution and then obtain a contradiction. If a solution does exist then by cancelling factors we can assume that m, n have no common factors. In particular m, n cannot both be even.

However n^2 is even because $2m^2$ is divisible by 2. So n must be even since the square of an odd number is odd.

Let us suppose that $n = 2k$ for some integer k. On substituting this value for n in our equation we get $4k^2 = 2m^2$ or $m^2 = 2k^2$. So by a repeat of the previous argument m is also even. This is the required contradiction and so $\sqrt{2}$ is irrational. □

We can construct the diagonal of a square geometrically but cannot measure it in any finite sequence of steps. Putting this in another way, the square root of 2 can be written down as a decimal to any finite number of places but the numbers never repeat or terminate. This result, over two thousand years old, is incredibly modern in spirit.

Exercise 1.18

A rectangle has the property that if you place two copies of it side by side the resulting rectangle of twice the area of the original has sides in the same ratio. Show that this ratio is $\sqrt{2}$.

The number π is of course the ratio of the length of the circumference of a circle to its diameter. This simple statement needs two qualifications. Firstly, what do we mean by the length of the circumference of a circle and how do we know it is true for all circles?

Consider a circle of unit diameter. We inscribe a polygon $ABC\ldots$ whose vertices lie on the circle and circumscribe a polygon $PQR\ldots$ whose edges touch the circle (Fig. 1.10).

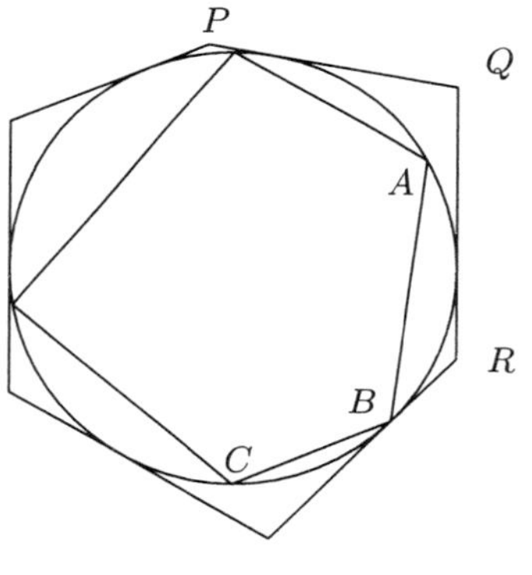

Fig. 1.10

Then we would expect the length of the circumference of the circle or perimeter to lie between those of the two polygons. In symbols we have

$$AB + BC + \cdots < \pi < PQ + QR + \cdots$$

However this is not at all easy to prove rigorously from scratch and I don't propose to do so here.

Now if we let the number of vertices of the polygons increase without bound and the length of the edges decrease to zero we would expect the inside polygonal perimeter to increase to π and the outside polygonal perimeter to decrease to π. In the notation of analysis

$$\overrightarrow{\lim}(AB + BC + \cdots) = \lim_{\rightarrow}(PQ + QR + \cdots) = \pi.$$

If we are happy to accept this then it is easy to see that the value of π as a ratio is independent of the circle chosen. We can choose the circles whose ratio we want to compare so that they are concentric. Now choose the two internal polygons so that one is the radial magnification of the other. By similar triangles (see Chapter 3) the edges and hence the perimeters of the two polygons are in the same ratio as the two diameters (see Fig. 1.11). We would expect that also to be true in the limit when we get π.

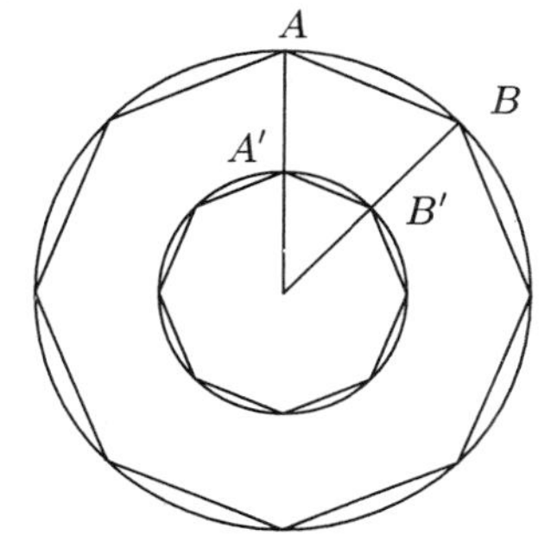

Fig. 1.11

We can use this and some geometry (which we will explain in later chapters) to get an interesting formula for π. Take a circle of unit diameter and centre O. Let $ADBE$ be points on the circle so that A is opposite B and DE are equidistant from B (Fig. 1.12). Then the line DE is at right angles to AB and we let them meet at C.

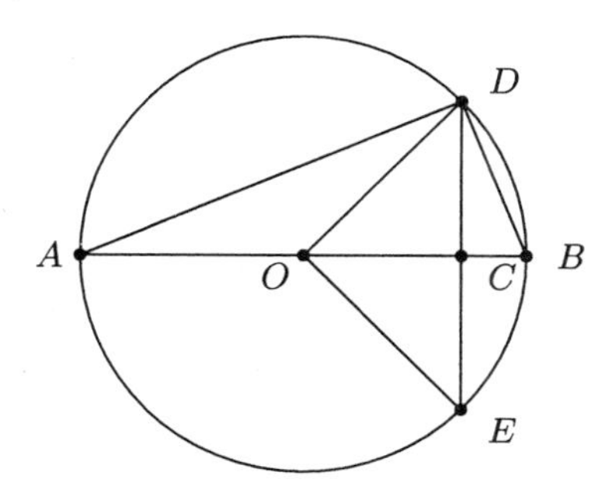

Fig. 1.12

Now consider the area of the right-angled triangle ADB

$$\triangle ADB = \frac{1}{2}AD \cdot BD = \frac{1}{2}AB \cdot CD = \frac{1}{2}CD.$$

By Pythagoras' theorem

$$AD^2 = AB^2 - BD^2 = 1 - BD^2.$$

Writing $CD = DE/2$ and eliminating the length AD we obtain the following quadratic equation for BD^2

$$4BD^4 - 4BD^2 + DE^2 = 0$$

with solution

$$BD^2 = \frac{4 \pm \sqrt{16 - 16DE^2}}{8} = \frac{1}{2} - \frac{1}{2}\sqrt{1 - DE^2}.$$

Note that we take the negative root because $BD < \frac{1}{2}$. Now suppose that DE is the side of an inscribed square. Then $DE = \frac{1}{2}\sqrt{2}$ and the perimeter is $2\sqrt{2}$. From our formula, BD which is the length of an edge of a regular inscribed octagon, is $\frac{1}{2}\sqrt{2-\sqrt{2}}$ whose perimeter is $4\sqrt{2-\sqrt{2}}$. Now repeat this to find the perimeter of a regular inscribed 16-gon, 32-gon etc. given by

$$8\sqrt{2-\sqrt{2+\sqrt{2}}}, \quad 16\sqrt{2-\sqrt{2+\sqrt{2+\sqrt{2}}}}\ldots$$

Repeating *ad infinitum* we get the following limit for π due to Viète

$$2^n \underbrace{\sqrt{2-\sqrt{2+\sqrt{2+\cdots+\sqrt{2}}}}}_{n \text{ roots}} \to \pi.$$

For $n = 6$ this gives $3.141277519\ldots$ which is not a particularly good approximation to $\pi = 3.14159265\ldots$

You can find an estimate of π by tightly and evenly wrapping a thread around a broom handle and drawing a line across the threads. Then an estimate of π can be obtained by taking the average distance between marks on the threads and dividing by the diameter of the broom handle.

There are two references in the Bible to the value $\pi = 3$, namely 1Kings, ch.7, ver.23 and 2Chronicles, ch.4, ver.2. There is an amusing story of how the Indiana State legislature was almost conned into passing a law making $\pi = 3$. This would have meant that wheels in this God-fearing state would have had to be hexagonal.

Other approximations taken for the number π in the past have been, in increasing accuracy, $\sqrt{10} = 3.162277660\ldots$, $\frac{22}{7} = 3.142857143\ldots$ and $\frac{355}{113} = 3.1415929\ldots$(!)

The following statement was found in a school geometry book and is given without comment.

$$\pi = 3.1415 \text{ cm}.$$

Let us finish with some more pretty formulæ for π. Only the last one has any practical value for evaluation.

$$\frac{\pi}{4} = 1 - \frac{1}{3} + \frac{1}{5} - \frac{1}{7} + \cdots$$

$$\frac{\pi^2}{6} = \frac{1}{1^2} + \frac{1}{2^2} + \frac{1}{3^2} + \frac{1}{4^2} + \cdots \qquad \textbf{(Euler)}$$

$$\frac{\pi}{2} = \frac{2}{1} \cdot \frac{2}{3} \cdot \frac{4}{3} \cdot \frac{4}{5} \cdot \frac{6}{5} \cdot \frac{6}{7} \cdots \qquad \textbf{(Wallis)}$$

$$\frac{\pi}{4} = \cfrac{1}{1 + \cfrac{1^2}{2 + \cfrac{3^2}{2 + \cfrac{5^2}{2 + \ddots}}}} \qquad \textbf{(Euler)}$$

$$\frac{\pi}{6} = \frac{1}{2} + \frac{1}{2} \cdot \frac{1}{3 \cdot 2^3} + \frac{1}{2} \cdot \frac{3}{4} \cdot \frac{1}{5 \cdot 2^5} + \cdots$$

The last series has

$$\frac{1}{2} \cdot \frac{3}{4} \cdots \frac{2n-1}{2n} \cdot \frac{1}{(2n+1)2^{2n+1}}$$

as its general term. Evaluating this series to 6 summands gives π to 4 decimal places.

To the ancient Greeks it was the height of taste to proportion their architecture in the *golden ratio*. Two sides of a rectangle are in proportion according to the golden ratio if, by cutting off a square on the smaller side, a rectangle results which is proportional to the first (Fig. 1.13).

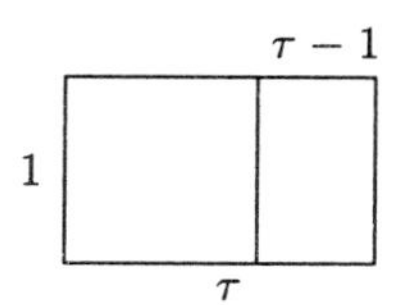

Fig. 1.13. The golden ratio

We can continue the sequence of removing smaller and smaller squares. If we insert quadrants into the squares we get an attractive spiral (Fig. 1.14.).

We can calculate the golden ratio as follows. Let the length of the longer side be τ and the length of the shorter side be 1. Then by hypothesis

$$\frac{\tau - 1}{1} = \frac{1}{\tau}.$$

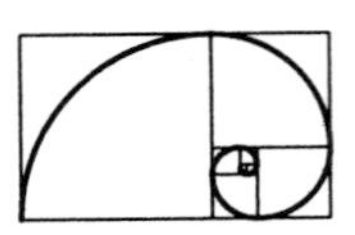

Fig. 1.14.

So τ satisfies the quadratic equation

$$\tau^2 - \tau - 1 = 0$$

and solving gives

$$\boldsymbol{\tau = \frac{1}{2} + \frac{\sqrt{5}}{2} = 1.618033988\ldots}$$

Exercise 1.19

Show that if we had labelled the shorter side σ and taken the longer side to be 1 then

$$\sigma = \frac{\sqrt{5}}{2} - \frac{1}{2} = \tau - 1 = \frac{1}{\tau} = 0.618033988\ldots$$

We shall see later when we come to consider regular polygons that the golden ratio is intimately connected with the regular pentagon and that

$$\tau = 2\cos\frac{\pi}{5}.$$

From the fundamental equation we get

$$\boldsymbol{\tau = 1 + \frac{1}{\tau} = 1 + \cfrac{1}{1 + \cfrac{1}{1 + \cfrac{1}{1 + \cfrac{1}{1 + \cfrac{1}{\ddots}}}}}}.$$

The successive convergents are

$$1 = \frac{1}{1}, \qquad 1 + \frac{1}{1} = \frac{2}{1}, \qquad 1 + \cfrac{1}{1 + \cfrac{1}{1}} = \frac{3}{2}, \qquad 1 + \cfrac{1}{1 + \cfrac{1}{1 + \cfrac{1}{1}}} = \frac{5}{3}, \ldots$$

We may write these as

$$\frac{f_n}{f_{n-1}} \text{ where } f_n = f_{n-1} + f_{n-2} \text{ and } f_1 = f_2 = 1.$$

The remarkable sequence (f_n) is called the *Fibonacci numbers.* These arise in all sorts of ways in nature where growth is concerned.

Exercise 1.20

Assume that we have one new-born potentially breeding pair of rabbits and that each month a pair of rabbits produce another pair of rabbits which become productive after 2 months. Assuming no deaths, show that after n months there are f_n pairs of rabbits.

A graphical way of seeing the Fibonacci numbers in terms of monotone paths is illustrated in Fig. 1.15.

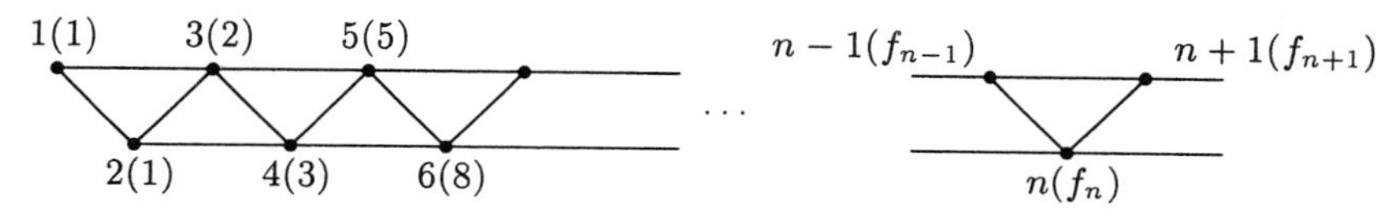

Fig. 1.15

The odd numbers are arranged along the top semi-infinite line and the even numbers are arranged along the bottom semi-infinite line. In brackets next to each number is the corresponding Fibonacci number. The whole object looks like the arm of a crane.

If $n > m$ a *path* from m to n is a sequence $m, m_1, \dots, m_k, n$ whose terms increase by 1 or 2. For example there are five paths between 3 and 7 namely

$$34567,\ 3457,\ 3467,\ 3567,\ 357.$$

Since any path from 1 to n passes either through $n-1$ or $n-2$ we see that the number of paths from 1 to n is the sum of the number of paths from 1 to $n-1$ plus the number of paths from 1 to $n-2$. Letting the number of paths from 1 to 1 be 1, we see that the number of paths from 1 to n is f_n if $n > 1$. More generally the number of paths from m to n is f_{n-m+1}.

Exercise 1.21

* Show, using the above graphical description that

$$f_{n+m} = f_n f_{m-1} + f_{n+1} f_m.$$

Exercise 1.22

* Prove Binet's formula

$$f_n = \frac{\tau^n - (-\tau)^{-n}}{\sqrt{5}}.$$

Exercise 1.23

* Show that the golden ratio $\tau = \sqrt{1+\sqrt{1+\sqrt{1+\sqrt{1+\cdots}}}}$.

The number $e = 2.7182818\ldots$, called *Euler's constant*, is the basis for natural logarithms and like the golden ratio τ also arises in situations of growth and decay. The value for e can be obtained quite quickly using the formula

$$e = 1 + \frac{1}{1!} + \frac{1}{2!} + \frac{1}{3!} + \frac{1}{4!} + \cdots$$

We will see the real power of e when we come to look at complex numbers.

Exercise 1.24

Let us write quite formally and ignoring convergence

$$\exp(x) = 1 + \frac{x}{1!} + \frac{x^2}{2!} + \frac{x^3}{3!} + \frac{x^4}{4!} + \cdots$$

Show by manipulation of series that

$$\exp(0) = 1, \ \exp(1) = e \text{ and } \exp(x+y) = \exp(x)\exp(y).$$

Hence we are justified in thinking of exp as a power and writing

$$\exp(x) = e^x.$$

Answers to Selected Questions in Chapter 1

1.1 563 and 1287. DCCXCVI and MCMXCIX (or MIM?).

1.2 (a) For associativity of addition see Fig. 1.16.

(b) The associative law for multiplication can be made convincing by looking at a cuboid of side n, m, p containing $(nm)p = n(mp)$ dots.

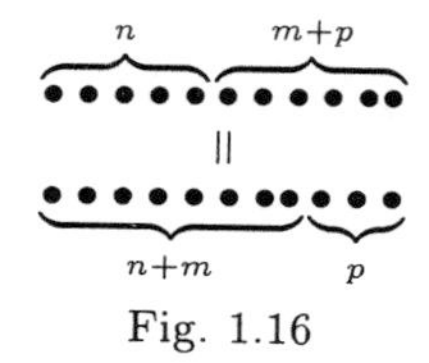

Fig. 1.16

(c) For distributivity see Fig. 1.17.

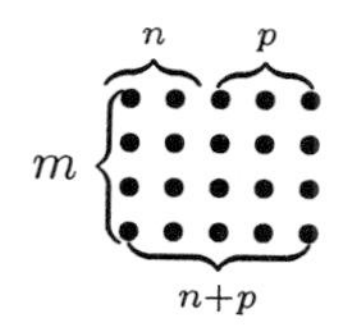

Fig. 1.17

1.3 See Fig. 1.18.

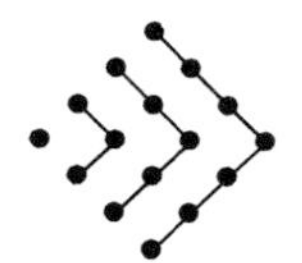

Fig. 1.18

1.5 Let $n = 755/5 = 151$. Then the base needs n^2 blocks, the next layer $(n-1)^2$ blocks and so on. From the formula $n(2n+1)(n+1)/6$ for the sum of the first n squares we get 1,159,076 blocks.

1.6 We have

$$n^4 - (n-1)^4 = 4n^3 - 6n^2 + 4n - 1.$$

Summing from 1 to n and using the formulæ already worked out we get

$$n^4 = 4\sum i^3 - n(n+1)(2n+1) + 2n(n+1) - n.$$

Solving for $\sum i^3$ gives the required result.

1.8 By expansion,

$$(1-x)(1-x^2)(1-x^3) = 1 - x - x^2 + x^4 + x^5 - x^6,$$

which is correct for the series up to powers of x less than 4. Since x^3 does not appear, we have verified that 3 is not a pentagonal number.

1.9

$$(4^3)^2 = 64^2 = 4096$$

whereas

$$4^{(3^2)} = 4^9 = 262144.$$

1.10 $2 - 3 = -1 \neq 1 = 3 - 2$ and $2 - (3 - 1) = 0 \neq -2 = (2 - 3) - 1$. The commutative law holds if and only if $p = q$ and the associative law holds if and only if $r = 0$.

1.11 (a) Yes, the equivalence class is the set of towns in a particular country.

(b) No, the transitive law fails.

(c) Yes: if a town has a railway station then its equivalence class if it lies in Europe is every town in Europe with a railway station and probably those of Africa and Asia too. I am not sure what happens in America. Of course a town without a railway station lies alone in an equivalence class with just one member.

1.12 We can rewrite the equation

$$\frac{1}{p} + \frac{1}{q} = \frac{1}{2}$$

as

$$(p - 2)(q - 2) = 4.$$

So $p - 2$, $q - 2$ are factors of 4, i.e. 1, 2 or 4. This gives solutions

$$(p, q) = (3, 6),\ (4, 4),\ (6, 3).$$

The inequality reduces to

$$(p - 2)(q - 2) < 4.$$

Solutions are

$$(p, q) = (1, \text{any}),\ (2, \text{any}),\ (3, 3),\ (3, 4),\ (3, 5),\ (4, 3),\ (5, 3),\ (\text{any}, 2),\ (\text{any}, 1).$$

Suppose

$$\frac{1}{p} + \frac{1}{q} + \frac{1}{r} = 1.$$

Then $r = 1$ is impossible since p, q are positive. If $r = 2$ then $\frac{1}{p} + \frac{1}{q} = \frac{1}{2}$ and the solutions are given above. If $r > 2$ then $\frac{1}{r} < \frac{1}{2}$ and $\frac{1}{p} + \frac{1}{q} > \frac{1}{2}$ and we get the other solutions above.

1.13 (a) $-710/457$, (b) $225/157$.

1.14

$$\frac{187}{57} = 3 + \cfrac{1}{3 + \cfrac{1}{1 + \cfrac{1}{1 + \cfrac{1}{3 + \frac{1}{2}}}}}, \qquad \frac{-19}{70} = -1 + \cfrac{1}{1 + \cfrac{1}{2 + \cfrac{1}{1 + \cfrac{1}{2 + \frac{1}{6}}}}}.$$

1.15

$$\frac{1}{s + t\sqrt{2}} = \frac{s}{s^2 - 2t^2} - \frac{t}{s^2 - 2t^2}\sqrt{2}.$$

However full marks for the formula can only be obtained by pointing out that $s^2 - 2t^2 \neq 0$ if $s + t\sqrt{2} \neq 0$.

1.16

$$\frac{1}{2 - 3\sqrt{5}} = -\frac{2}{41} - \frac{3}{41}\sqrt{5}, \quad \frac{1}{2 - 3\sqrt{-5}} = \frac{2}{49} + \frac{3}{49}\sqrt{-5}.$$

1.17 Solving the quadratic equation

$$x = 2 + \frac{1}{1 + \frac{1}{x}} = 2 + \frac{x}{1 + x}$$

gives a positive root $x = \sqrt{3} + 1$.

Let $x = \sqrt{5} + 2$, then

$$x^2 - 4x - 1 = 0 \text{ or } x = 4 + \frac{1}{x} = 4 + \cfrac{1}{4 + \cfrac{1}{4 + \ddots}}.$$

1.18 Suppose the longer side (with suitable units) is x and the shorter side is 1. Then the new rectangle has sides 2 and x. Hence if these are in the same ratio then $x/2 = 1/x$ so $x^2 = 2$ or $x = \sqrt{2}$. The metric measurement of paper has these proportions. For example A4 paper is 297mm × 210mm. Squaring the ratio gives $(297/210)^2 = 2.000204\cdots$, almost 2. Two A4s placed side by side gives A3 which is 420mm × 297mm.

1.19 The new equation is

$$\frac{1}{\sigma} = \frac{\sigma}{1 - \sigma} \text{ or } \sigma^2 + \sigma - 1 = 0$$

which gives the answer.

1.21 Any path from 1 to $n+m$ either passes through the edge from n to $n+2$ or passes through $n+1$ but not both. The number of paths of the first kind is $f_n f_{m+n-(n+2)+1} = f_n f_{m-1}$ and the number of paths of the second kind is $f_{n+1} f_{m+n-(n+1)+1} = f_{n+1} f_m$. This proves the result.

1.22 Let $g_n = \frac{\tau^n - (-\tau)^{-n}}{\sqrt{5}}$. If we can show that

$$g_n = g_{n-1} + g_{n-2} \text{ and } g_1 = g_2 = 1$$

then clearly $g_n = f_n$. However this is an easy algebraic manipulation using the fact that τ and $(-\tau)^{-1}$ are the two roots of $x^2 - x - 1 = 0$.

1.23 Let $t = \sqrt{1+\sqrt{1+\sqrt{1+\sqrt{1+\cdots}}}}$. Squaring once yields $t^2 = 1 + \sqrt{1+\sqrt{1+\sqrt{1+\cdots}}} = 1+t$. So t is the positive root of $t^2 - t - 1 = 0$, that is τ.

2
Coordinate Geometry

Cogito ergo sum
Descartes

2.1 Coordinates

Suppose we have a portion of a flat plane, a map say. Points in the plane are given by two real numbers called coordinates. Consider the pirate's instruction for where the treasure is buried.
From Point Origin, walk 12 paces east and then 7 paces north.
The place where the treasure is buried is now coded by the pair of numbers $(12, 7)$.

Now suppose that a further more accurate pirate's map is discovered which gives the following instructions.
From Point Origin, walk 12 paces east and then 7 paces north. Now dig down 6 feet.
If we take a pace as 3 feet then the place where the treasure is buried can be coded by the triple $(12, 7, -2)$.

The idea of using coordinates to specify points in space is popularly attributed to René Descartes. The use of such coordinates to specify lines, conics, etc. is sometimes called cartesian geometry in consequence. Descartes did indeed recognise its importance and helped to popularise it. However coordinates had been in use since the 10th century and Pierre de Fermat wrote a book on coordinate geometry independently of Descartes.

Whatever the rights and wrongs of history there is no doubt that coordinate geometry is an immensely powerful mathematical tool. For a start it allowed the idea of replacing geometry by algebra. Secondly, combined with the calculus, which it surely helped to create, mathematicians were able to calculate curve lengths, areas etc. which would have been unthinkable before. On the negative side the use of coordinates did tend to put mathematicians into a euclidean mind set so that the idea of a non-euclidean geometry was that much harder to accept. Moreover, as we shall see in the next chapter some results are much easier to prove using global pre-coordinate geometry arguments.

In this chapter we shall look at how to calculate length and area with cartesian coordinates, define angles, polar coordinates and describe lines, planes and their generalisations.

2.2 $\mathbb{R}^n$, the Space of Coordinates

Let us denote by $\mathbb{R}^n$ the set of all ordered n-tuples of real numbers such as $(x_1, x_2, \ldots, x_n)$.

Then $\mathbb{R}^n$ is shorthand for $\mathbb{R} \times \cdots \times \mathbb{R}$ (n times).

We call $\mathbb{R}^n$ *euclidean* space of dimension n, so the real line $\mathbb{R}$ is $\mathbb{R}^1$, the plane is $\mathbb{R}^2$, three-dimensional space is $\mathbb{R}^3$ and so on.

Points in $\mathbb{R}^n$ will be denoted by capital letters such as $X = (x_1, x_2, \ldots, x_n)$. The n-tuple $(x_1, x_2, \ldots, x_n)$ is sometimes called the *position vector* of the point X and the numbers $x_1, x_2, \ldots, x_n$ are called the (cartesian) *coordinates* of X. The zero point or *origin* is always denoted by $O = (0, 0, \ldots, 0)$.

In so far as it is possible we shall adopt the convention that a point, represented by a capital letter, will have coordinates represented by the corresponding lower case letters. So Y is the point $Y = (y_1, y_2, \ldots, y_n)$ and so on.

However we will usually have to suspend this convention when considering the special case of the plane, $\mathbb{R}^2$, since the long-held tradition is that general points in the plane have an x-coordinate, the first one, and a y-coordinate, the second one. The same goes for $\mathbb{R}^3$ where the general point is (x, y, z).

We represent $\mathbb{R}^2$ as points on graph paper with an infinitely fine mesh. The set of points of the form $(x, 0)$ is called the $\boldsymbol{x}$**-axis** and the set of points of the form $(0, y)$ is called the $\boldsymbol{y}$**-axis**. We always represent these axes at right angles, with the x-axis ordered from left to right and the y-axis ordered upwards.

If $y = f(x)$ is a function of x, we can use the rectangular grid structure of

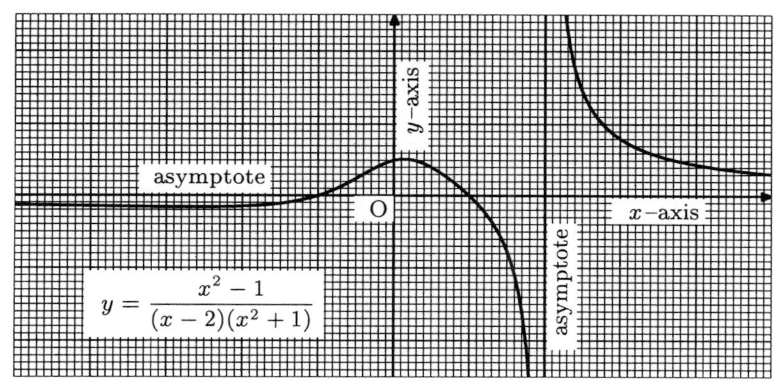

Fig. 2.1 The coordinate plane and the graph of $(x^2 - 1)/(x - 2)(x^2 + 1)$

cartesian coordinates to draw the *graph* of the function f by plotting all points $(x, f(x))$. See Fig. 2.1 for an example.

An *asymptote* is a line which the graph *tends to* at infinity. In the example above both the x-axis and the line $x = 2$ are asymptotes.

Exercise 2.1

All mathematics students *must* be able to sketch the graphs of the standard functions $\sin, \cos, \tan, \exp$ or $\log$. If you cannot do so, learn how immediately. Sketch the graphs of
(a) $2\cos(x - 5\pi/4)$,
(b) $\log(x) + 4$,
(c) $x(x^2 + 1)/(x^2 - 1)$,
(d) $\sin(1/x)$,
(e) $x(1 - \exp(-x^{-m}))$ for $m = 1,\ 2$ and 3
indicating all asymptotes.

Returning now to the general situation in $\mathbb{R}^n$, points can be added and multiplied by real numbers (sometimes called scalars) as follows. Let

$$X = (x_1, x_2, \ldots, x_n) \text{ and } Y = (y_1, y_2, \ldots, y_n)$$

be points and a be a real number. Then the sum $X + Y$ is defined by

$$\boldsymbol{X + Y = (x_1 + y_1, x_2 + y_2, \ldots, x_n + y_n)}$$

and the product aX is defined by

$$\boldsymbol{aX = (ax_1, ax_2, \ldots, ax_n).}$$

For example $3(2,-1,2)-(1,-3,2)=(6-1,-3+3,6-2)=(5,0,4)$.

The geometric interpretation of these operations is as follows. If O, X, Y lie in a plane p then $X+Y$ is the far point of a parallelogram in p defined by O, X, Y (see Fig. 2.2).

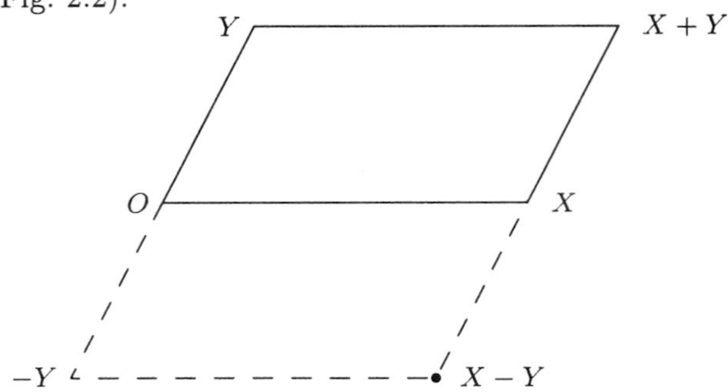

Fig. 2.2 The points $X+Y$ and $X-Y$

The point $(-1)Y$ is written $-Y$. The point $X+(-Y)$ is usually written $X-Y$ and its position in the plane p is illustrated in Fig. 2.2.

Multiplication by a real number changes the scale of measurements. So for example if the units are feet then division by 3.28083999... changes the units to metres. Any multiplication by a fixed (non-zero) number is called a *scale change* transformation (Fig. 2.3).

Exercise 2.2

Find the scale change which takes $(-2,1,-3)$ to $(1,-0.5,1.5)$.

If the number is negative then multiplication reverses direction. In particular multiplication by -1 can be thought of as a reflection in O. If the space is the plane $\mathbb{R}^2$ then multiplication by -1 is a rotation in the plane, through 180° about O. In this case the transformation is sometimes called a *half turn* about O. Half turns may be performed about any point in the plane.

If P is kept fixed and X varies over $\mathbb{R}^n$ then the resulting transformation $X \to X+P$ is called a *translation* of $\mathbb{R}^n$ by P. An oriented segment such as the one from X, its *tail*, to $X+P$, its *head*, is called a *vector*. Vectors are said to be *based* at their tail. Vectors based at the origin, O, can and will be identified with their head. Let us denote by $\overrightarrow{PQ}$ the vector from P to Q. So $\overrightarrow{PQ}$ is based at P and is parallel to $Q-P$. (If we ignore base points, parallel vectors are equal and we will constantly equate $\overrightarrow{PQ}$ with $Q-P$.) Translations take vectors to parallel vectors. So $X \to X+P$ takes $\overrightarrow{OQ}$ (which we can think of as the position vector Q) to $\overrightarrow{PR}$ where $R=P+Q$ (Fig. 2.4).

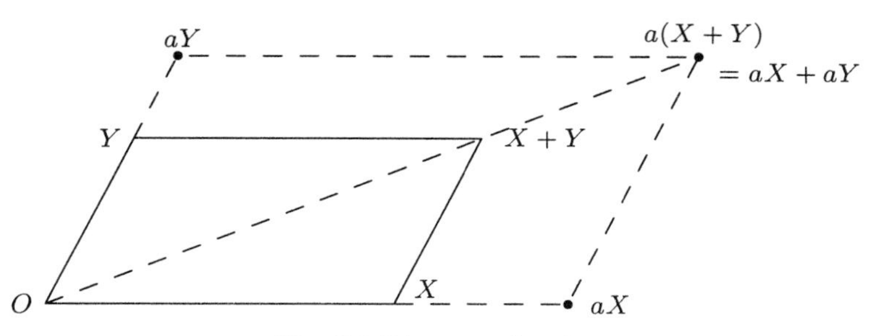

Fig. 2.3 Changes of scale

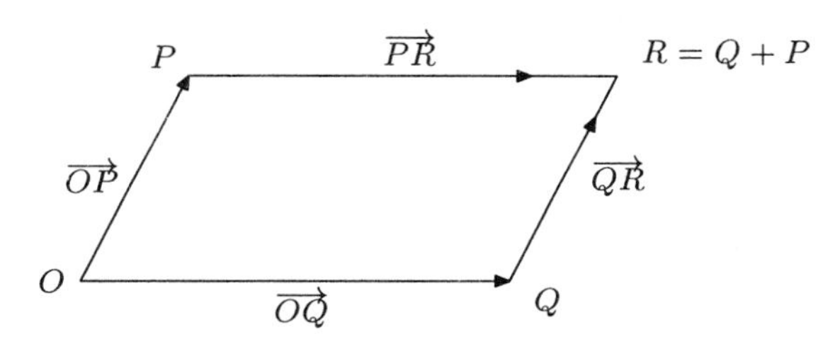

Fig. 2.4 Translating the vector $\overrightarrow{OQ}$ by P

Transformations of $\mathbb{R}^n$ such as the ones above are examples of homeomorphisms. That is they are 1–1 correspondences which are continuous and have a continuous inverse transformation. In this book we shall only consider homeomorphisms which have some geometric property such as preserving angles or length. Transformations which preserve length are called *congruences* or *isometries*. Examples of congruences of the plane are translations, reflections and rotations including half turns. Later on we shall meet another kind called a glide reflection.

Congruences which keep the origin fixed are linear and are represented by orthogonal matrices.

The properties enjoyed by the two operations of addition and multiplication by a real number are listed below where X, Y, Z are points and a, b are real numbers

$$\begin{aligned} (X+Y)+Z &= X+(Y+Z) \qquad & a(X+Y) &= aX+aY \\ X+O &= X & (a+b)X &= aX+bX \\ X+(-X) &= O & (ab)X &= a(bX) \\ X+Y &= Y+X & 1X &= X. \end{aligned}$$

A set with the properties above is called a *vector space* (over the reals).

Recall from Chapter 1 that properties such as those above have technical names. So $(X+Y)+Z = X+(Y+Z)$ is called associativity of addition (Fig. 2.5) and $X+Y = Y+X$ is called commutativity of addition.

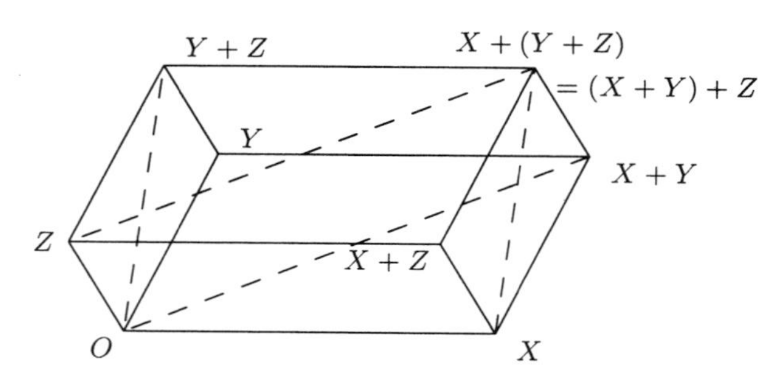

Fig. 2.5 Associativity in action

Exercise 2.3

Let T, S be two translations. Show that if one is performed after the other then the result is again a translation and they can be performed in either order; in symbols, $TS = ST$ is another translation. On what properties of vector addition does this depend?

Translations form a *group*.That is, the product of two translations, in symbols TS, is their repeated application and the result is again a translation. The associativity law holds, $(TS)U = T(SU)$. There is an identity translation, E, which moves nothing and every translation has an inverse translation, T^{-1}, such that $TT^{-1} = T^{-1}T = E$. Exercise 2.3 shows that the product in a group of translations is commutative or abelian. This is not true in general for other groups.

Exercise 2.4

What is the inverse translation of $T(X) = X + P$?

There is not space in this book to describe the theory of vector spaces. However there are important definitions and consequent theorems which will be useful in what follows. We will give the definitions and theorem without proof. However this section can be skipped on a first reading.

Let $\{A_1, \ldots, A_k\}$ be a finite set of points in $\mathbb{R}^n$. Then a sum such as $P = x_1A_1 + \cdots + x_kA_k$ is called a *linear combination* of $\{A_1, \ldots, A_k\}$. The set $\{A_1, \ldots, A_k\}$ is called *linearly independent* if the origin, O, can only be written in one way as a linear combination. That is, in the above, $x_1 = \cdots = x_k = 0$. The set $\{A_1, \ldots, A_k\}$ is called *generating* if any point in $\mathbb{R}^n$ can be written as a linear combination. If a set is both linearly independent and generating then it is called a *basis*.

The important properties of bases are:

- Any linearly independent set can be included in a basis. □

- Any generating set contains a basis. □

- Any two bases have the same number of elements. □

The number of elements of a basis is called the *dimension* of the vector space. For $\mathbb{R}^n$ the dimension is n.

2.3 The Line through Two Points

Let X, Y be distinct points. Then the line from O to $X - Y$ is parallel to the line from Y to X. So a point on the line from Y to X is of the form

$$\boldsymbol{P = Y + a(X - Y) = aX + (1 - a)Y}$$

where a is any real number (Fig. 2.6). Putting $a = 0, 1$ recovers Y, X respectively. If $0 < a < 1$ then P lies on the finite segment between Y and X. In particular

$$\boldsymbol{\frac{X + Y}{2}}$$

is the *midpoint* between X and Y. If $a < 0$ then P lies on the Y side of the line and if $a > 1$ then P lies on the X side.

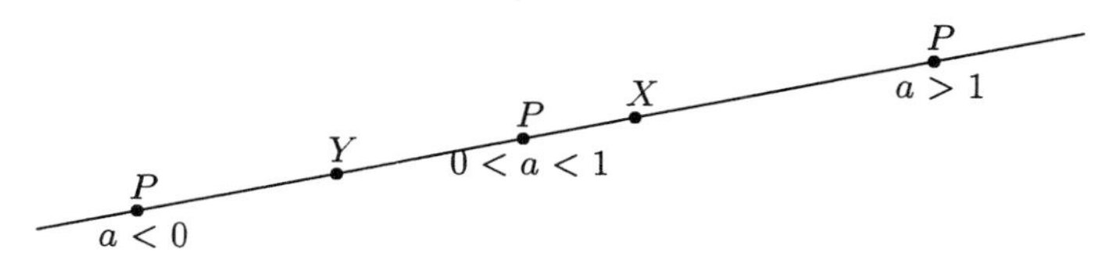

Fig. 2.6 Position of $P = aX + (1 - a)Y$

A balanced or homogeneous way of writing the general point of the line is

$$\boldsymbol{P = aX + bY \text{ where } a + b = 1.}$$

Exercise 2.5

Find the point A on the line through X and Y such that X is halfway between A and Y.

Exercise 2.6

Let H_P denote the half turn about the point P (that is reflection in the point P). Write down an expression for $H_P(X)$. Show that $H_P^{-1} = H_P$. If H_Q is another half turn, show that their product $H_P H_Q$ is a translation. Deduce that $H_P H_Q = H_Q H_P$ if and only if $P = Q$.

Exercise 2.7

Let H_P, H_Q, H_R be half turns about P, Q, R respectively. Show that $H_P H_Q = H_Q H_R$ if and only if Q is the midpoint of P, R.

The number a is an example of a *parameter* and the formula $P = aX + (1-a)Y$ is called a *parametric* representation of the general point P on the line through X and Y.

The function $a \to aX + (1-a)Y$ is a homeomorphism from the real line $\mathbb{R}$ to a general line. Of course this is not the only such homeomorphism, $a \to a^3X + (1-a^3)Y$ will also do, but it is the most natural for two reasons. Firstly it is affine (first order in a) and secondly it respects distance if we scale the line by the distance between X and Y.

By an abuse of notation, the line through X and Y will often be called XY. This is the same notation which we will use for the distance between X and Y and if possible the product of X and Y but will not lead to confusion if the context is kept in mind.

2.4 The Plane Containing Three Points

Let X, Y, Z be three points not lying in a line. Then they determine a plane, XYZ, containing all three. Since the vectors $X - Z$ and $Y - Z$ span a linear subspace of dimension two, parallel to this plane, the general point is

$$\boldsymbol{P = Z + a(X - Z) + b(Y - Z) = aX + bY + (1 - a - b)Z}$$

where a, b are any real numbers. A balanced form is

$$\boldsymbol{P = aX + bY + cZ \text{ where } a + b + c = 1.}$$

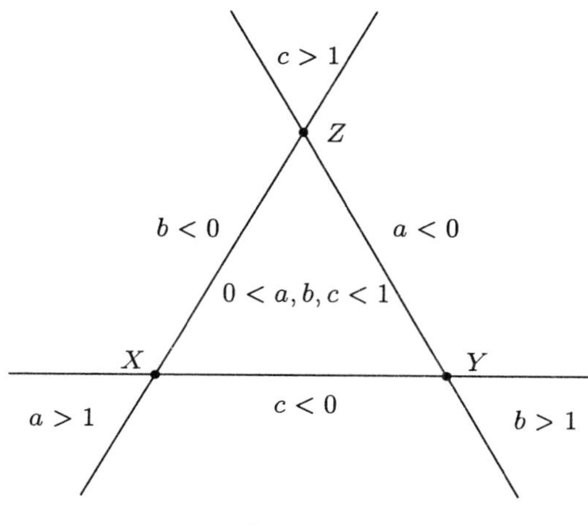

Fig. 2.7

If any of the parameters are zero then the point lies on one of the lines through two points, for example, if $c = 0$, then P lies on the line XY. Figure 2.7 shows the range of possibilities for the parameters and the regions they define.

Exercise 2.8

In Fig. 2.7 describe the set of points for which $a = 1$.

It doesn't take a genius to see that there is a pattern developing here. If we let $X_1, X_2, \dots, X_k$ be points in $\mathbb{R}^n$ then we can define the generalised plane containing these points to be the set of points $P = a_1X_1 + a_2X_2 + \dots + a_kX_k$ where $a_1 + a_2 + \dots + a_k = 1$. The dimension d is defined by the dimension of the vector space spanned by $X_1 - X_k, X_2 - X_k, \dots, X_{k-1} - X_k$ and in that case the generalised plane is called a *d-plane.* A 1-plane is a line and a 2-plane just a plane. Sometimes a 3-plane is called a solid. If $d = n - 1$ (codimension 1) then the d-plane is called a hyperplane.

If all the parameters $a_1, a_2, \dots, a_k$ are positive then the point $P = a_1X_1 + a_2X_2 + \dots + a_kX_k$ is said to belong to the *convex hull* of $X_1, X_2, \dots, X_k$.

Exercise 2.9

Find the dimension d of the d-plane containing the points $(0, 1, 1)$, $(2, 0, -1)$ and $(1, -1, 1)$. Repeat for the points $(0, 1, 1)$, $(2, 0, -1)$ and $(0, 3, -1)$.

Exercise 2.10

What solid in $\mathbb{R}^3$ is the convex hull of
(a) the eight points $(\pm1, \pm1, \pm1)$,
(b) the four points $(1,1,1), (-1,-1,1), (1,-1,-1), (-1,1,-1)$,
(c) the six points $(\pm1,0,0), (0,\pm1,0), (0,0,\pm1)$?

2.5 Distance and Angle

In man's original view of the world, as we find it among primitives, space and time have a very precarious existence. They become "fixed" concepts only in the course of his mental development, thanks largely to the introduction of measurement

C.G. Jung

So far we have only considered properties which come from the fact that $\mathbb{R}^n$ is a vector space. These define the so called *affine* geometry of $\mathbb{R}^n$. That is the geometry of $\mathbb{R}^n$ which doesn't depend on measurement. In order for $\mathbb{R}^n$ to merit the name euclidean we must look at metric properties.

The first of these is distance. Let $X = (x_1, x_2, \ldots, x_n)$ be the position vector of a point X. Then $|X|$, pronounced *mod* X, is defined by

$$|\boldsymbol{X}| = \sqrt{\boldsymbol{x_1^2 + x_2^2 + \cdots + x_n^2}}.$$

Thinking of this as an extension of the familiar formula for the length of a hypotenuse given by Pythagoras' theorem means that $|X|$ is the distance from the point X to the origin O. It follows by a translation of the origin to Y that the distance XY between two points $X = (x_1, x_2, \ldots, x_n)$ and $Y = (y_1, y_2, \ldots, y_n)$ is $|X - Y|$, i.e.

$$\boldsymbol{XY = |X - Y| = \sqrt{(x_1 - y_1)^2 + (x_2 - y_2)^2 + \cdots + (x_n - y_n)^2}}.$$

The function $|X|$ respects positive scale changes so $|aX| = |a||X|$.

If we let $P = aX + (1-a)Y$ then

$$\frac{YP}{YX} = \frac{|aX + (1-a)Y - Y|}{|X - Y|} = \frac{|aX - aY|}{|X - Y|} = |a|.$$

In other words we have:

■ The distance from $aX + (1-a)Y$ to Y along the line between X and Y is $|a|$ scaled by the distance between X and Y. □

Also, distance satisfies

$$XY = 0 \Leftrightarrow X = Y, \text{ otherwise } XY > 0$$
$$XY = YX$$
$$XY \leq XZ + ZY.$$

The first property follows because a sum of squares of (real) numbers can only be zero if each element in the sum is zero. The second property is a direct consequence of the definition. The last one, called the *triangle inequality* is easy to justify geometrically. It says that the side of a triangle is less than the sum of the other two sides. On the other hand, an algebraic proof of this fact is quite tricky and will require some initial groundwork.

The set of points in $\mathbb{R}^n$ which are a fixed distance r from a point C is called a hypersphere of dimension $n-1$ with centre C and radius r. If $n = 2$ it is a circle and if $n = 3$ it is a sphere. We will have more to say about these in Chapter 8.

Let $X = (x_1, x_2, \ldots, x_n)$ and $Y = (y_1, y_2, \ldots, y_n)$ be points. Define the *dot product* $X \cdot Y$ by the rule

$$\boldsymbol{X \cdot Y = x_1 y_1 + x_2 y_2 + \cdots + x_n y_n}.$$

Then $X \cdot Y$ is a real number or scalar associated to every pair of points. We can easily check that the dot product satisfies

$$X \cdot X = |X|^2$$
$$X \cdot Y = Y \cdot X$$
$$(aX) \cdot Y = X \cdot (aY) = a(X \cdot Y)$$
$$(X + Y) \cdot Z = X \cdot Z + Y \cdot Z.$$

In addition the dot product satisfies the *Cauchy–Schwarz* inequality which is sufficiently important to justify a title and a proof.

■ (Cauchy–Schwarz) For any points X, Y in $\mathbb{R}^n$ we have

$$\boldsymbol{|X \cdot Y| \leq |X||Y|}.$$

Proof If x is any real number then

$$\begin{aligned} |X + xY|^2 &= (X + xY) \cdot (X + xY) \\ &= X \cdot X + 2xX \cdot Y + x^2 Y \cdot Y \\ &= |X|^2 + 2xX \cdot Y + x^2 |Y|^2. \end{aligned}$$

A quadratic in x say $ax^2 + bx + c$ is never negative if its *discriminant*, $b^2 - 4ac$, is not positive. But the above quadratic in x is never negative. By looking at its discriminant it follows that

$$(X \cdot Y)^2 - |X|^2|Y|^2 \leq 0.$$

Rearranging and taking square roots gives the result. □

Exercise 2.11

Show that $|X \cdot Y| = |X||Y|$ if and only if one of X, Y is a scalar multiple of the other.

▮ (Minkowski) For any points X, Y in $\mathbb{R}^n$ we have

$$|\boldsymbol{X} + \boldsymbol{Y}| \leq |\boldsymbol{X}| + |\boldsymbol{Y}|.$$

Proof Putting $x = 1$ in the quadratic above and using Cauchy–Schwarz gives

$$\begin{aligned} |X + Y|^2 &= |X|^2 + 2X \cdot Y + |Y|^2 \\ &\leq |X|^2 + 2|X||Y| + |Y|^2 \\ &= (|X| + |Y|)^2. \end{aligned}$$

Taking square roots gives the result. □

▮ (The triangle inequality) For any points X, Y, Z in $\mathbb{R}^n$ we have

$$\boldsymbol{XY} \leq \boldsymbol{XZ} + \boldsymbol{ZY}.$$

Proof If we replace X by $X - Z$ and Y by $Z - Y$ in Minkowski's inequality we get

$$|\boldsymbol{X} - \boldsymbol{Y}| \leq |\boldsymbol{X} - \boldsymbol{Z}| + |\boldsymbol{Z} - \boldsymbol{Y}|$$

which is what we want. □

Consider now the definition of angle. Let X and Y be non-zero points in $\mathbb{R}^n$. We want to define the angle $\angle XOY$ between OX and OY. From the Cauchy–Schwarz inequality we have

$$-1 \leq \frac{X \cdot Y}{|X||Y|} \leq 1.$$

So there is a unique real number $\theta = \angle XOY$ where $0 \leq \theta \leq \pi$ such that

$$\cos\boldsymbol{\theta} = \frac{\boldsymbol{X} \cdot \boldsymbol{Y}}{|\boldsymbol{X}||\boldsymbol{Y}|}.$$

We call θ the *absolute* or *unoriented* angle $\angle XOY$. We say that $\angle XOY$ is the angle *subtended* by the segment XY at O. We can rewrite the equation above as

$$\boldsymbol{X} \cdot \boldsymbol{Y} = |\boldsymbol{X}||\boldsymbol{Y}| \cos\boldsymbol{\theta}$$

which is true even if X or Y are O. The definition of angle given is in terms of *radian* measure. So half a complete turn is π radians.

In everyday life, angles are measured by dividing a complete turn into 360 degrees. To convert to degrees we replace θ in radians by $180 \times \theta/\pi$ where $\pi = 3.14159\ldots$. Degrees are subdivided into 60 *minutes of arc* and each minute is further divided into 60 *seconds* of arc. The notation used can best be illustrated by an example: so $25°31'49''$ is $25 + 31/60 + 49/3600 = 25.530278\ldots$ degrees.

The number of degrees in a complete turn is almost the number of days in a year. This is for historical and astronomical reasons. In fact astronomers sometimes measure angle by time, dividing a complete turn into 24 hours (the passing of a day). So one hour is $15°$, one minute is $15'$ and one second is $15''$.

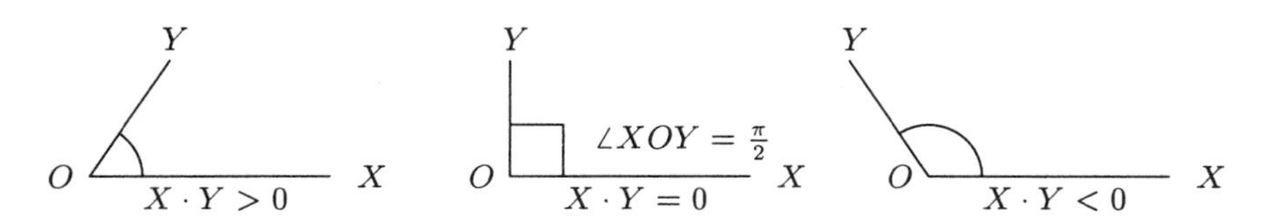

Fig. 2.8 Ranges of $\angle XOY$

If $\theta = \pi/2$ or $90°$ then $X \cdot Y = 0$ and OX is said to be at *right angles* to OY (Fig. 2.8). Alternative phrases are OX is *orthogonal* or *perpendicular* to OY. Right angles are indicated in diagrams by a little square at the point where they occur.

Exercise 2.12

Let $X = (1, 1, 2), Y = (2, 2, 1)$. Find the angle $\angle XOY$. Verify that the vector $\overrightarrow{OX}$ is at right angles to the vector $\overrightarrow{XY}$.

This definition of angle is fine as far as it goes. However for the geometry of the plane ($n = 2$) we need a more refined definition which allows addition and includes negative angles. Consider the following example in the plane.

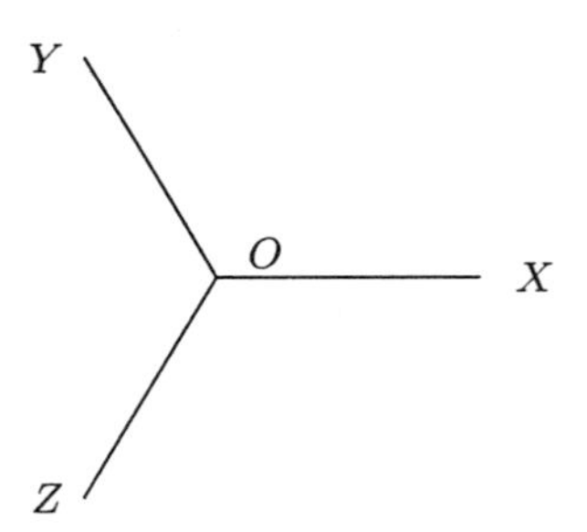

Fig. 2.9

Let $\angle XOY = \angle YOZ = 2\pi/3$ (Fig. 2.9). We would like $\angle XOY + \angle YOZ = \angle XOZ$ and this is impossible for absolute angles. So let us define an (oriented) *angle* to be represented by a real number, say $26\pi/5$, but with the condition that angles which differ by an integer multiple of 2π are deemed to be equal. So $26\pi/5 = 16\pi/5 = 6\pi/5 = -4\pi/5 = \cdots$. These numbers are said to be elements of $\mathbb{R}$ *modulo* 2π. With this definition we can add and subtract angles provided that we always take the result modulo 2π.

Taking angles modulo 2π is an example of an equivalence relation. We say that the real numbers θ and ϕ define the same angle if $\theta - \phi$ is an integer multiple of 2π. A representative of an angle equivalence class is usually taken in the range $0 \leq \theta < 2\pi$ although some authors prefer $-\pi \leq \theta < \pi$.

The set of angles forms a group under addition. This group together with $\mathbb{R}$ under addition are the simplest examples of *Lie groups*.

We can use this new definition of angle in $\mathbb{R}^2$ because the plane is naturally oriented. The definition goes as follows. Let X, Y be non-origin points in $\mathbb{R}^2$. Because our definition is independent of the length of X, Y we will assume they both have length 1. Then there is a unique rotation *in a positive direction* which takes OX to OY. We will take the positive direction to be anticlockwise. Suppose that the rotation, which we think of as a continuous motion, traces out a distance θ on the circle of radius 1 with centre O. Then θ, taken modulo 2π, is our definition of *oriented angle* or just angle if no confusion can arise in $\mathbb{R}^2$. We will always take oriented angles as the default definition from now on unless circumstances dictate otherwise.

By a sleight of hand we have introduced an undefined concept, namely the length of a circle arc. We have already defined the length of a complete arc as a limit and a subarc is defined proportionally to the angle subtended at the centre. Any one bothered by this should read a book on elementary analysis.

An oriented angle α, $-\pi < \alpha \leq \pi$ can be converted to an absolute angle by the rule

$$\alpha \to \begin{cases} \alpha, & \text{if } \alpha > 0; \\ -\alpha, & \text{otherwise.} \end{cases}$$

So absolute angles stand in relation to oriented angles in a similar way as absolute real numbers $|a|$ stand to all (positive and negative) real numbers a.

Angles measured in radians are useful because they give the values of circular arc lengths by the formula $r \times \psi$ where r is the radius and ψ is the value of the angle subtended at the centre by the circular arc in radians. Consequently radians are sometimes called *circular measure* (Fig. 2.10).

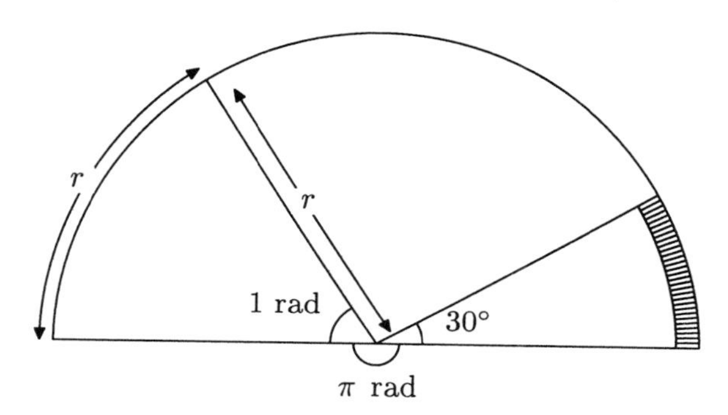

Fig. 2.10 Radians and degrees

Exercise 2.13

Let $A = (1, 0), B = (-1, -1)$. Find the oriented angle $\angle AOB$ in radians and degrees.

A moment's thought should convince you that the oriented angle now satisfies,

$$\angle XOY = -\angle YOX$$
$$\angle XOY = \angle XOZ + \angle ZOY.$$

Moreover the following properties are also easy to see.

Adjacent angles: The points X, O, Z are collinear if and only if $\angle XOY + \angle YOZ = \pi$. (Three or more points are said to be *collinear* if they lie on a line.)

Opposite angles: Suppose the points X, O, Z are collinear. Then Y, O, W are collinear if and only if $\angle YOZ = \angle WOX$.

Exercise 2.14

The points A, O, E are collinear. If OB bisects $\angle AOC$ and OD bisects $\angle COE$ show that $\angle BOD$ is a right angle.

Exercise 2.15

The lines BA and DE are parallel. If $\angle ABC = \alpha$ and $\angle CDE = \beta$ find $\angle DCB$.

2.6 Polar Coordinates

It follows from the above discussion that any point $X = (x, y)$ in the plane other than the origin has a unique representation in the form

$$X = (x, y) = (r\cos\theta, r\sin\theta) \qquad (r > 0).$$

The pair (r, θ) define the *polar* coordinates of X.

The polar radius coordinate is given by

$$r = \sqrt{x^2 + y^2}.$$

The polar angle coordinate satisfies $\tan\theta = y/x$ but this is only sufficient to define θ as an absolute angle. The full definition is

$$\theta = \begin{cases} \tan^{-1} y/x & \text{if } x > 0; \\ \pi + \tan^{-1} y/x & \text{if } x < 0. \end{cases}$$

If $x = 0$ then $\theta = \pi/2$ if $y > 0$ and $\theta = -\pi/2$ if $y < 0$. As usual we consider the values of the function $\tan^{-1}$ to lie in the range $-\pi/2 < \tan^{-1} < \pi/2$. Of course θ is not defined for the origin $(0, 0)$.

Exercise 2.16

Show that the polar angle coordinates for the two points with cartesian coordinates $(1, 1), (-1, -1)$ both satisfy $\tan\theta = 1$. Find the polar coordinates for both points.

Exercise 2.17

Find the polar coordinates of the points with cartesian coordinates $(1, \sqrt{3})$ and $(-\sqrt{3}, -1)$.

Polar coordinates can be used to plot curves of the form $r = r(\theta)$ just as in the cartesian case. Indeed sometimes it is much more natural to think of a curve in polar coordinates than in cartesian coordinates, especially if the curve comes back to where it started as in a loop. For example the polar curve $r = a$ where a is a positive constant is the circle $x^2 + y^2 = a^2$. To find the cartesian form of

$r^2 \cos 2\theta = a^2$ we note that $\cos 2\theta = \cos^2\theta - \sin^2\theta$ so $r^2\cos^2\theta - r^2\sin^2\theta = a^2$ and the curve is the *rectangular hyperbola* $x^2 - y^2 = a^2$.

The cartesian form of the *lemniscate* $r^2 = 2a^2\cos 2\theta$ is $(x^2+y^2)^2 = 2a^2(x^2 - y^2)$ which is much more complicated. The plot of $r = 2 + \cos\theta$ is shown in Fig. 2.11.

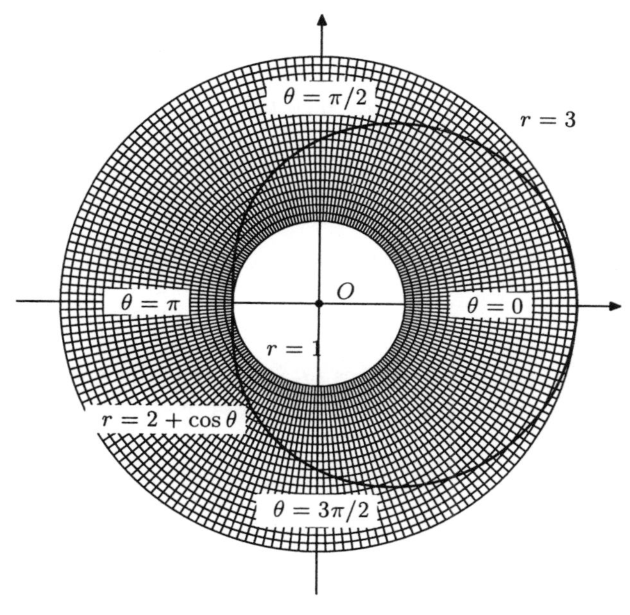

Fig. 2.11 $r = 2 + \cos\theta$

Exercise 2.18

Show that the polar equation $r = 2a\cos\theta$ represents a circle and find its centre and radius.

Exercise 2.19

Find the polar equation of the line with cartesian equation $ax + by = c$.

When plotting with polar coordinates it is sometimes useful to maintain the fiction that r can be negative and let $(-r, \theta)$ represent the same point as $(r, \pi + \theta)$. For example it is more convenient to let $\theta = \psi$ represent the whole line $y = x\tan\psi$ otherwise we would have had to include the equation $\theta = \psi + \pi$.

Exercise 2.20

Identify the following curves: $r \sin\theta = 1$, $r(r - 2\cos\theta) = 1$ and $r(\cos\theta + \sin\theta) = 1$.

Exercise 2.21

Plot the curves $r = 1 + \cos\theta$ and $r = 1/2 + \cos\theta$.

Curves of the form $r = a + b\cos\theta$ are called *limaçons* from the French for snails. Their shape depends on the relative magnitudes of a, b.

Exercise 2.22

Convert the curve $x^4 + y^4 = 1$ to polar form and plot it.

Exercise 2.23

Plot the lemniscate $r^2 = 2a^2 \cos 2\theta$. (Note that this equation only makes sense when $\cos 2\theta \geq 0$, so $-\pi/4 \leq \theta \leq \pi/4$ or $3\pi/4 \leq \theta \leq 5\pi/4$.)

Exercise 2.24

* Show that for any point on the lemniscate, the product of its distances from $(a, 0)$ and $(-a, 0)$ is a^2.

A function f is said to have *period* p if $f(x+p) = f(x)$ for all x. More definitely, *the* period is the smallest such positive p. The period of cos and sin is 2π; the period of tan and $\cos 2\theta$ is π. If the period of the function $f(\theta)$ is a fraction of 2π then the polar plot of $r = f(\theta)$ will repeat symmetrically according to that fraction.

Exercise 2.25

Plot the guitar shape $r = 2 + \cos 2\theta$.

The *rosettes* are of the form $r = \cos(n\theta)$. They have n petals if n is odd and $2n$ petals if n is even.

Exercise 2.26

Plot the rosettes $r = \cos 3\theta$ and $r = \cos 4\theta$.

If the period of $f(\theta)$ is a multiple of 2π then the polar plot of $r = f(\theta)$ may not return to its starting point after one revolution. Accordingly we may have to rotate a number of times (the multiple of 2π) in order to do this. For example $2 + \cos(\theta/2)$ has period 4π and its plot is shown in Fig. 2.12.

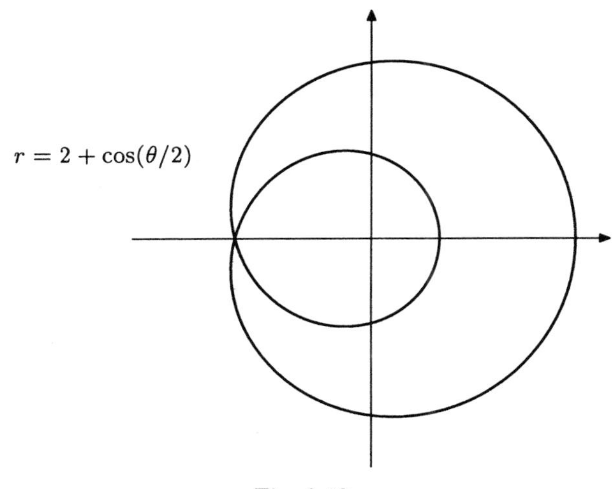

Fig. 2.12

Exercise 2.27

Plot the curves $r = 2 + \cos(3\theta/2)$ and $r = 2 + \cos(4\theta/5)$.

Exercise 2.28

The identity $\cos\theta = 2\cos^2(\theta/2) - 1$ seems to suggest that cos has period 2π *and* 4π. What is wrong with this argument?

If the function $f(\theta)$ has no period at all then its curve will wind round and round like a spiral. Figure 2.13 shows the plot of the equiangular spiral $r = ae^{\mu\theta}$. This spiral is called equiangular because the tangent at the point P makes a constant angle ψ with the radius vector OP. This fixed angle is given by $\cot\psi = \mu$.

Exercise 2.29

Plot the spirals $r = \theta$ (the archimedian spiral) and $r = \theta/(\theta + 1)$.

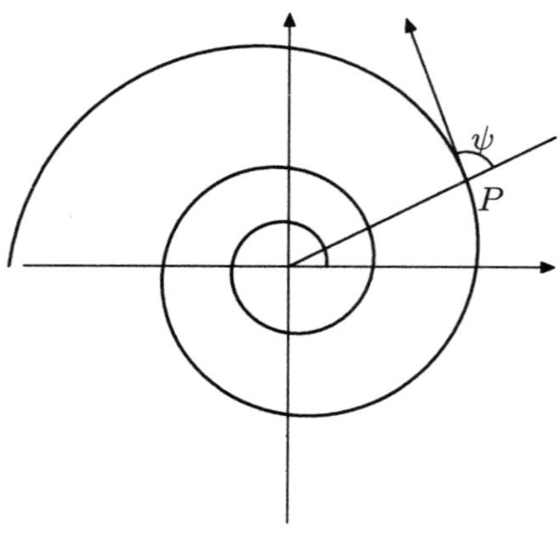

Fig. 2.13

There is something curious going on here because $r = f(\theta)$ is supposed to be a function of angle and this is only true if f has period 2π or a fraction of 2π. What is happening for functions of period a multiple of 2π is that they have been "lifted" to another function of angle which "covers" the original. A function with no period lifts to a function of the reals $\mathbb{R}$. This point will be covered in more detail when we come to consider de Moivre's theorem in Chapter 4.

2.7 Area

We now consider area. Since this potentially opens up a whole complicated (but interesting) field of study which goes well beyond the scope of this book we will only consider the simplest cases, which we will treat naïvely.

The area of a rectangle, being a picture of multiplication, is ab where a and b are the lengths of the (non-parallel) sides. To deduce the area of a triangle OXY consider Fig. 2.14.

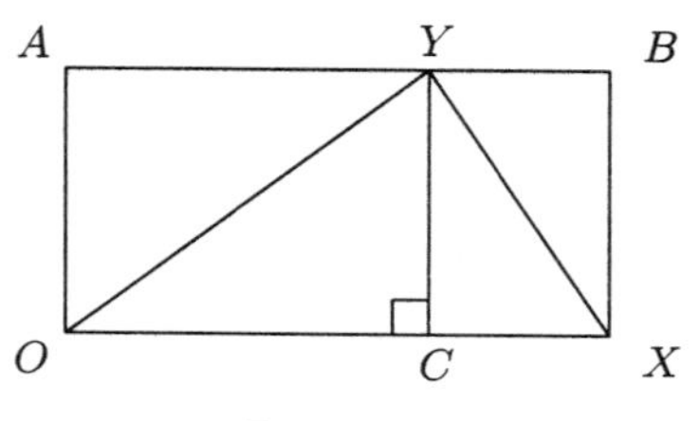

Fig. 2.14

The base OX is also the length of one side of the rectangle $OXBA$. The other length is $OA = CY$, the length of the perpendicular from Y to the base.

Since the area of the rectangle is clearly twice the area of the triangle we get the well-known formula,

$$\text{area of triangle} = \text{half base} \times \text{perpendicular height.}$$

Then the area of a more complicated body, say a polygon, is defined to be the sum of the areas of the triangles into which it can be decomposed. We leave to one side any considerations as to whether this is well defined (it is) and continue with our fingers crossed.

■ **The area of the triangle XOY is**

$$\boldsymbol{\triangle XOY = \frac{\sqrt{|X|^2|Y|^2 - (X \cdot Y)^2}}{2}.}$$

Proof Let θ be the absolute angle $\angle XOY$. Then using our formula, we have

$$\begin{aligned} \text{area} =& \frac{1}{2}OX \times \text{ height} = \frac{1}{2}|X||Y|\sin\theta \\ =& \frac{1}{2}|X||Y|\sqrt{1 - \left(\frac{X \cdot Y}{|X||Y|}\right)^2} = \frac{\sqrt{|X|^2|Y|^2 - (X \cdot Y)^2}}{2}. \end{aligned}$$

□

Exercise 2.30

Let $X = (1, 0, -2, 4)$ and $Y = (2, -2, 0, 5)$. Find
(a) the distance of X from O,
(b) $|2X - 3Y|$,
(c) the absolute angle $\angle XOY$,
(d) the area $\triangle XOY$.

Exercise 2.31

By translating the point Z to the origin show that the area of a general triangle XYZ is

$$\triangle XYZ = \frac{\sqrt{|X - Z|^2|Y - Z|^2 - ((X - Z) \cdot (Y - Z))^2}}{2}.$$

(You may assume that the area of a triangle is preserved under translation.)

Exercise 2.32

Suppose the points $X = (x_1, y_1), Y = (x_2, y_2), Z = (x_3, y_3)$ lie in the plane. Show that the formula for the area above reduces to

$$\frac{1}{2}(x_1y_2 - x_2y_1 + x_2y_3 - x_3y_2 + x_3y_1 - x_1y_3).$$

This can be written as half of a 3×3 determinant

$$\frac{1}{2}\begin{vmatrix} x_1 & y_1 & 1 \\ x_2 & y_2 & 1 \\ x_3 & y_3 & 1 \end{vmatrix}.$$

(You may have to change the sign if you want a positive area.)

Exercise 2.33

Find the area of the triangle in $\mathbb{R}^2$ with vertices $(0, 2), (1, -3), (-2, 0)$.

Exercise 2.34

Show that the area of the parallelogram in the plane with vertices whose coordinates are $(0, 0), (a, b), (c, d), (a + c, b + d)$ is the value of the determinant

$$ad - bc = \begin{vmatrix} a & b \\ c & d \end{vmatrix}$$

up to sign.

The area of a circle of radius r is given by the famous formula πr^2. You can get a feel for why this is true as follows. Cut the circle up into n equal sectors and pack them like sardines. The case $n = 16$ is illustrated in Fig. 2.15.

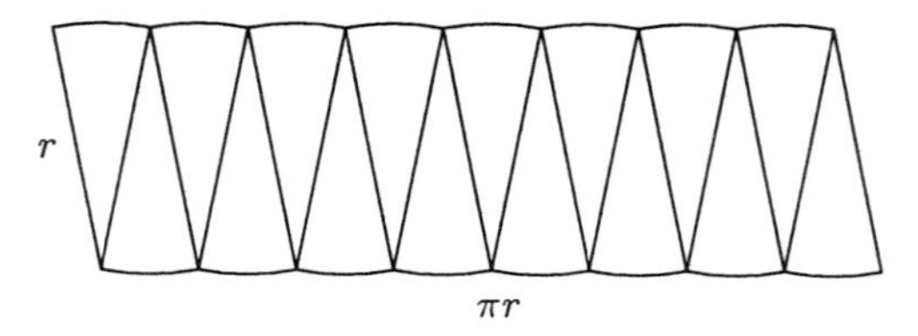

Fig. 2.15

If n is large enough then they will approximately form a rectangle with sides r and πr. As n tends to ∞ then the shape will tend to a rectangle giving the area $r \times \pi r = \pi r^2$.

2.8 Hyperplanes

A *hyperplane* in $\mathbb{R}^n$ is the collection of points, $X = (x_1, x_2, \ldots, x_n)$, which satisfy a linear equation of the form

$$a_1x_1 + a_2x_2 + \cdots + a_nx_n = b.$$

If $n = 2$ a hyperplane is a line and so the general equation of a line in the plane is

$$ax + by = c$$

where the numbers a, b are not both zero.

Through two points in $\mathbb{R}^n$ there is a unique line. Let $P = (-1, 5)$ and $Q = (0, 3)$ be two points in $\mathbb{R}^2$ and let ℓ be the unique line through P and Q. All points (x, y) on ℓ are given by the equation $2x + y = 3$.

We can find the equation of the line ℓ by a method due to Cramer. Suppose that $P = (x_1, y_1)$ and $Q = (x_2, y_2)$. Write the coordinates of the points in an array as follows

$$\begin{vmatrix} x_1 & y_1 & 1 \\ x_2 & y_2 & 1 \end{vmatrix}.$$

The coefficients of the equation of the line are given by the three 2×2 determinants with alternating sign obtained by successively deleting the columns of the array. For the example above the array is

$$\begin{vmatrix} -1 & 5 & 1 \\ 0 & 3 & 1 \end{vmatrix}$$

and the determinants with sign are

$$\begin{vmatrix} 5 & 1 \\ 3 & 1 \end{vmatrix} = 2, \quad -\begin{vmatrix} -1 & 1 \\ 0 & 1 \end{vmatrix} = 1, \quad \begin{vmatrix} -1 & 5 \\ 0 & 3 \end{vmatrix} = -3.$$

So the equation of the line is

$$2x + y - 3 = 0.$$

If you are curious as to why this works, note that if a point (x, y) lies on the line defined by the two points $(x_1, y_1), (x_2, y_2)$ then the area of the resulting triangle is zero. So from the formula above for the area of a triangle we have

$$\begin{vmatrix} x & y & 1 \\ x_1 & y_1 & 1 \\ x_2 & y_2 & 1 \end{vmatrix} = 0.$$

The equation of the line is obtained by expanding this determinant using the top row.

If you don't like this method we will describe an alternative in terms of slope soon.

We can find the intersection of two lines by solving a simple simultaneous linear equation. We can adopt Cramer's method again to do this. Take the two lines $2x - y + 2 = 0$ and $x + 2y = 3$. The array is

$$\begin{vmatrix} 2 & -1 & 2 \\ 1 & 2 & -3 \end{vmatrix}$$

which has determinants (with sign), $-1, 8, 5$. So their point of intersection is

$$(\frac{-1}{5}, \frac{8}{5})$$

as may be easily verified.

Exercise 2.35

Find the intersection of the lines $x - y + 2 = 0$ and $2x + 2y = 3$

Exercise 2.36

Find the intersection of the lines $y = m_1 x + k_1$ and $y = m_2 x + k_2$. What happens if $m_1 = m_2$?

Two distinct lines either meet in a point or are disjoint when we say they are *parallel.*

If $b \neq 0$ we can write the general equation of the line in the form

$$y = mx + \ell$$

where $m = -a/b$ and $\ell = c/b$.

If $b = 0$ then the equation of the line is of the form $x = k$ where $k = c/a$ is constant. This is a line parallel to the y-axis and perpendicular to the x-axis which it meets in the point $(k, 0)$.

Let $y = mx + k$ be a line. Then we call m its *slope.* There is a unique angle α such that $0 \leq \alpha < \pi$ and $\tan\alpha = m$ which we call the angle of slope (with respect to the x-axis) of the line. Note that lines are not oriented in general so the angle of slope is an absolute angle and is only defined modulo π.

If $x = c$ is the equation of the line then we put $\alpha = \pi/2$ and $m = \infty$.

If (x_1, y_1) and (x_2, y_2) are two points on a line ℓ then the slope of ℓ is given by the formula

$$m = \frac{y_1 - y_2}{x_1 - x_2}.$$

This simple formula can be used to give a quick answer to the equation of a line through two points. Returning to the example earlier, a general point (x, y)

on the line ℓ through $P = (-1, 5)$ and $Q = (0, 3)$ can be used to calculate the slope in two different ways. So

$$\frac{y-5}{x-(-1)} = \frac{3-5}{0-(-1)}$$

which reduces to the answer, $2x + y = 3$.

Let $y = m_1 x$ and $y = m_2 x$ be two lines passing through the origin. The angle between them is α where

$$\tan\alpha = \frac{m_1 - m_2}{1 + m_1 m_2}.$$

This is easily seen using the addition formula for tan. Since $\tan(\pi/2) = +\infty$ we have:

■ Lines $y = m_1 x + k_1$ and $y = m_2 x + k_2$ are at right angles if and only if

$$m_1 m_2 = -1.$$ □

Exercise 2.37

Show that the line through (x_1, y_1) which is perpendicular to the line through (x_2, y_2) and (x_3, y_3) has equation

$$y = \left(\frac{x_3 - x_2}{y_2 - y_3}\right) x + \frac{x_1 x_2 - x_1 x_3 - y_1 y_3 + y_1 y_2}{y_2 - y_3}$$

if $y_2 \neq y_3$. Find the equation if $y_2 = y_3$.

If $n = 3$ a hyperplane is just an ordinary plane in space and is given by an equation of the form:

$$ax + by + cz = d.$$

Three non-collinear points determine a plane. For example to determine the equation of the plane through $(1, 2, 3), (0, 1, -2), (2, 0, 5)$ we assume that it is of the form $ax + by + cz = d$. Then the coefficients satisfy

$$\begin{aligned} a + 2b + 3c &= d \\ b - 2c &= d \\ 2a \qquad + 5c &= d. \end{aligned}$$

We solve this by the method of Gaussian elimination. Roughly speaking we try to make the system of equations into upper triangular form as far as we can so that coefficients below the NW–SE diagonal are zero. This is done using elementary row operations. If you have not encountered Gaussian elimination before then you should look up any decent book on linear algebra.

We change the system above by $R_3 \to R_3 - 2R_1$ to get

$$\begin{aligned} a + 2b + 3c &= d \\ b - 2c &= d \\ -4b - c &= -d. \end{aligned}$$

Here the instruction $R_3 \to R_3 - 2R_1$ means replace row 3 by the old row 3 with twice row 1 removed. A further change $R_3 \to R_3 + 4R_2$ gives

$$\begin{aligned} a + 2b + 3c &= d \\ b - 2c &= d \\ -9c &= 3d. \end{aligned}$$

This is the required upper triangular form. We can now solve for a, b, c by working upwards. We can give any convenient value to d say $d = 3$. Then $c = -1$ and $b = 3 - 2 = 1$ and $a = 3 - 2 + 3 = 4$. So the equation of the plane is $4x + y - z = 3$. Any other choice of d other than zero would have given the same answer after cancellation.

Exercise 2.38

It is known that the four points $(1, 1, 2, 2)$, $(-1, 0, 1, -3)$, $(0, 1, 1, -1)$, $(2, 1, -1, 1)$ lie on a unique hyperplane in $\mathbb{R}^4$. Find the equation of this hyperplane.

The intersection of two or more planes in $\mathbb{R}^3$ can be found by Gaussian elimination in a similar way. In general you would expect the intersection of two planes in $\mathbb{R}^3$ to be a line lying in both and the intersection of three planes in $\mathbb{R}^3$ to be a single point. However it is possible for two planes in $\mathbb{R}^3$ to be parallel and hence to have empty intersection and it is possible for three planes in $\mathbb{R}^3$ to have a line in common or to have empty intersection. The possible configurations of three distinct planes in $\mathbb{R}^3$ are given by Fig. 2.16.

The examples are **a** 3 planes meeting in a (triple) point; **b** planes which meet in pairs in 3 parallel lines; **c** 3 planes, 2 of which are parallel; **d** 3 parallel planes; and **e** 3 planes through 1 line.

Exercise 2.39

Find the intersection of the following three sets of three planes in $\mathbb{R}^3$ and

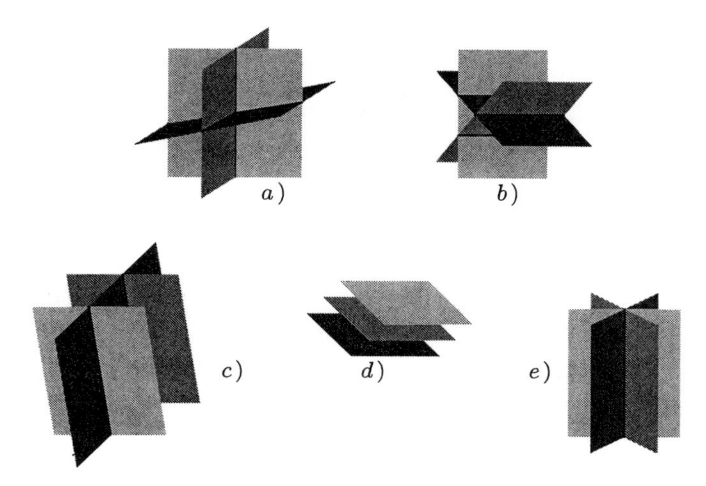

Fig. 2.16 Configurations of planes in space

determine what kind of configuration they form.

$$\text{(i)}\quad \begin{aligned} x - 4y + z &= -10 \\ 4x - y + 2z &= -4 \\ 2x + 2y - 3z &= 15 \end{aligned} \qquad \text{(ii)}\quad \begin{aligned} x + 2y - 3z &= -1 \\ 3x - y + 2z &= 7 \\ 5x + 3y - 4z &= 2 \end{aligned}$$

$$\text{(iii)}\quad \begin{aligned} x + y + z &= 0 \\ 4x + y + 3z &= 3 \\ 2x - y + z &= 3 \end{aligned}$$

2.9 Angles between Hyperplanes and Nearest Points to Hyperplanes

A hyperplane given by the equation $a_1x_1 + a_2x_2 + \cdots + a_nx_n = b$ can be written in the more compact form,

$$\boldsymbol{A} \cdot \boldsymbol{X} = \boldsymbol{b}$$

where $X = (x_1, \ldots, x_n)$ and $A = (a_1, \ldots, a_n)$ is non-zero.

The line through O and A is perpendicular to the hyperplane because if X_1 and X_2 lie in the hyperplane then $A \cdot (X_1 - X_2) = b - b = 0$. So A is a *normal* to the hyperplane. This allows us to formulate answers to two questions.

(1) What is the angle between two hyperplanes?

■ Let $A_1 \cdot X = b_1$ and $A_2 \cdot X = b_2$ be two hyperplanes. Then the angle between them is the same as the angle between A_1 and A_2. That is

$$\text{angle} = \cos^{-1}\left(\frac{A_1 \cdot A_2}{|A_1||A_2|}\right). \qquad \square$$

Exercise 2.40

Find the angle between the planes $x + z = 1$ and $x - y = 0$.

(2) What is the nearest point from the origin to a given hyperplane and what is its distance from the origin?

Clearly the nearest point is the intersection of the line through O and A with the hyperplane. Suppose that the point is cA, then it satisfies $A \cdot (cA) = b$. So $c = b/|A|^2$ and we have the following.

■ The nearest point from the origin O to the hyperplane $A.X = b$ is

$$\textbf{nearest point} = \frac{\boldsymbol{b}}{|\boldsymbol{A}|^2}\boldsymbol{A}$$

and its distance from the origin is

$$\textbf{distance} = \frac{|\boldsymbol{b}|}{|\boldsymbol{A}|}. \qquad \square$$

We can generalise this by translating the origin to the point P by the transformation $X \to X' = X - P$ when the new equation of the hyperplane becomes $A \cdot X' + A \cdot P = b$ and so by the above formula the nearest point in X' coordinates is $\frac{b - A \cdot P}{|A|^2}A$. Translating back to X coordinates we have the following.

■ The nearest point on the hyperplane $A.X = b$ to a point P is

$$\textbf{nearest point} = \frac{\boldsymbol{b} - \boldsymbol{A} \cdot \boldsymbol{P}}{|\boldsymbol{A}|^2}\boldsymbol{A} + \boldsymbol{P}$$

and its distance from P is

$$\textbf{distance} = \frac{|\boldsymbol{b} - \boldsymbol{A} \cdot \boldsymbol{P}|}{|\boldsymbol{A}|}. \qquad \square$$

For example in $\mathbb{R}^3$, suppose we want to find the nearest point on the plane $2x - 3z = 2$ to the point $(1, -1, 2)$. Then $A = (2, 0, -3)$ and $b = 2$ so the nearest point is

$$\frac{2 - (2 - 6)}{2^2 + (-3)^2}(2, 0, -3) + (1, -1, 2) = (\frac{25}{13}, -1, \frac{8}{13})$$

and the distance apart is $\frac{6}{\sqrt{13}}$.

Exercise 2.41

In $\mathbb{R}^2$ find the nearest point P on the line ℓ given by $x - 3y = 2$ to the origin O and find $|P|$.

Exercise 2.42

Let the line ℓ be given by the equation $ax + by = c$. Show that the nearest point to ℓ from an arbitrary point (p, q) is

$$\left(\frac{ac + b^2p - abq}{a^2 + b^2}, \frac{bc + a^2q - abp}{a^2 + b^2}\right)$$

and the distance is

$$\frac{|c - ap - bq|}{\sqrt{a^2 + b^2}}.$$

Answers to Selected Questions in Chapter 2

2.2 Multiplication by $-1/2$.

2.3 $TS(X) = T(X + Q) = (X + Q) + P = X + (Q + P)$. The associativity law shows that the composition is a translation. Now the commutivity law $Q + P = P + Q$ shows that $TS(X) = ST(X)$.

2.4 $T^{-1}(X) = X - P$.

2.5 $A = 2X - Y$.

2.6 The point P is halfway between $H_P(X)$ and X so $P = (H_P(X) + X)/2$. Solving gives $H_P(X) = 2P - X$. The half turn H_P is its own inverse because H_P^2 is the identity. $H_PH_Q(X) = H_P(2Q - X) = 2P - (2Q - X) = X + 2(P - Q)$, a translation through $2(P - Q)$. Similarly $H_QH_P(X) = X + 2(Q - P)$. These are equal if and only if $P - Q = 0$.

2.7 $H_PH_Q(X) = H_QH_R(X)$ if and only if $X + 2(P - Q) = X + 2(Q - R)$. That is $Q = (P + R)/2$.

2.8 The line through X parallel to YZ.

2.9 In the first case $d = 2$ and is the plane $4x + 2y + 3z = 5$. In the second case $d = 1$ (a line) with general point $(1 + t, 1 - 2t, t), t \in \mathbb{R}$.

2.10 (a) a cube, (b) a tetrahedron, (c) an octahedron.

2.11 We may as well assume that X, Y are non-zero. This means that if θ is the angle between them

$$|X||Y|\cos\theta = \pm|X||Y|.$$

So $\cos\theta = \pm 1$ or $\theta = 0$ or π. Hence X, O, Y are collinear which is the same as saying that one of X, Y is a scalar multiple of the other.

2.12 Calculations show that $X \cdot Y = 6$, $|X|^2 = 6$ and $|Y|^2 = 9$. So $\cos\theta = \frac{6}{\sqrt{6}\cdot 3} = \frac{\sqrt{6}}{3}$. So $\theta = 35.26\ldots°$. The vector $\overrightarrow{XY} = (2-1, 2-1, 1-2) = (1, 1, -1)$ and $\overrightarrow{XY} \cdot \overrightarrow{OX} = 1 + 1 - 2 = 0$.

2.13 $\angle AOB = 225° = 5\pi/4$ rad.

2.14 $\angle AOB + \angle BOC + \angle COD + \angle DOE = \pi$, by hypothesis $2\angle BOC + 2\angle COD = \pi$ so $\angle BOD = \pi/2$.

2.15 $\angle DCB = \alpha + \beta$.

2.16 $\tan\theta = 1/1 = (-1)/(-1)$, $r = \sqrt{2}$ in both cases, $\theta = \pi/4$ and $\theta = 5\pi/4$.

2.17 $(2, \pi/3)$, $(2, 7\pi/6)$.

2.18 We have $r^2 = 2ar\cos\theta$ or $x^2 + y^2 = 2ax$. This can be rewritten as $(x - a)^2 + y^2 = a^2$ which is a circle centre $(a, 0)$ with radius $|a|$.

2.19 This equation can be written as

$$ar\cos\theta + br\sin\theta = c \qquad \text{or} \qquad r = \frac{c}{a\cos\theta + b\sin\theta}.$$

2.20 The first equation is the line $y = 1$, the second converts to $(x-1)^2 + y^2 = 2$ in cartesian coordinates which is a circle centre $(1, 0)$ with radius $\sqrt{2}$ and the third is the line $x + y = 1$.

2.21 Fig. 2.17 gives the two plots with that of $r = 2 + \cos\theta$ for comparison.

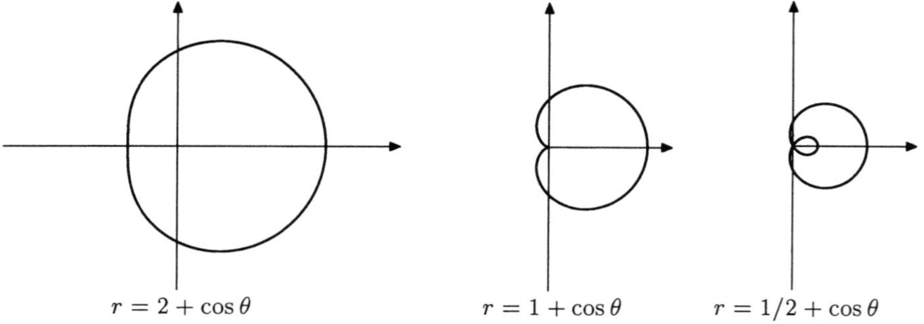

Fig. 2.17

2.22 In polar form the equation is

$$r = \left(\frac{1}{\cos^4 \theta + \sin^4 \theta} \right)^{\frac{1}{4}}.$$

The curve looks like a square with rounded corners (Fig. 2.18).

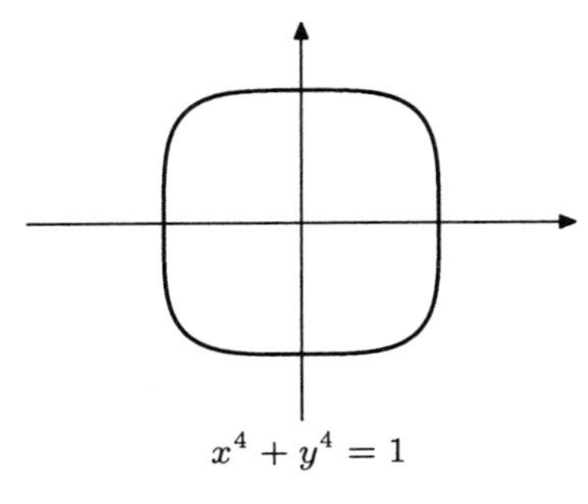

$x^4 + y^4 = 1$

Fig. 2.18.

2.23 See Fig. 2.19.

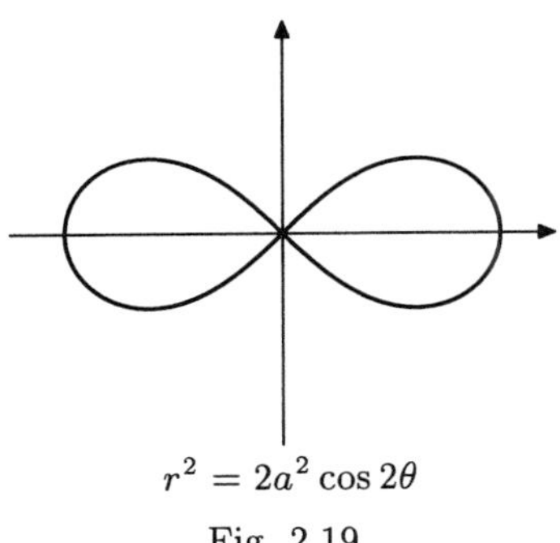

$r^2 = 2a^2 \cos 2\theta$

Fig. 2.19

2.24 The square distance from a point (x, y) on the lemniscate to $(a, 0)$ is

$$(x - a)^2 + y^2 = a^2(2 \cos 2\theta + 1) - 2ax.$$

Similarly the square distance to $(-a, 0)$ is

$$(x + a)^2 + y^2 = a^2(2 \cos 2\theta + 1) + 2ax.$$

The product, which is a difference of two squares, is

$$a^4(2\cos 2\theta + 1)^2 - 4a^2x^2 = a^4(2\cos 2\theta + 1)^2 - 4a^2 2a^2 \cos 2\theta \cos^2 \theta.$$

This is equal, after some trigonometric manipulation, to a^4.

2.25 See Fig. 2.20.

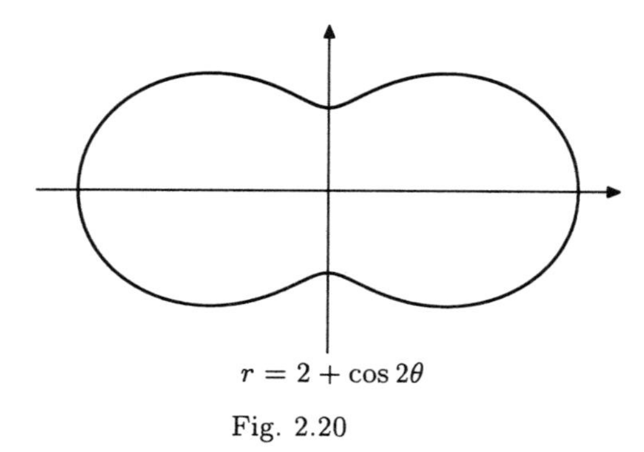

Fig. 2.20

2.26 See Fig. 2.21.

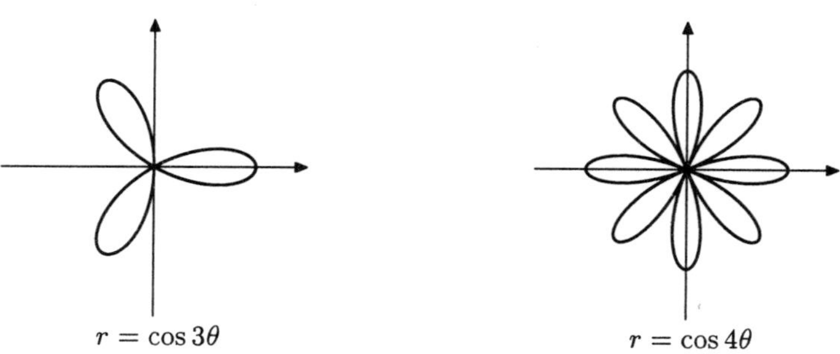

Fig. 2.21

2.27 See Fig. 2.22.

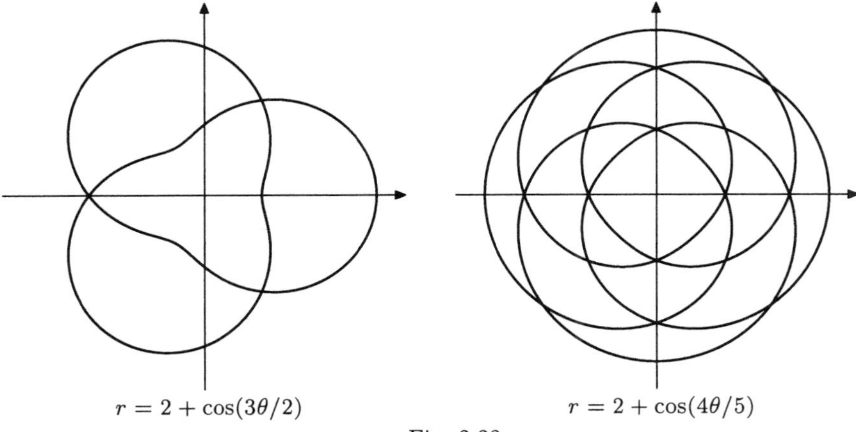

$r = 2 + \cos(3\theta/2)$ $r = 2 + \cos(4\theta/5)$

Fig. 2.22

2.28 $\cos(\theta/2)$ does indeed have period 4π but $\cos^2(\theta/2)$ has period 2π.

2.29 See Fig. 2.23.

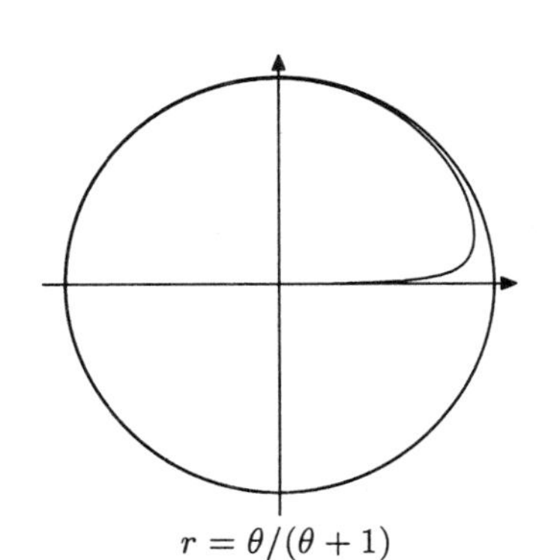

$r = \theta$ $r = \theta/(\theta + 1)$

Fig. 2.23

2.30 (a) $\sqrt{1+4+16} = \sqrt{21} = 4.5825757\ldots$

(b) $2X - 3Y = (-4, 6, -4, -7)$ so $|2X - 3Y| = \sqrt{16+36+16+49} = \sqrt{117} = 10.816654\ldots$.

(c) Let $\angle XOY = \theta$ then $\cos\theta = (2{+}20)/(\sqrt{21}\sqrt{4+4+25}) = 0.83578089\ldots$ Hence $\theta = 33.310066° = 0.58137032$ rad.

(d) Using the formula

$$\Delta XOY = \frac{\sqrt{|X|^2|Y|^2 - (X \cdot Y)^2}}{2}$$

we have that the area is

$$\frac{\sqrt{21 \times 33 - 22^2}}{2} = 7.228\ldots$$

2.33

$$\frac{1}{2}\begin{vmatrix} 0 & 2 & 1 \\ 1 & -3 & 1 \\ -2 & 0 & 1 \end{vmatrix} = -6$$

so the numerical value of the area is 6.

2.34 The area is twice the area of the triangle with vertices $(0,0), (a,b), (c,d)$. That is

$$\begin{vmatrix} a & b & 1 \\ c & d & 1 \\ 0 & 0 & 1 \end{vmatrix} = \begin{vmatrix} a & b \\ c & d \end{vmatrix}.$$

2.35 $x = -1/4$, $y = 7/4$.

2.36 $x = (k_1 - k_2)/(m_2 - m_1)$, $y = (m_2k_1 - m_1k_2)/(m_2 - m_1)$. If $m_1 = m_2$ the lines are parallel.

2.38 $2x_1 + x_3 - x_4 = 2$.

2.39 (i) is of type (a), they all meet in the point $(1, 2, -3)$, (ii) is of type (b) because solving the linear equations

$$\begin{array}{cccc} x & +2y & -3z & = -1 \\ 3x & -y & +2z & = 7 \end{array}$$

gives a general point on $p \cap q$ as $(-t/7 + 13/7, 11t/7 - 10/7, t)$ where $t \in \mathbb{R}$ is arbitrary. So $p \cap q$ is a line through $(13/7, -10/7, 0)$ in the direction $(-1, 11, 7)$. Similarly $q \cap r$ and $r \cap p$ are lines all in the direction $(-1, 11, 7)$. So $p \cap q \cap r = \emptyset$; and (iii) is of type (e), they all meet in the line $(1 - 2t/3, -1 - t/3, t), t \in \mathbb{R}$.

2.40 The unit normals to the planes are $\frac{1}{\sqrt{2}}(1, 0, 1)$ and $\frac{1}{\sqrt{2}}(1, -1, 0)$. So the cosine of the angle between them is $1/2$, that is the angle is $60°$.

2.41 $P = (1/5, -3/5)$ and $|P| = \sqrt{10}/5 = 2/\sqrt{10}$.

3
The Geometry of the Euclidean Plane

3.1 The Life of Euclid

Let no one come to our school, who has
not first learned the elements of Euclid
Notice on the door of Greek Philosophical Schools

We know very little about Euclid, who wrote the *Elements of Geometry*, one of the most famous and influential books of mathematics ever. Most of what we know is due to Proclus (AD 410–485). He wrote, "This man lived at the time of the first Ptolemy, *a Greek King*. For Archimedes, who came immediately after the first, makes mention of Euclid: and, further, they say that Ptolemy once asked him if there was in geometry any shorter way than that of the elements, and he answered that there was no royal road to geometry."

Since Proclus also mentions that Euclid was younger than the first pupils of Plato but older than Eratosthenes and Archimedes we can deduce from this that Euclid flourished about 300 BC. This date agrees well with the fact that Ptolemy reigned from 306 to 283 BC.

A story told about Euclid is that a pupil once asked him, "But what shall I get by learning these things?" Euclid responded by telling his slave, "Give him threepence, since he must make gain out of what he learns." A doubtful fact about Euclid perhaps, but a good story.

3.2 The Euclidean Axioms for the Plane

A point is that which has no part. A line is breadthless length.
Definitions by Euclid

The most essential idea to have come down to us from Euclid is not the geometry itself, beautiful and important as it is, but the idea of deducing true results from a number of self-evident propositions or axioms. The idea of proof in mathematics is logical: deduce a result from one already known.

Actually this is not the way mathematicians work, which is to leap ahead by intuition and fill in any gaps in the proof later.

In our more sophisticated times we now know that the euclidean axioms were incomplete. For example, in the first book of Euclid, an equilateral triangle is constructed by drawing two circles. But how do we know that they meet? We also know that however many axioms we write down, there are true statements which cannot be deduced from them. Saddest of all for those who would like to claim euclidean geometry as unique is the existence of other, or non-euclidean geometries.

We can see from the definitions quoted above that Euclid had great difficulty with points and lines. His descriptions are more poetic statements about meanings of words than true definitions. Nowadays we would consider points and lines as undefined concepts which have an incidence property. So, points belong to or lie on lines and lines pass through points. Here are the euclidean axioms for the plane.

- Through any two points is a unique line.
- A finite part of a line may be infinitely extended.
- There exist circles with any centre and any radius.
- All right angles are equal to one another.
- Let ℓ be a line and P be a point not on ℓ. Then all lines through P meet ℓ, except just one which is parallel to ℓ.

Euclid's fifth proposition, perfect in conception (5).
Crossword clue

Actually the last axiom (the fifth) is another, equivalent to Euclid's and is due to Playfair. For many years mathematicians tried to prove that the fifth axiom, which has a quite different feel to the other four, could be deduced from the previous axioms. The discovery of non-euclidean geometries, which satisfy the first four axioms but not the last, precludes this.

These axioms are easily seen to be true in $\mathbb{R}^2$ using the methods of coordinate geometry. Conversely it is fun to start from these axioms and see where they lead us. We will not get the so-called euclidean plane because, as we saw above, they are insufficient. However, here is a taste of this programme.

■ Two distinct lines meet in at most one point.

Proof Let P and Q be two points containing two lines. Then by axiom 1 these lines are equal. □

If the line ℓ_1 is parallel to ℓ_2 we write $\ell_1 || \ell_2$.

■ Let ℓ_1, ℓ_2, ℓ_3 be three lines with $\ell_1 || \ell_2$ and $\ell_2 || \ell_3$. Then $\ell_1 || \ell_3$.

Proof We can assume that all three lines are distinct. Suppose that ℓ_1 is not parallel to ℓ_3 and let the point P be common to both ℓ_1 and ℓ_3. Then by hypothesis P cannot lie in ℓ_2. By the fifth axiom there is a unique line through P parallel to ℓ_2. So $\ell_1 = \ell_3$, which is a contradiction. □

3.3 Angles and Lines

We know from the last chapter how to define angles between lines. Here are some basic facts about angles and lines which will be useful to quote.

■ **Adjacent angles:** Let three lines meet at O, making angles α, β as shown in Fig. 3.1. Then the points X, O, Z are collinear if and only if $\alpha + \beta = \pi$. □

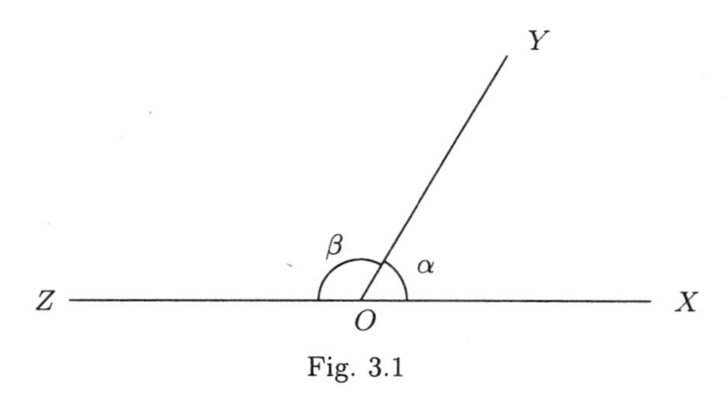

Fig. 3.1

The pair α, β are then called adjacent angles.

■ **Opposite angles:** Consider the angles α, β in Fig. 3.2 where the points Y, O, W are collinear. Then the points X, O, Z are collinear if and only if $\alpha = \beta$. □

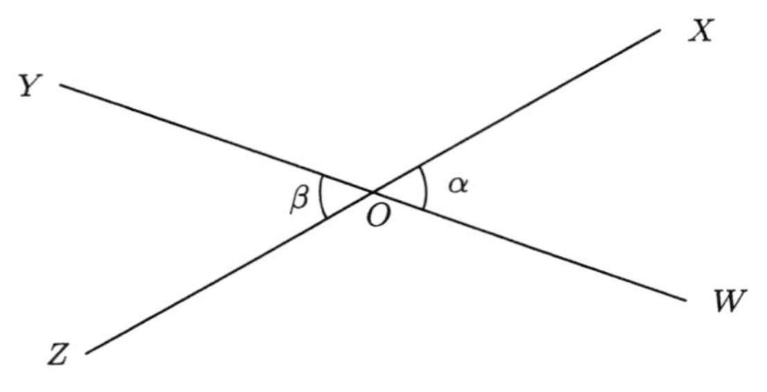

Fig. 3.2

The pair α, β are then called opposite angles.

Exercise 3.1

Consider Fig. 3.3 where the points Y, O, W are collinear.

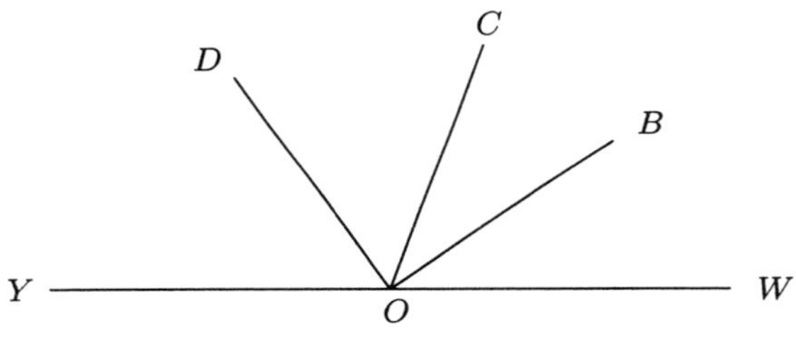

Fig. 3.3

Suppose that OB bisects $\angle WOC$ and OD bisects $\angle COY$. Show that $\angle BOD$ is a right angle.

Alternate and corresponding angles: The lines ℓ_1 and ℓ_2 are cut by a further line making the angles shown in Fig. 3.4. Then the lines ℓ_1 and ℓ_2 are parallel if and only if $\alpha = \beta$ or $\alpha = \gamma$. □

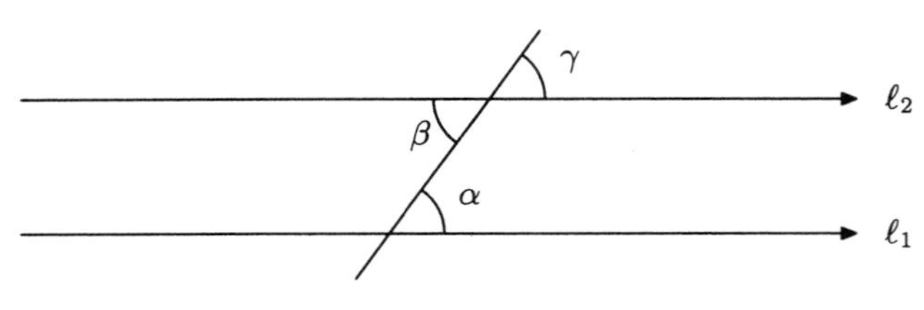

Fig. 3.4

The pair α, β are then called alternate angles and the pair α, γ are called corresponding angles.

3.4 Some Basic Facts about Triangles

... a bunch of dead Greek geeks like Pythagoras, Archimedes and Hypotenuse.
Character from Doonsbury strip cartoon

A triangle is defined by three non-collinear points, its *vertices*. Or, dually, it is defined by three lines which are not concurrent or parallel. (Three or more lines are said to be *concurrent* if they all have a point in common.) If A, B, C are the vertices of a triangle then the corresponding interior angles are denoted by α, β, γ. We sometimes say that α is the angle *subtended* by BC at the point A.

The corresponding exterior angles are denoted by $\hat{\alpha}, \hat{\beta}, \hat{\gamma}$. Clearly $\hat{\alpha} = \pi - \alpha$ etc. Fig. 3.5 illustrates the interior and exterior angles of a triangle.

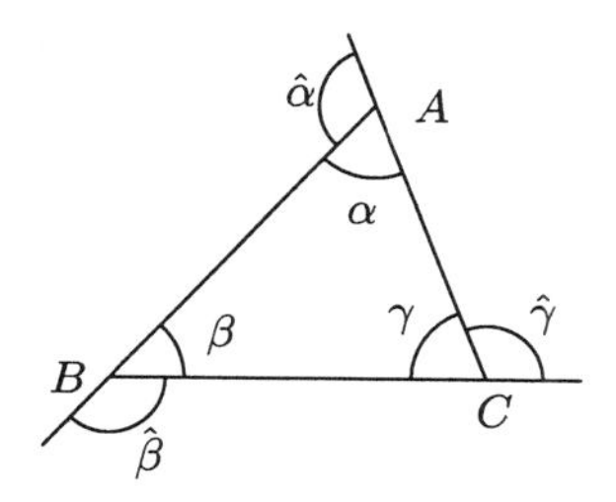

Fig. 3.5

As usual, angles in a triangle or other polygon are measured anticlockwise. There is a choice of extension to indicate the exterior angles but both choices give the same answer. As far as is possible we will chose the extension so that an anticlockwise rotation is *towards* the triangle.

Triangles are divided into types by their angles.

scalene	all three angles distinct
isosceles	two angles equal
equilateral	all three angles equal
right-angled	one angle equal to $\pi/2$

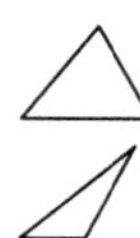

acute-angled all three angles less than $\pi/2$

obtuse-angled one angle greater than $\pi/2$

The first three names actually mean unequal sides, two equal sides and three equal sides respectively. But as we will see below the above definitions involving angles are equivalent.

The side of an isosceles triangle abutting the two equal angles is called the *base* of the isosceles triangle. The side of a right angled triangle opposite the right angle is called its *hypotenuse*.

■ An exterior angle of a triangle is the sum of the opposite interior angles. In symbols

$$\hat{\alpha} = \beta + \gamma.$$

The sum of the interior angles of a triangle is two right angles. In symbols

$$\boldsymbol{\alpha + \beta + \gamma = \pi.}$$

Proof If we divide the exterior angle at C by a line parallel to BA then we see that $\hat{\gamma}$ is made up of an alternate angle α and a corresponding angle β.

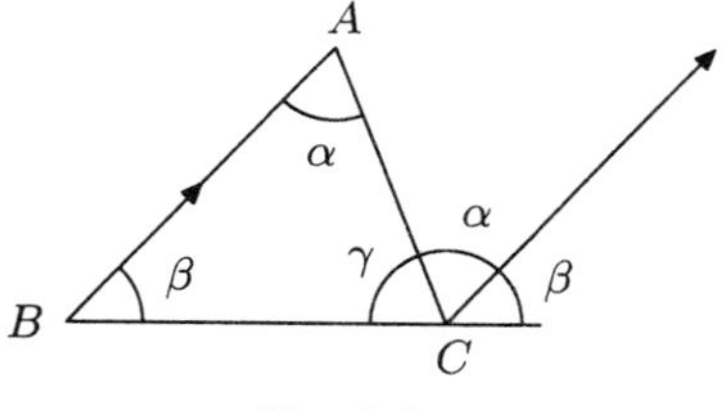

Fig. 3.6

The equation $\alpha + \beta + \gamma = \pi$ can be seen from Fig. 3.6. □

Notice that this was proved by constructing the unique line through C parallel to BA. In non-euclidean geometries this will not be possible and in fact for such triangles the sum of the interior angles is not two right angles.

Exercise 3.2

Show that the sum of the exterior angles of a triangle is 2π.

Exercise 3.3

Show that if two of the corresponding angles of two triangles are equal then so is the third.

Exercise 3.4

In a right-angled triangle show that
(i) the right angle is greatest;
(ii) the sum of the remaining two angles is a right angle; and
(iii) each of these angles is acute.

Exercise 3.5

Show that at least two of the angles in a triangle are acute.

Exercise 3.6

From a given point off a given line, show that exactly one straight line can be drawn perpendicular to the given line.

Exercise 3.7

The angles of a triangle are π/p, π/q and π/r where p, q, r are integers. What values can p, q, r take?

Exercise 3.8

The external angles of a triangle are $\hat{\alpha}$, $\hat{\beta}$ and $\hat{\gamma}$. Show that if $\hat{\alpha}+\hat{\beta} = 3\hat{\gamma}$ then the triangle is right-angled.

Exercise 3.9

* Let $\boldsymbol{\Delta}$ be the set of points (x, y, z) in $\mathbb{R}^3$ such that x, y, z are the side lengths of a euclidean triangle. If $\boldsymbol{p}$ is the plane $z = c$, describe $\boldsymbol{\Delta} \cap \boldsymbol{p}$ and its subsets of equilateral, isosceles, right-angled, acute and obtuse triangles.

3.5 General Polygons

A polygon with p *vertices* or a *p-gon* is defined by p points (the vertices), $V_1, V_2, \ldots, V_p$. The *edges* are the line segments V_1V_2, $V_2V_3, \ldots, V_{p-1}V_p$, V_pV_1.

Here are two examples: an irregular convex pentagon and a pentangle or pentagram (Fig. 3.7).

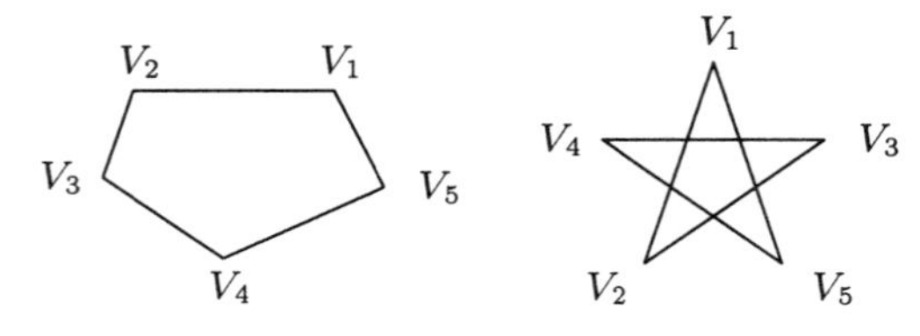

Fig. 3.7

A polygon is *convex* if the line interval drawn from a point on an edge to a point on any other edge is contained inside the polygon. A polygon, such as the pentagram, which intersects itself is called *singular*. All convex polygons are non-singular although the reverse is not true. We shall generally restrict ourselves to convex polygons except for interesting special cases like the pentagram.

A *p-polygon* has a special name if p is small. We have already met triangles, $p = 3$ and pentagons, $p = 5$. Other names are *quadrilaterals*, $p = 4$; *hexagons*, $p = 6$; *heptagons* or *septagons*, $p = 7$; *octagons*, $p = 8$; *nonagons*, $p = 9$; and *decagons*, $p = 10$.

Quadrilaterals are divided into a number of types as follows.

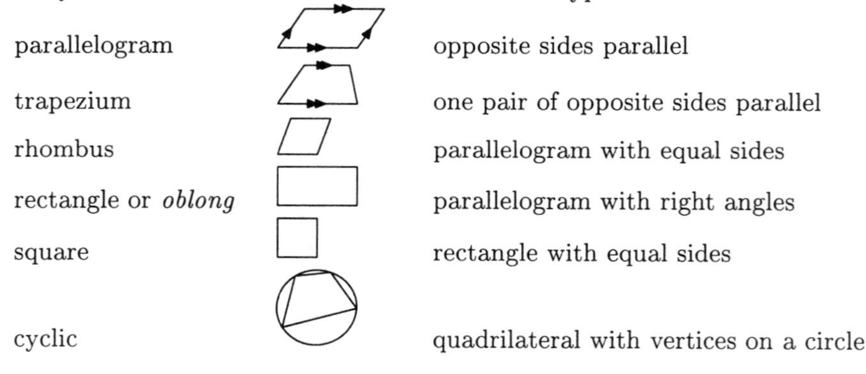

Exercise 3.10

How would you describe a cyclic trapezium which is not a rectangle?

Exercise 3.11

* Draw a Venn diagram to indicate boolean relationships between the various types of quadrilateral.

A polygon is *regular* if the vertices are regularly distributed around a circle. The Schläfli symbol for a regular p-gon is $\{p\}$. Fig. 3.8 gives $\{p\}$'s for $p = 3$ up to 10. Note that a regular quadrilateral is a square.

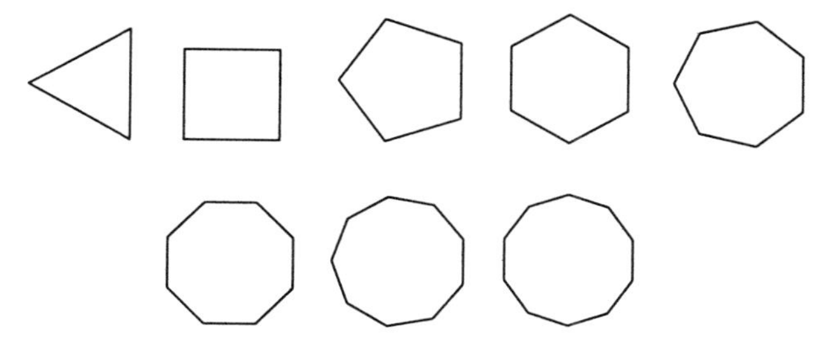

Fig. 3.8

The pentagon is the shape of the US military building. The borders of France roughly form a hexagon, as do the cells of a honeycomb. The heptagon is the shape of British 20p and 50p pieces. It is also the smallest regular polygon which cannot be constructed with a ruler and compass only. (See the next chapter.)

A *star* p-gon also has its vertices situated regularly around a circle, but the edges jump along a fixed number q of vertices. The Schläfli symbol is $\{\frac{p}{q}\}$. So for example the symbol for the pentangle is $\{\frac{5}{2}\}$. Some examples of star polygons are given in Fig. 3.9.

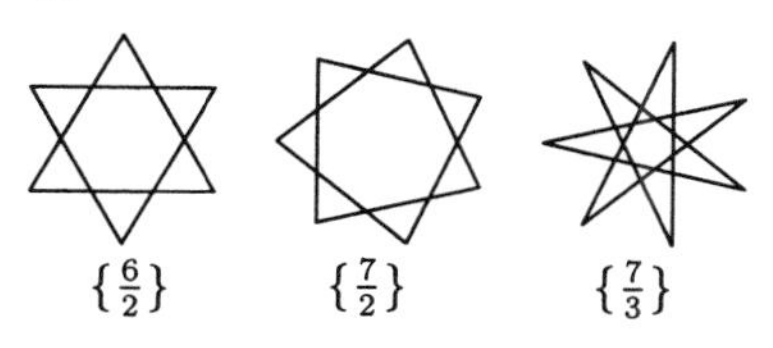

Fig. 3.9

The pentangle, as oriented above, is the star on the US flag and is a traditional symbol for a person (head, two arms, two legs). An inverted pentangle ⛧ is a symbol for the devil (two horns, two ears, a beard). The Star of David, $\{\frac{6}{2}\}$, is the star on the Israeli flag.

Exercise 3.12

Draw the octagram $\{\frac{8}{3}\}$ and the decagram $\{\frac{10}{3}\}$.

Formulæ for the sums of the interior and exterior angles of a polygon can be given. For example in the pentagon illustrated in Fig. 3.10 the internal angles are $\alpha_1,\ \alpha_2,\ \alpha_3,\ \alpha_4,\ \alpha_5,$ and the exterior ones are $\hat{\alpha}_1,\ \hat{\alpha}_2,\ \hat{\alpha}_3,\ \hat{\alpha}_4,\ \hat{\alpha}_5$.

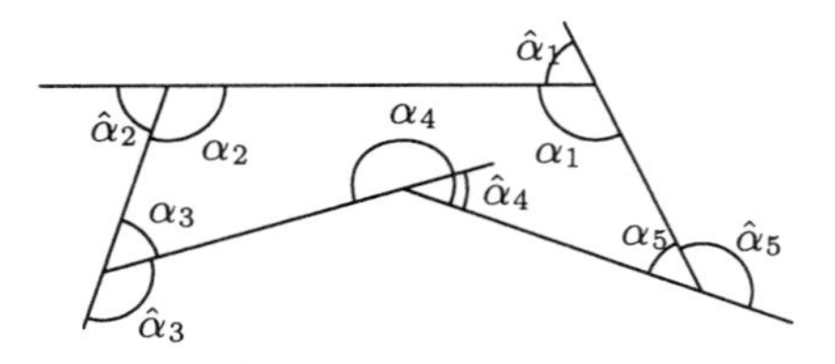

Fig. 3.10

Then

$$\hat\alpha_1 + \hat\alpha_2 + \hat\alpha_3 + \hat\alpha_4 + \hat\alpha_5 = 2\pi$$

and

$$\alpha_1 + \alpha_2 + \alpha_3 + \alpha_4 + \alpha_5 = 3\pi.$$

Note that $\hat\alpha_4$ is negative.

To get the general result we need the following generalisation of convex. A polygon is called *star convex* from an interior point O if the segment from O to any point on the polygon does not meet the polygon elsewhere.

■ Let $V_1, \ldots, V_p$ be the vertices of a p-gon star convex from O. Suppose that the interior angles are $\alpha_1, \ldots, \alpha_p$ and the exterior angles are $\hat\alpha_1, \ldots, \hat\alpha_p$. Then

$$\sum \hat\alpha_i = 2\pi$$

and

$$\sum \alpha_i = (p-2)\pi.$$

Proof Break up the interior of the polygon into the p triangles

$$OV_1V_2, OV_2V_3, \ldots, OV_pV_1.$$

Then the sum of the angles in the p triangles is πp which is $s - 2\pi$ where $s = \sum \alpha_i$. The formula for $\sum \hat\alpha_i$ follows from the identity $\hat\alpha_i = \pi - \alpha_i$. □

Exercise 3.13

Find the interior angle of a regular pentagon and a pentangle.

Exercise 3.14

Show that the interior angle of a $\{p\}$ is $(1 - \frac{2}{p})\pi$.

3.6 Congruences and Similarities

The objects of the euclidean plane (triangles, squares, circles, etc.) are the basic building blocks of euclidean geometry. The transformations of the plane which preserve their shape are *congruences/isometries* (distances are preserved) or *similarities* (combinations of isometries and scale changes). For example if someone told you that they had a triangle with sides 3, 4 and 5 centimetres then you could draw a copy of their triangle and the two triangles would be congruent or isometric. Moreover both triangles would be right-angled and as far as euclidean geometry is concerned they would be equivalent. Now suppose they had said that the triangle had sides 3, 4 and 5 inches long and you were not sure what an inch was but you thought it was about 2.5 centimetres. Then you could draw a triangle with sides 7.5, 10 and 12.5 centimetres. The two triangles would not be congruent any more but they would be similar. That is the length of their sides would be in proportion. Again as far as euclidean geometry is concerned they would be equivalent since size is only a matter of choosing units.

In the first case there is a euclidean isometry taking one to the other and in the second case a similarity or scale change taking one to the other.

An isometry can either be *direct* (preserving orientation) or *indirect* (reversing orientation). Direct isometries of the euclidean plane (other than the identity) are either *translations* or *rotations*. The effect of these on the line segment AB are illustrated in Fig. 3.11.

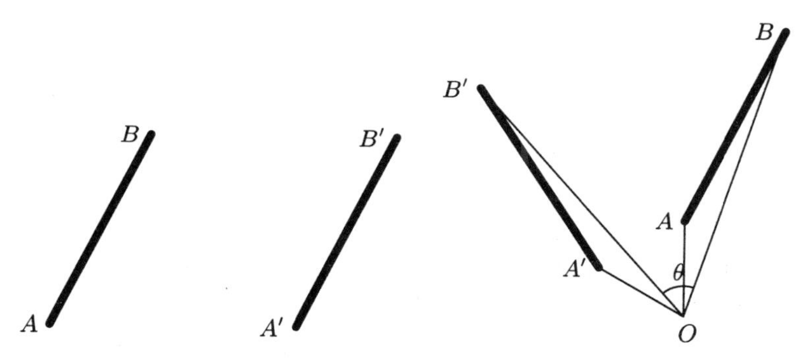

Fig. 3.11 Translation through $\overrightarrow{AA'} = \overrightarrow{BB'}$ Rotation about O through $\theta = \angle BOB'$

We have met translations before in Chapter 2. A rotation about O through θ keeps O fixed and sends A to A' where $OA = OA'$ and $\angle AOA' = \theta$.

An indirect isometry of the euclidean plane is either a *reflection* in a line or a combination of a reflection in a line followed by a translation parallel to the line. This last kind of isometry is sometimes called a *glide reflection*. Their effect on a line segment AB is illustrated in Fig. 3.12.

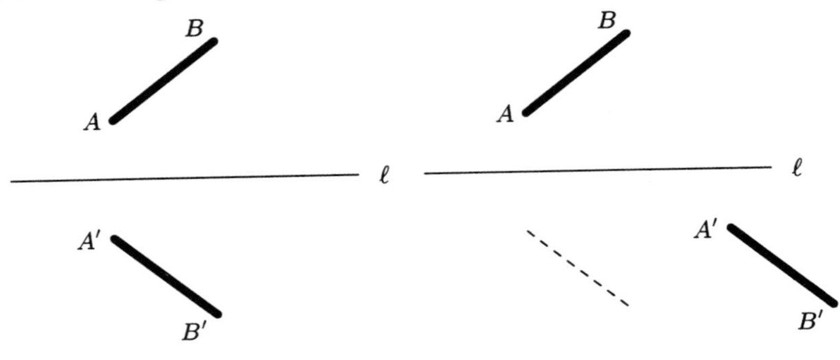

Fig. 3.12

However a more common effect of a glide reflection is illustrated by footprints in the sand (Fig. 3.13).

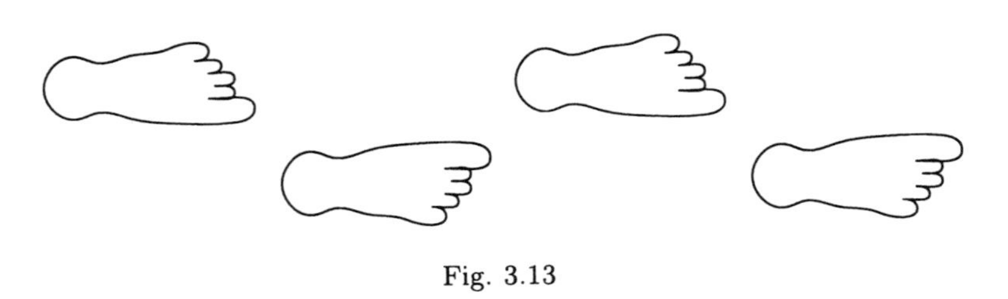

Fig. 3.13

Exercise 3.15

Discuss how you might prove the assertions above.

A similarity is a euclidean transformation. The set of all euclidean transformations forms the euclidean group. Euclidean geometry is the set of properties invariant under the action of the euclidean group according to Klein's definition of a geometry.

There are a number of conditions which determine if two triangles are congruent (related by a congruence).

▮ **SSS** Three corresponding sides are equal.

▮ **SAS** Two corresponding sides and the included angle between the sides are equal.

▮ **SS⊥** Both triangles are right-angled and two corresponding sides are equal.

■ **AAS** Corresponding angles and one corresponding side are equal.

Proof We will only prove the first of these conditions using the traditional methods of euclidean geometry. The idea is to construct a triangle with the given side lengths and show that the result is unique up to isometry.

Let BC be one side of the triangle and draw two circles; one having centre B and radius BA, the other having centre C and radius CA. Because we are told that a triangle with these side lengths exists then we know that the circles intersect in two points A and A' say. (You can see at this point how important it is to have an axiom on the intersection of circles.)

One triangle ABC is the reflection of the triangle $A'BC$ in the line BC but otherwise it is unique since the original line segment BC could have coincided with any other of the same length. □

Exercise 3.16

Show that the adjective included is necessary in **SAS** by constructing two non-congruent triangles with two equal sides and one equal angle. What happens if the triangle contains a right angle?

■ If two triangles have corresponding angles equal then their sides are in proportion. So if the triangles are ABC and $A'B'C'$ then

$$\frac{AB}{A'B'} = \frac{BC}{B'C'} = \frac{CA}{C'A'}.$$

Proof By an isometry move ABC so that $A = A'$ and AB lies on the line $A'B'$ and in the same direction. Make a scale change so that A stays fixed and B' becomes B. All the triangle sides change in proportion. Moreover ABC and $A'B'C'$ are now congruent by **ASA**. □

We say that two triangles with equal angles are *similar* and write **AAA** for the condition above.

Note that any given condition involving two equal angles of triangles is the same as three equal angles since their sum is the constant π.

■ Consider the triangles ABC and $AB'C'$ illustrated in Fig. 3.14 where B' lies in the segment AB and C' lies in the segment AC. Then ABC and $AB'C'$ are similar if and only if $BC||B'C'$.

Proof If the triangles are similar then $\angle AB'C' = \angle ABC$ and $BC||B'C'$ by corresponding angles. The converse is just as easy. □

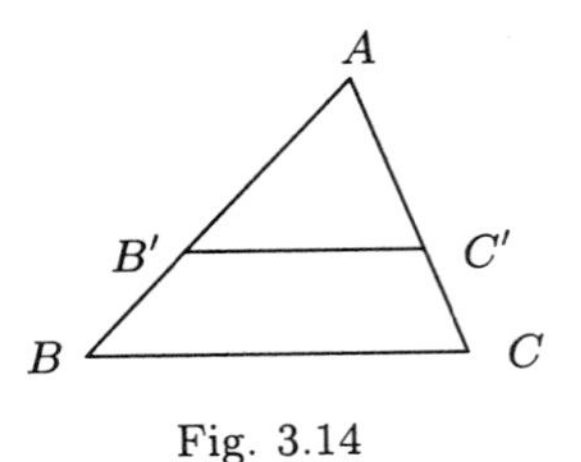

Fig. 3.14

3.7 Isosceles Triangles

It might seem extravagant to occupy a whole section with the properties of a special kind of triangle, but isosceles triangles are so useful and keep cropping up in so many situations that this treatment is justified. We start off by showing, among other things, that having two equal sides is an equivalent definition.

Let ABC be an isosceles triangle with $\beta = \gamma$ and let A' be the midpoint of BC (Fig. 3.15).

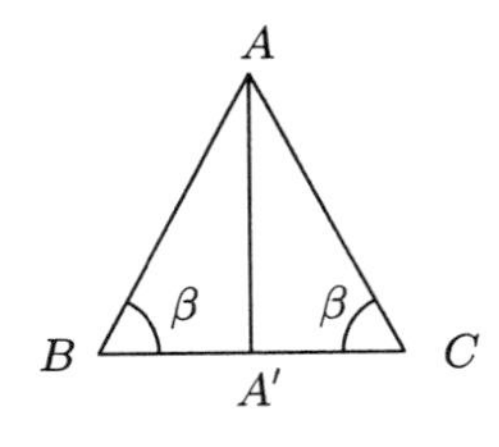

Fig. 3.15

■ In the isosceles triangle in Fig. 3.15, the following holds:

$$AB = AC \text{ and } AA' \perp BC.$$

Moreover for any triangle, if any one of the two above conditions hold then ABC is isosceles.

Proof Consider the line, ℓ, through A' perpendicular to BC. Suppose that ℓ meets AB at X. Then the triangles $XA'B$ and $XA'C$ are congruent by **SAS** or **SS⊥** (the included angle is $\pi/2$.) So $\angle XBA' = \angle XCA'$ and hence $X = A$.

In other words, the triangles $AA'B$ and $AA'C$ are congruent and so $AB = AC$ and $AA' \perp BC$.

The converse is even easier because if say $AB = AC$ then the triangles $AA'B$ and $AA'C$ are congruent by **SSS** and if $AA' \perp BC$ they are congruent by **SS**$\perp$.

□

■ If two sides of a triangle are unequal then the greater side has the greater angle opposite.

Proof

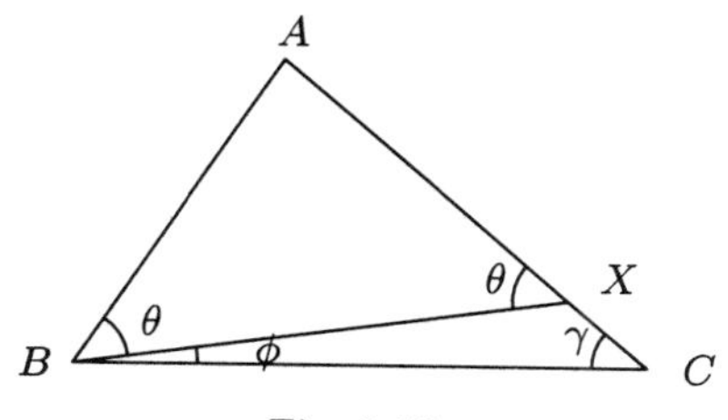

Fig. 3.16

Suppose in the triangle ABC in Fig. 3.16, $AC > AB$. Then let X be on AC so that $AB = AX$. Using the isosceles triangle ABX we see that $\angle AXB = \angle ABX = \theta$ say. Let $\angle XBC = \phi$, then $\phi + \gamma = \theta$ and so $\theta > \gamma$. A fortiori $\angle ABC = \theta + \phi > \gamma = \angle ACB$.

□

■ **The triangle inequality:** *For any triangle ABC, the inequality $BA + AC > BC$ holds.*

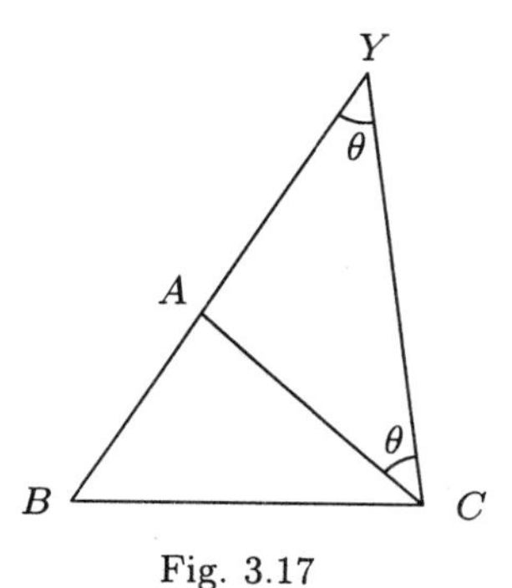

Fig. 3.17

Proof Let Y be a point on the line BA so that $AY = AC$ (see Fig. 3.17). Then $BY = BA + AC$. Because ACY is isosceles, $\angle AYC = \angle ACY < \angle BCY$, so $BC < BY = BA + AC$.

□

Isosceles triangles can be used by sailors to estimate their distance in a boat from an object, see Fig. 3.18.

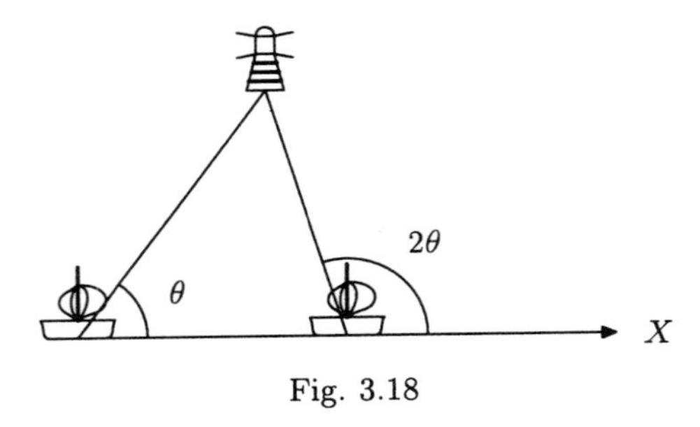

Fig. 3.18

The idea is to sail on a fixed bearing towards X say until the angle made with the object is doubled. Then the distance travelled is the distance you are now away.

3.8 Circles

A circle is the set of points which are a fixed distance from a fixed point. The fixed point is called the centre and the fixed distance is called the radius (plural, radii). The diameter is twice the radius. By an abuse of notation, a radius, at a point X of the circle is also the name for any segment with endpoints consisting of the centre and X. A diameter also means a segment joining two points on the circle which passes through the centre. A chord is any segment joining two points on the circle. A tangent at a point T on the circle is a line through T which is orthogonal to the radius at T. See Fig. 3.19.

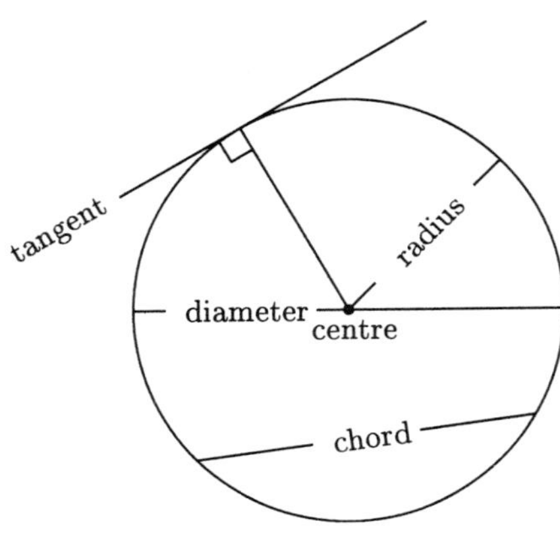

Fig. 3.19

The points of the plane not on the circle are divided into the inside, where the distance from the centre is less than the radius, and the outside where it is greater. Two radii of a circle determine an isosceles triangle. This simple fact will be a constant leitmotiv in the discussions ahead.

■ A line in the plane, if it meets a circle at all, meets it either in two points and so defines a chord or is a tangent and meets it in one point.

Proof Let T be the nearest point on the line to the centre O of the circle. If OT is bigger than the radius then the line will not meet the circle. If OT is equal to the radius then T lies on the circle and the line is a tangent which only meets the circle at T. If OT is less than the radius then there will be two points on the line, A, B say, equally spaced on either side of T for which the distances OA and OB are equal to the radius and hence lie on the circle. □

■ Let A be any point outside of the circle. Then there are two tangents from A to the circle. Moreover these are bisected by the line through A and the centre. Conversely any point O on the angle bisector of two lines through A is the centre of a unique circle tangent to the two lines.

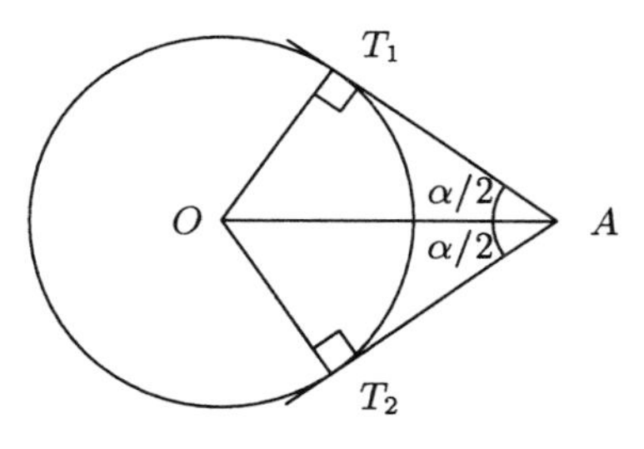

Fig. 3.20

Proof Construct two congruent right-angled triangles OAT_1 and OAT_2 with OA as common hypotenuse and $OT_1 = OT_2$ equal to the radius (Fig. 3.21). Then AT_1 and AT_2 are the required tangents and the angles $\angle OAT_1$ and $\angle OAT_2$ are equal. Conversely if O lies on the angle bisector of two lines ℓ_1 and ℓ_2, let T_1, T_2 be the nearest points to O on these lines. Then comparing the two congruent right-angled triangles OAT_1 and OAT_2 we see that $OT_1 = OT_2$ and there is a unique circle with this radius which has ℓ_1 and ℓ_2 as tangents. □

The following results illustrate the usefulness of isosceles triangles made up of radii.

■ The perpendicular bisector of any chord of a circle passes through the centre of that circle.

Proof Let the chord be AB and the centre O. Then by properties of the isosceles triangle AOB the perpendicular bisector of AB passes through O. □

■ Given three non-collinear points there is a unique circle through them.

Proof

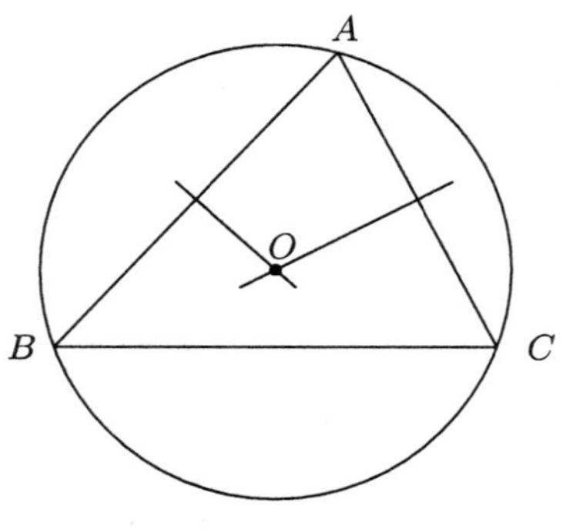

Fig. 3.21

Let the three non-collinear points be A, B, and C. The perpendicular bisectors of AB and AC are not parallel (otherwise A, B, C would be collinear) and hence meet in a unique point O, the centre of the circle through A, B, C (Fig. 3.21). □

Note that O also lies on the perpendicular bisector of BC.

■ Two distinct circles meet in at most two points. □

Exercise 3.17

Find the centre and radius of the circle through the points $A = (-1, 0)$, $B = (0, 1)$ and $C = (1, 1)$.

■ Two chords of a circle are equidistant from the centre if and only if they are equal in length.

Proof A chord of a circle determines an isosceles triangle with apex at the centre of the circle. Clearly these are congruent if and only if the chords are equal. □

■ Let A, A', B, C be points on a circle with centre O. Suppose that A, A' are on opposite sides of the chord BC and that the chord subtends angles α, α' at A, A'. Then

$$\alpha + \alpha' = \pi.$$

If the centre O is on the same side of the chord as A, then the angle subtended by the chord at O is 2α.

Proof

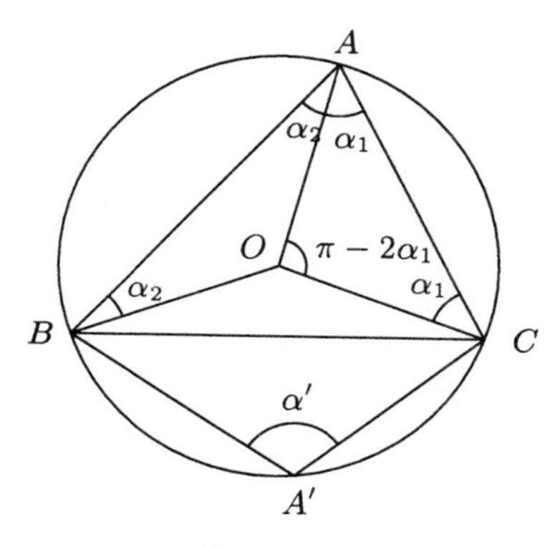

Fig. 3.22

Fig. 3.22 shows lots of isosceles triangles with sides made out of radii, OCA and OAB to name but two. Now $\angle BAC = \alpha_1 + \alpha_2 = \alpha$ and $\angle COA = \pi - 2\alpha_1$. Similarly $\angle AOB = \pi - 2\alpha_2$. So the angle subtended by BC at O is

$$\angle BOC = 2\pi - (\pi - 2\alpha_1) - (\pi - 2\alpha_2) = 2(\alpha_1 + \alpha_2) = 2\alpha.$$

The *external* angle is $\angle COB = 2\pi - 2\alpha$ (note order) and so the angle subtended on the opposite side at A' is half that, i.e. $\alpha' = \pi - \alpha$. □

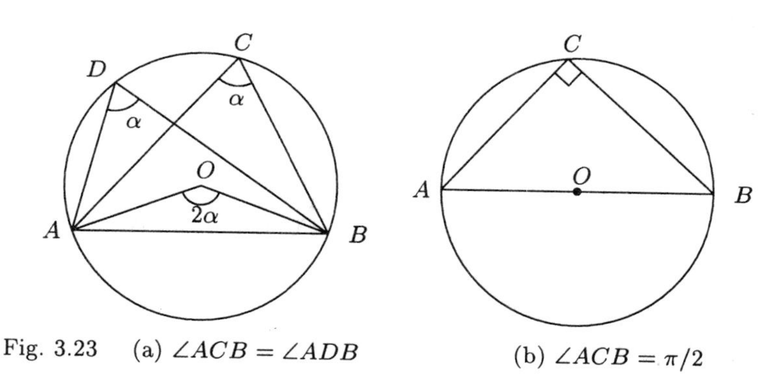

Fig. 3.23 (a) $\angle ACB = \angle ADB$ (b) $\angle ACB = \pi/2$

■ The angle subtended on one side of a circle by a chord is constant and in particular if the chord is a diameter then the angle subtended at the circle is $\pi/2$ (Fig. 3.23). □

■ A quadrilateral is cyclic if and only if opposite angles add to π (Fig. 3.24). □

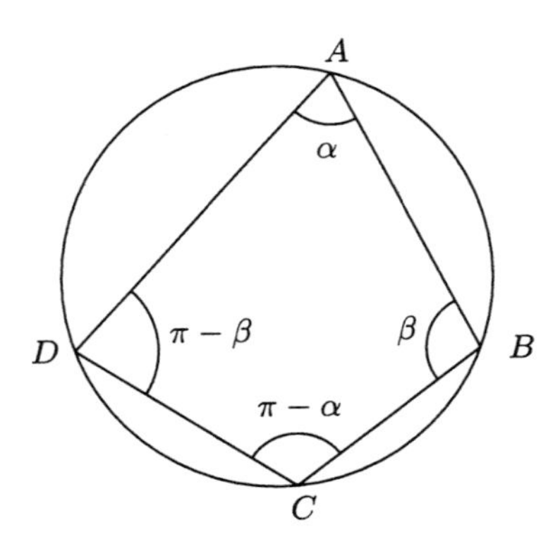

Fig. 3.24 Angles in a cyclic quadrilateral

Exercise 3.18

Let $ABYX$ be a cyclic quadrilateral and let the diagonals XB and YA meet at N.
If AB is parallel to XY, prove that $\angle ANX = 2\angle ABX$ and $NX = NY$.
If XA and YB when extended outside the circle meet at K and if AB is parallel to XY, prove that $KX = KY$.

Exercise 3.19

Show that a quadrilateral is a parallelogram if and only if opposite angles are equal.

Exercise 3.20

Show that a quadrilateral is a parallelogram if and only if each diagonal bisects the other.

Exercise 3.21

Recall that a rhombus is a parallelogram with equal sides. Show that a parallelogram is a rhombus if and only if the diagonals bisect the corner angles and if and only if they meet at right angles.

3.9 Triangles and their Centres

Triangles have a number of naturally occurring lines of symmetry. These come in sets of three lines. By a natural miracle they are concurrent. The points where they meet constitute the various centres of a triangle. Their definitions and properties will now be considered.

The Circumcentre O

We have already seen that three non-collinear points A, B, C lie on a unique circle. This circle is called the *circumcircle* of the triangle ABC. The centre O of the circumcircle is called the *circumcentre* of the triangle ABC.

The perpendicular bisectors of the sides AB, BC, CA all meet at O.

The Centroid G

Let A, B, C be the vertices of a triangle. The midpoints of AB, BC, CA will be denoted by C', A', B'. The lines AA', BB', CC' are called the *medians* of the triangle (Fig. 3.25).

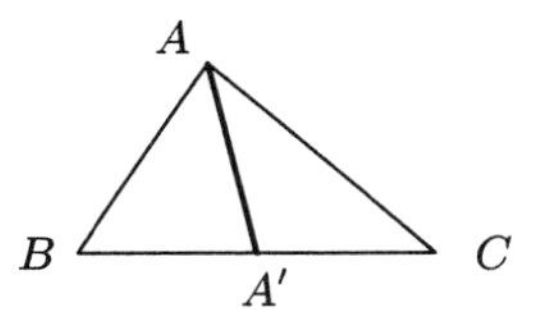

Fig. 3.25 A median

■ The medians of a triangle are concurrent and meet at the point G called the *centroid* of a triangle (Fig. 3.26).

Proof It is convenient to prove this using vector methods. The midpoint A' is given by the formula $A' = (B + C)/2$ with similar formulæ for the other midpoints. It follows that the point G given by

$$G = (A + B + C)/3$$

can be written as $G = A/3 + (B + C)/2 \times 2/3$ and so lies on the median AA'. By symmetry it lies on the other medians which are therefore concurrent at G.

□

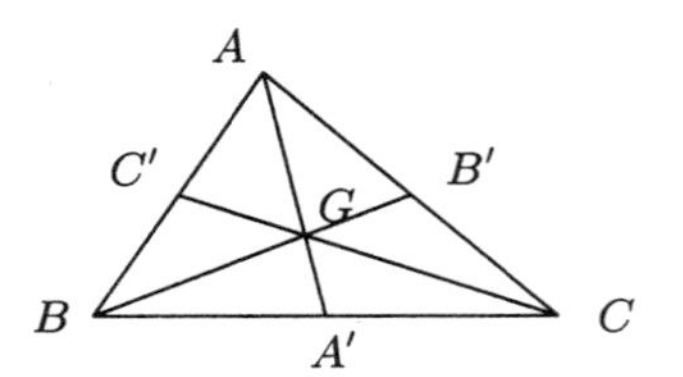

Fig. 3.26 The centroid of a triangle

Other names for the centroid are barycentre or centre of gravity. This is because it is the centre of gravity of three equal masses placed at A, B, C. It is also the centre of gravity of a uniform flat triangular plate with vertices A, B, C.

■ The centroid of a triangle lies one third up the medians. □

Exercise 3.22

A triangle ABC has medians AA', BB' and CC'. The coordinates of ABC are $(0,1), (t,0)$ and $(s,0)$ respectively. Find the coordinates of A', B', C' and the centroid G.

The Orthocentre H

Let A, B, C be the vertices of a triangle. The point D is the nearest point of the line BC to the point A. It is the point where a line through A perpendicular to BC meets BC and is called the *foot* of the perpendicular or *altitude* AD. The points E, F are defined similarly. They are the feet of the altitudes BE, CF respectively.

■ The altitudes AD, BE, CF of a triangle are concurrent at a point H called the *orthocentre*.

Proof Again vector methods are used. Take the circumcentre O to be the origin. Then the vectors A, B, C all have the same length, equal to the radius of the circumcircle. Let $H = A + B + C$. Then we have

$$\begin{aligned}(H - A) \cdot (B - C) &= (B + C) \cdot (B - C) \\ &= B \cdot B - C \cdot C \\ &= |B|^2 - |C|^2 \\ &= 0.\end{aligned}$$

So the line HA is perpendicular to the line BC and hence H lies on the altitude through A. By symmetry H lies on the other altitudes which are therefore concurrent at H. □

Note that if the angles at B or at C are not acute then the point D will lie outside of the interval BC. This is indicated in the right side of the Fig. 3.27.

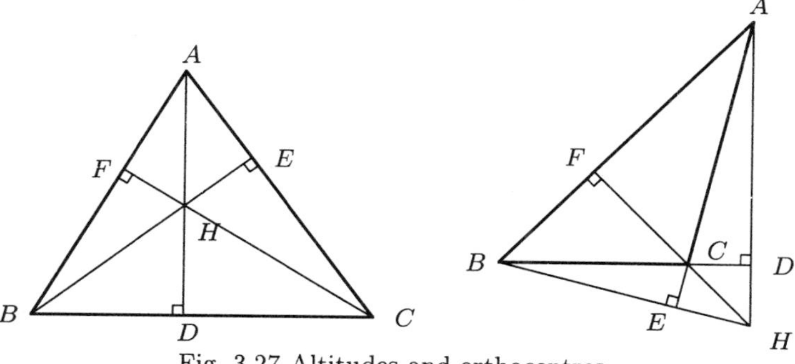

Fig. 3.27 Altitudes and orthocentres

A glance at Fig. 3.27 will reveal that A is the orthocentre of the triangle BHC. Similarly B, C are the orthocentres of the triangle CHA, AHB, respectively.

Exercise 3.23

The coordinates of ABC are $(0,1),(t,0)$ and $(s,0)$ respectively. Let the feet of perpendiculars from A, B, C be at D, E, F respectively and let the orthocentre be at H. Find the coordinates of D, E, F, H.

■ The circumcentre O, the centroid G and the orthocentre H all lie on a line called the *Euler line* (Fig. 3.28). The centroid G is a third of the way along OH.

Proof With O as origin, $H = A + B + C$. Since $G = (A + B + C)/3$, this is what we want. □

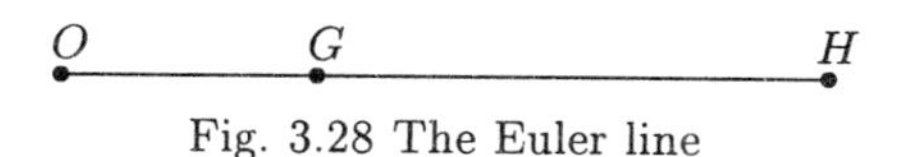

Fig. 3.28 The Euler line

Exercise 3.24

The coordinates of ABC are $(0,1),(t,0)$ and $(s,0)$ respectively. Show that the coordinates of the circumcentre O are $(\frac{s+t}{2}, \frac{1+st}{2})$ and verify that O, G, H are collinear, where G is the centroid and H is the orthocentre.

The Incircle, Outcircles and their Centres, I, I_A, I_B, I_C

This section will repeatedly use the following symmetry property of the circle, namely, if AT and AS are two tangents to a circle from a point A outside then the angle bisector of $\angle TAS$ passes through the centre of the circle. Moreover given any point X on the angle bisector there is a unique circle with centre at that point X which is tangent to the two lines AT and AS.

The largest circle which can be inscribed inside the triangle ABC is called the *incircle*. Its centre is denoted by I and called the *incentre*. Since AB and AC are tangents to the incircle from A it follows that the incentre lies on the angle bisector of $\angle BAC$. We therefore have:

■ The (interior) angle bisectors of a triangle ABC are concurrent and meet at the incentre I. □

Exercise 3.25

Show that the incentre I is the unique point on the bisector of $\angle BAC$ such that $\angle BIC = \pi/2 + \alpha/2$.

Let the bisector of the external angle at A meet the bisector of the internal angle at B in the point I_B. Then because I_B lies on the bisector of the external angle at A there is a unique circle, outside of the triangle with centre I_B, which is tangent to the lines BA and AC. But I_B also lies on the bisector of the internal angle at B and so, in addition, this circle is tangent to the line BC (Fig. 3.29).

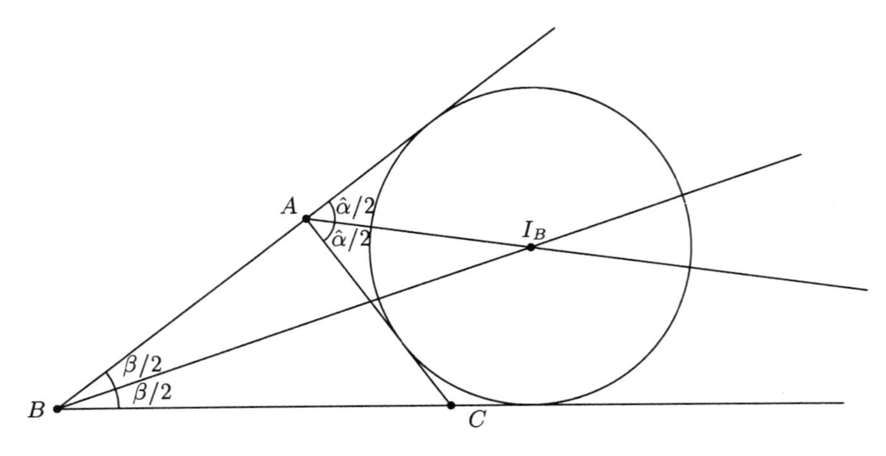

Fig. 3.29

The resulting circle is one of the three *outcircles* of the triangle ABC (Fig. 3.30).

Their centres, called *outcentres*, are denoted by I_A, I_B and I_C. Note that since I lies on the internal bisector of the angle at A the points A, I and I_A are collinear. Similarly the points B, I, I_B and C, I, I_C are collinear.

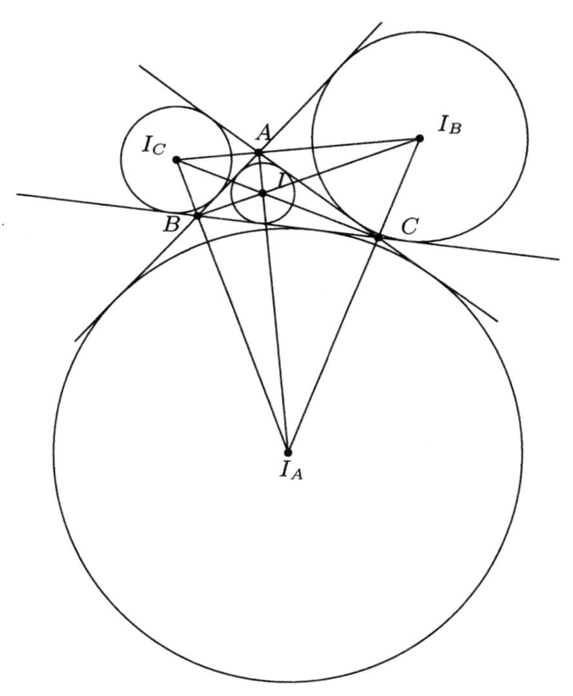

Fig. 3.30 Incircle and outcircles of a triangle

Exercise 3.26

Show that I is the orthocentre of the triangle $I_A I_B I_C$.

3.10 Metric Properties of Triangles

Associated to any triangle are various numerical quantities such as the lengths of the sides, the radii of various circles, trigonometric functions of the angles and so on. These are all related by various formulæ which we will develop in this section. Since a triangle is determined by the length of its sides, written $a = BC, b = CA$ and $c = AB$, we shall attempt to relate everything to these quantities. The first formula is the famous *cosine* formula. Recall that α, β, γ denote the internal angles of a triangle with vertices A, B, C. Then

$$\cos\alpha = \frac{b^2 + c^2 - a^2}{2bc}.$$

To prove this we use vector methods. Assume that $A = O$ is the origin. Then

$$\begin{aligned} a^2 &= (B - C) \cdot (B - C) \\ &= B \cdot B + C \cdot C - 2B \cdot C \\ &= b^2 + c^2 - 2bc\cos\alpha. \end{aligned}$$

Now a little algebraic juggling gives the cosine formula.

Exercise 3.27

When does the quadratic equation in x

$$x^2 - 2bx\cos\alpha + b^2 - a^2 = 0$$

have equal roots? Interpret the answer geometrically.

We can get a *sine* formula from the cosine formula as follows:

$$\begin{aligned} \sin^2\alpha &= 1 - \cos^2\alpha \\ &= \frac{-a^4 - b^4 - c^4 + 2a^2b^2 + 2b^2c^2 + 2c^2a^2}{4b^2c^2} \\ &= \frac{(a+b+c)(-a+b+c)(a-b+c)(a+b-c)}{4b^2c^2}. \end{aligned}$$

Let Δ denote the area of the triangle. We can find Δ using the formula "area of a triangle equals half base times height". Let the base be the side AC and the height $c\sin\alpha$. So

$$\begin{aligned} \Delta &= \frac{1}{2}bc\sin\alpha \\ &= \frac{1}{4}\sqrt{(a+b+c)(-a+b+c)(a-b+c)(a+b-c)} \\ &= \sqrt{s(s-a)(s-b)(s-c)}. \end{aligned}$$

Where

$$s = \frac{a+b+c}{2}$$

is the *semiperimeter*. This is usually called Heron's formula.

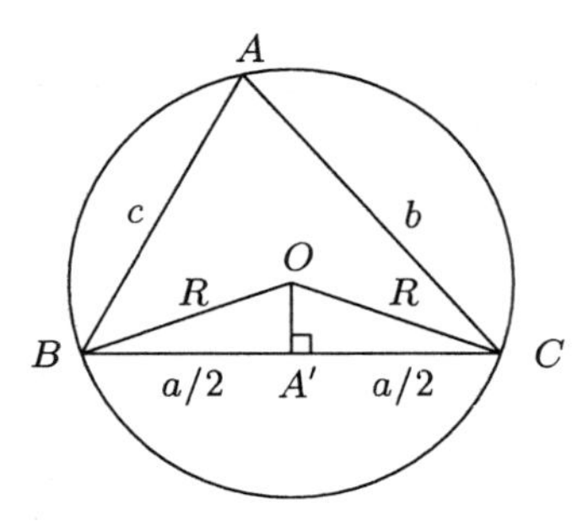

Fig. 3.31 The circumcircle

Let R denote the radius of the circumcircle (Fig. 3.31). As usual O denotes its centre and so by properties of the circle, the angle $\angle BOC$ is 2α or twice the angle $\angle BAC = \alpha$. It follows that the perpendicular bisector OA' divides the triangle BOC into two congruent triangles each having angle α at O. Whence

$$R \sin \alpha = BA' = \frac{1}{2}a$$

and so

$$\mathbf{2R = \frac{a}{\sin \alpha} = \frac{b}{\sin \beta} = \frac{c}{\sin \gamma}}.$$

Exercise 3.28

** Let Δ denote the area of the triangle ABC and let Δ' denote the area of the triangle $I_A I_B I_C$ of outcentres. Show that

$$\frac{\Delta'}{2\Delta} = \frac{a+b+c}{a\cos\alpha + b\cos\beta + c\cos\gamma}.$$

Exercise 3.29

** If R is the radius of the circumcircle, r is the radius of the incircle and r_A, r_B, r_C are the radii of the outcircles, show that

$$4R = r_A + r_B + r_C - r.$$

3.11 Three Surprising (and Beautiful) Theorems

Plane euclidean geometry abounds with connections between its various parts. Some of these are not so surprising but some connections are literally wonderful.

We have met some connections, for example the Euler line, already. In this section three theorems about the triangle are exhibited. All are relatively recent inventions but all would have been understood and appreciated by the ancient Greek geometers.

The Nine-Point Circle

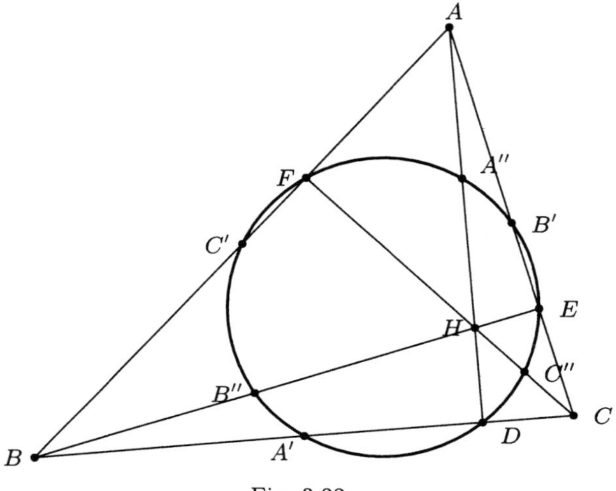

Fig. 3.32

If the sign of a surprising result in plane euclidean geometry is the inability to draw a convincing diagram freehand then the nine-point circle discovered by Euler must take some sort of prize. Somehow my attempts at a blackboard drawing of the circle in front of a class of sceptical undergraduates never seem to be convex let alone oval or even vaguely circular. The theorem that nine easily defined points of a triangle are concyclic is the first of our surprising and beautiful theorems.

■ **Nine-point circle:** Let H denote the orthocentre of a triangle ABC and let A'', B'', C'' be the midpoints of AH, BH, CH respectively. Then these three points along with the six points, A', B', C', D, E, F, all having their usual meanings, lie on a circle (Fig. 3.32).

Proof The points C', B' are midpoints of edges in the triangle ABC and so $C'B'$ is parallel to BC. Similarly using the triangles BHC, CHA and BHA we see that $B''C''$ is parallel to BC (and hence to $C'B'$) and $B''C'$, $C''B'$ are both parallel to AH. Since AH is perpendicular to BC we deduce that $C'B'C''B''$ is a rectangle (Fig. 3.33). This means that there is a circle with $C'C''$ and $B'B''$ as diameters. By symmetry $A'A''$ is another diameter. The angle $\angle ADA'$ is a right angle so D also lies on this circle as does E and F. □

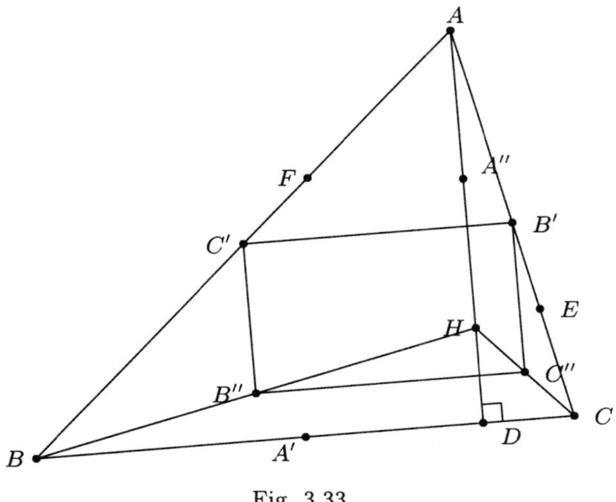

Fig. 3.33

■ The centre, N, of the nine-point circle is the midpoint of the circumcentre, O, and the orthocentre, H (and so lies on the Euler line). The radius of the nine-point circle is half the circumradius R.

Proof Take the circumcentre O as origin. Then $H = A + B + C$ and $C'' = (H + C)/2 = (A + B + 2C)/2$. So

$$\begin{aligned} N &= \frac{C' + C''}{2} \\ &= \frac{1}{2}\left(\frac{A+B}{2} + \frac{A+B+2C}{2}\right) \\ &= \frac{A+B+C}{2}. \end{aligned}$$

The radius of the nine-point circle is the length of NC' which is $|(A + B + C)/2 - (A + B)/2| = |C|/2 = R/2$. □

Exercise 3.30

The coordinates of ABC are $(0, 1), (t, 0)$ and $(s, 0)$ respectively. Show that the coordinates of the nine-point centre N are $(\frac{s+t}{4}, \frac{1-st}{4})$.

Exercise 3.31

** Show that the nine-point circles of $AB'C'$ and $A'B'C'$ are tangent at the midpoint of $B'C'$.

Exercise 3.32

** Show that the nine-point circle touches the incircle and the three outcircles (**Feuerbach's theorem**).

Exercise 3.33

* In a triangle ABC let O, the circumcentre, be the origin. If X lies on the circumcircle show that $(A + B + C + X)/2$ lies on the nine-point circle.

Napoleon's Theorem

We start this section with an extremal problem:
Fermat's problem *In an acute-angled triangle ABC, find a point P whose distances from A, B, C have the smallest possible sum.*

Take an arbitrary point P' in the interior of ABC. Our task is to minimise $AP' + BP' + CP'$. This is achieved by the following construction. Rotate the triangle BAP' through $60°$ about B to obtain the triangle BZP'' (Fig. 3.34). Then the triangles BZA and $BP''P'$ are equilateral. So it is easy to see that the length of the polygonal path $ZP'' + P''P' + P'C$ is equal to the quantity that we want to minimise, $AP' + BP' + CP'$.

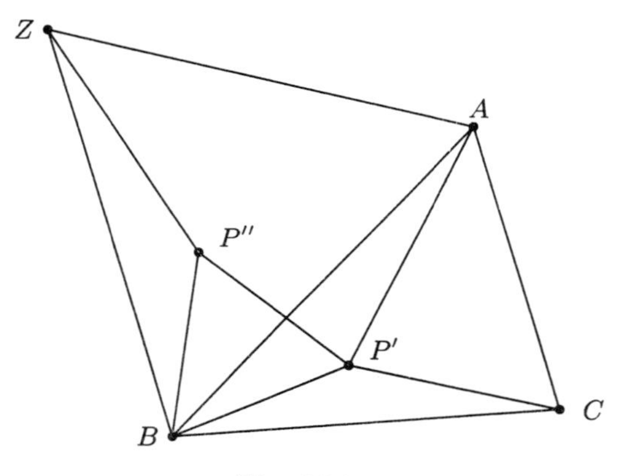

Fig. 3.34

But this length is minimal when the points $ZP''P'C$ are collinear. Let us assume that this is the case when the point $P' = P$. Then

$$\angle BPC = 180° - \angle BPP'' = 120°.$$

Similarly

$$\angle CPA = \angle APB = 120°.$$

So the desired point P for which $AP + BP + CP$ is minimal, is the point for which each of the sides BC, CA and AB subtends an angle of $120°$. This point P is called the *Fermat point* and it is easy to see from the construction that it is unique.

■ **Napoleon's Theorem** Let ABZ, BCX and CAY be equilateral triangles constructed on the outside edges of a triangle ABC (Fig. 3.35). Then the segments AX, BY and CZ are equal in length, pass through the Fermat point P and are inclined at $120°$. Moreover the circumcircles ABZ, BCX and CAY all pass through P and their centres form a fourth equilateral triangle.

Proof The first conclusion follows from the construction of P and symmetry. For the second conclusion we know that $\angle APB = 120°$ and so the points $APBZ$ are concyclic. Similarly, so are the points $BPCX$ and $APCY$.

Finally, consider the lines joining pairs of centres. They are the perpendicular bisectors of the common chords AP, BP and CP which are inclined at $120°$. Hence they make angle of $60°$ with each other and so form an equilateral triangle. □

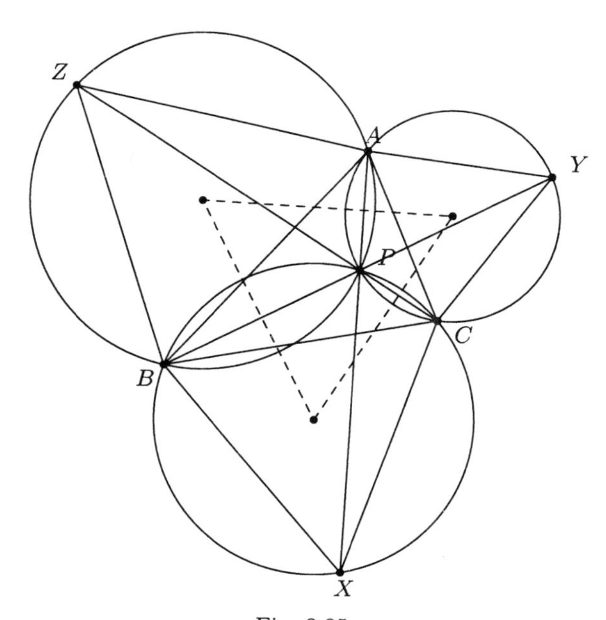

Fig. 3.35

Morley's Theorem

The third and final example of surprising and beautiful theorems also has an equilateral triangle as a theme. Morley's theorem says that the angle trisectors of a triangle meet in an equilateral triangle. This theorem, as easy to state as it is hard to prove, was not found until nearly the 20th century. Why did it take so long to be discovered? Perhaps the reason is because of the appearance of angle trisectors in the hypothesis. A general angle cannot be trisected with ruler and compass only. Nevertheless many so-called constructions were presented from the fringes of mathematics and so angle trisection got a bad name.

■ Let ABC be a triangle and let U, V, W be points inside so that the lines AV, AW, BW, BU and CU, CV divide the angles at A, B and C, respectively, into three equal parts (Fig. 3.36). Then UVW is an equilateral triangle.

Proof The proof is indirect. The equilateral triangle, UVW, is given and then the triangle ABC is constructed. Let the angles, α, β, γ, of a triangle have their usual meaning. Define the angles x, y, z by

$$x = 60^\circ - \alpha/3,\ y = 60^\circ - \beta/3,\ z = 60^\circ - \gamma/3.$$

On the outside edges of UVW erect isosceles triangles VWU', WUV' and UVW' with base angles, x, y and z respectively. Now extend $W'V$ and $V'W$ to meet at A, $U'W$ and $W'U$ to meet at B, and $V'U$ and $U'V$ to meet at C.

Several angles in the diagram can now be calculated. For example $\angle U'WA = z$ since

$$y + 60^\circ + x + z = 180^\circ.$$

Similarly $\angle AVU' = y$ and hence

$$\angle WAV = 180^\circ - 2x - y - z = \alpha/3.$$

Likewise $\angle UBW = \beta/3$ and $\angle VCU = \gamma/3$. The line UU' bisects the angle $\angle WU'V$ and it is easily calculated that $\angle BUC$ is $90^\circ + \angle WU'V/2$. This is the condition for U to be the incentre of the triangle $U'BC$. Hence UC bisects $\angle U'CB$ and UB bisects $\angle CBU'$. It now easily follows that the lines AV, AW, BW, BU and CU, CV divide the angles at A, B and C, respectively, into three equal parts.

Since the angles at A, B and C are α, β and γ respectively we see that the triangle ABC is similar to any given triangle and the theorem is proved. □

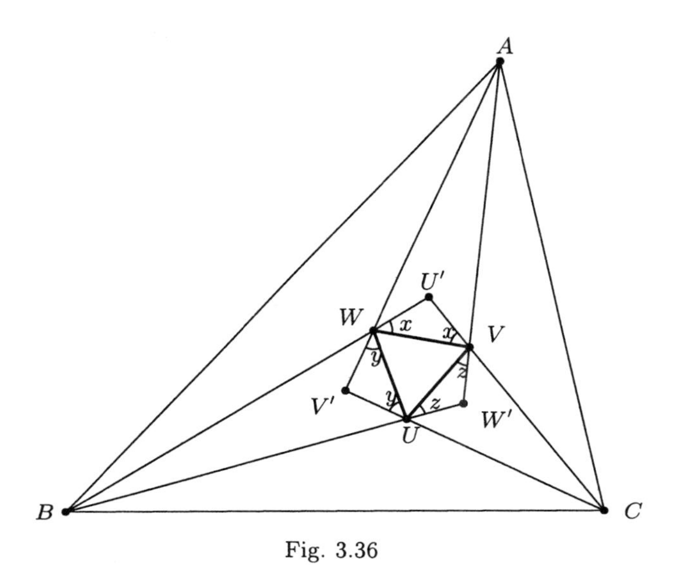

Fig. 3.36

Answers to Selected Questions in Chapter 3

3.1 $\angle WOB + \angle BOC + \angle COD + \angle DOY = \pi$ but $\angle WOB = \angle BOC$ and $\angle COD = \angle DOY$ so $\angle BOD = \angle BOC + \angle COD = \pi/2$.

3.2 $\hat{\alpha} = \pi - \alpha$ etc. so

$$\hat{\alpha} + \hat{\beta} + \hat{\gamma} = 3\pi - (\alpha + \beta + \gamma) = 2\pi.$$

3.6 Let the point be P and the line ℓ. By the fifth postulate there is a line ℓ' through P parallel to ℓ. Let ℓ'' be the line through P perpendicular to ℓ'. Then ℓ'' is perpendicular to ℓ by alternate angles.

3.7 The sum of the angles of a triangle is π so

$$\frac{1}{p} + \frac{1}{q} + \frac{1}{r} = 1.$$

The value $r = 1$ is impossible since p, q are positive. Suppose $r = 2$ then

$$\frac{1}{p} + \frac{1}{q} = \frac{1}{2}.$$

We can rewrite this equation as

$$(p-2)(q-2) = 4.$$

So $p-2,\ q-2$ are factors of 4, i.e. 1, 2 or 4. This gives solutions

$$(p,q) = (3,6),\ (4,4),\ (6,3)$$

and so the only solutions involving 2 are up to permutation

$$(p,q,r) = (3,6,2),\ (4,4,2).$$

If $r > 2$ then $\frac{1}{r} < \frac{1}{2}$ and $\frac{1}{p} + \frac{1}{q} > \frac{1}{2}$ and we get the inequality

$$(p-2)(q-2) < 4.$$

The positive solutions are

$$(p,q) = (1,1),\ (2,\text{any}),\ (3,3),\ (3,4),\ (3,5),\ (4,3),\ (5,3),\ (\text{any},2).$$

The only new solution with r an integer is

$$(p,q,r) = (3,3,3).$$

3.9 The intersection of $\boldsymbol{\Delta}$ with the plane $z = c$ is the semi-infinite open strip bounded by the lines $x + y = c$, $x - y = c$ and $y - x = c$. The isosceles triangles lie on the lines $x = y$, $x = c$, $y = c$ where they meet $\boldsymbol{\Delta}$. The equilateral triangle with side c is the point (c,c), right-angled triangles lie on the intersection of $\boldsymbol{\Delta}$ with the circle $x^2 + y^2 = c^2$ and the hyperbolæ $x^2 - y^2 = c^2$ and $y^2 - x^2 = c^2$ and the acute triangles lie within the tri-cuspular region bounded by the right-angled triangles. The obtuse triangles correspond to everything else. The general situation is the cone from $(0,0,0)$.

3.10 The four-sided region of an isosceles triangle cut off by a line parallel to the base.

3.12 See Fig. 3.37.

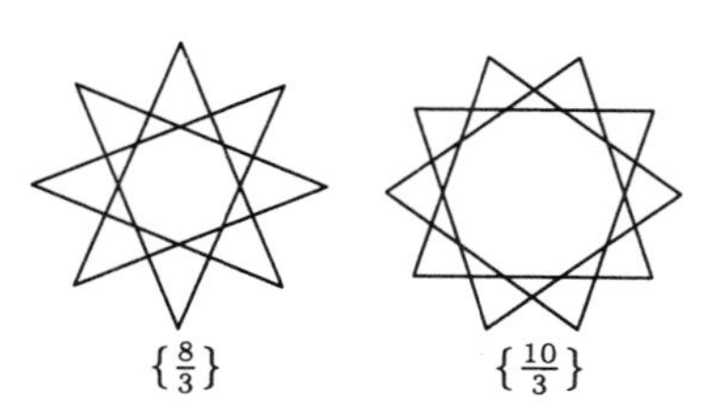

Fig. 3.37

3.13 $3\pi/5$ and $\pi/5$.

3.14 Let O be the centre of the $\{p\}$. Then the edges divide the whole angle at O into p angles of $2\pi/p$ radians. The angles at each vertex can now be seen to be $(1-2/p)\pi$.

3.17 The perpendicular bisector of AB has equation $x+y=0$ and the perpendicular bisector of BC has equation $x=1/2$. These meet at the point $(1/2,-1/2)$ which is $\sqrt{10}/2$ from each.

3.18 Consider Fig. 3.38 and use the two theorems:
Alternate angles cut by a transversal of two parallel lines are equal;
Angles subtended by the same chord of a circle are equal;
and the properties of isosceles triangles.

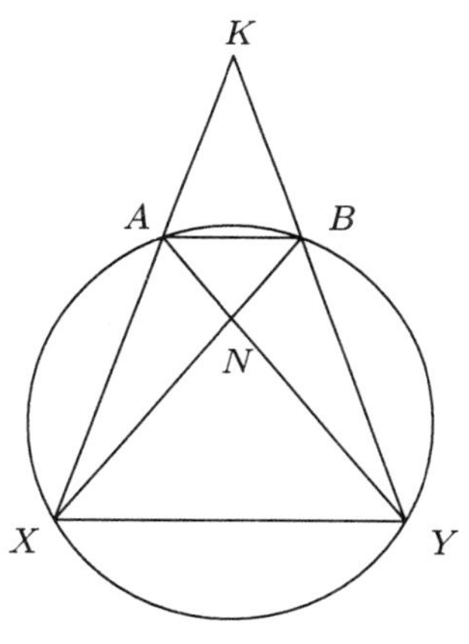

Fig. 3.38

3.19 If $ABCD$ is a parallelogram, then the angles at A and C likewise at B and D are clearly equal. Conversely if $ABCD$ is a quadrilateral and the angle α at A is equal to the angle at C and the angle β at B is equal to the angle at D then $2\alpha+2\beta=2\pi$. So $\alpha+\beta=\pi$ and by alternate and adjacent angles AB is parallel to DC. Similarly DA is parallel to CB. So $ABCD$ is a parallelogram.

3.20 The easiest way to a solution is to consider the points as vectors. Let $OACB$ be a quadrilateral. The midpoint of AB is $(A+B)/2$. The midpoint of OC is $C/2$. These are equal if and only if $C=A+B$, i.e. $OACB$ is a parallelogram.

3.21 As in the solution above, consider the points as vectors. Let the vertices of a parallelogram be $OACB$ where $C=A+B$. Then the cosine of $\angle AOC$ is $A\cdot(A+B)/(|A||A+B|)$ and the cosine of $\angle BOC$ is $B\cdot(A+B)/(|B||A+B|)$.

These are equal if and only if $(A{\cdot}B+A{\cdot}A)|B| = (A{\cdot}B+B{\cdot}B)|A|$. Similarly the other diagonal bisects if and only if $(B{\cdot}B-A{\cdot}B)|A| = (A{\cdot}A-A{\cdot}B)|B|$. Taking this from the previous equation yields $|A| = |B|$.

The line AB is perpendicular to OC if and only if $(A - B) \cdot (A + B) = A \cdot A - B \cdot B = O$. This again yields $|A| = |B|$.

3.22 $A' = (\frac{s+t}{2}, 0)$, $B' = (\frac{s}{2}, \frac{1}{2})$, $C' = (\frac{t}{2}, \frac{1}{2})$, $G = (\frac{s+t}{3}, \frac{1}{3})$.

3.23 D=$(0, 0)$, $E = (s\frac{1+st}{1+s^2}, s\frac{s-t}{1+s^2})$, $F = (t\frac{1+st}{1+t^2}, t\frac{t-s}{1+t^2})$, $H = (0, -st)$.

3.24 The perpendicular bisector of BC is the line $x = (s+t)/2$. The midpoint of AB is $(t/2, 1/2)$ and its slope is $-1/t$. It follows that the perpendicular bisector of AB has equation

$$\frac{y - 1/2}{x - t/2} = t.$$

Substituting $x = (s+t)/2$ gives $(\frac{s+t}{2}, \frac{1+st}{2})$ for the coordinates of O.

Exercise 3.22 gives the coordinates of G and H: OGH are collinear because $G - O$ is a multiple of $H - O$. Alternatively check that the determinant

$$\begin{vmatrix} (s+t)/2 & (1+st)/2 & 1 \\ (s+t)/3 & 1/3 & 1 \\ 0 & -st & 1 \end{vmatrix}$$

is zero.

3.25 Because of the angle-bisecting properties of I we have $\angle BIC = \pi - \beta/2 - \gamma/2 = \pi/2 + \alpha/2$ since $\alpha + \beta + \gamma = \pi$.

3.26 The angles $\angle IAB$ and $\angle IAC$ are equal by common tangents. The angles between the tangents through A to the two outcircles centre I_B and I_C are bisected by the line between the two centres. So IA bisects the line $I_B A I_C$. Hence A is the foot of the perpendicular from I_A to $I_B I_C$. Similar results hold for B and C. So the common point of these perpendiculars, I, is the orthocentre of $I_A I_B I_C$.

3.27 Equal roots occur when $a = b \sin\alpha$, that is when the triangle ABC is right-angled at B.

3.30 The nine-point centre N is the midpoint of OH and, by the results of Exercises 3.23 and 3.24, its coordinates are $((s+t)/2+0, (1+st)/2-st)/2 = (\frac{s+t}{4}, \frac{1-st}{4})$.

4
The Geometry of Complex Numbers

As far as I know, Cardano was the first to introduce complex numbers $a + \sqrt{-b}$ into algebra, but he had serious misgivings about it

van der Waerden

One can view the development of numbers as generated by the need to find solutions to more and more complicated equations. So the definition and use of negative integers was motivated by equations such as $x + 1 = 0$, rational numbers by equations such as $2x - 1 = 0$ and so on. Complex numbers were needed to find a solution to $x^2 + 1 = 0$, that is $\sqrt{-1}$. Each such advance in the use of numbers met some resistance from the current mathematical community. The use of complex numbers was no exception. Even the first user, Cardano in about 1539, who needed them to make sense of the solution of cubic equations was reluctant.

Two hundred years later Euler visualised complex numbers as points in the plane. In fact, the introduction of $\sqrt{-1}$ leads to a multiplication on elements of $\mathbb{R}^2$. This multiplication is important because it is distance preserving, as we shall see. The only other multiplications on $\mathbb{R}^n$ which are distance preserving occur when $n = 1$ (real number multiplication), $n = 4$ (quaternionic multiplication) and $n = 8$ (octonionic multiplication). We will consider these last two multiplications in Chapter 9.

4.1 What is $\sqrt{-1}$?

There is no real solution to the equation $x^2 = -1$. However we can easily construct a matrix solution. Let

$$\mathbf{I} = \begin{pmatrix} 0 & 1 \\ -1 & 0 \end{pmatrix} \text{ and } \mathbf{E} = \begin{pmatrix} 1 & 0 \\ 0 & 1 \end{pmatrix}.$$

Then we can easily check that $\mathbf{I}^2 = -\mathbf{E}$. If we think of $\mathbf{E}$ as the matrix equivalent of unity then $\mathbf{I}$ is a solution of the first equation.

Let us see if we can find a geometric interpretation. The matrix $\mathbf{I}$ defines a linear map by $X \to X\mathbf{I}$ or

$$(x, y) \to (x, y) \begin{pmatrix} 0 & 1 \\ -1 & 0 \end{pmatrix} = (-y, x).$$

This is a rotation of $\mathbb{R}^2$ about O through $\pi/2$. So its square, $X \to X\mathbf{I}^2$, is a rotation about O through π. That is the square of $\mathbf{I}$ is multiplication by -1.

We identify the real number 1 with $\mathbf{E}$ and with the point $(1, 0)$. Let

$$i = 1\mathbf{I} = (1, 0) \begin{pmatrix} 0 & 1 \\ -1 & 0 \end{pmatrix} = (0, 1).$$

Then

$$i^2 = (1\mathbf{I})(1\mathbf{I}) = 1\mathbf{I}^2 = -1\mathbf{E} = -1.$$

So we have found a solution to $\sqrt{-1}$ as $i = (0, 1)$ in $\mathbb{R}^2$.

A general point $Z = (x, y)$ in the plane $\mathbb{R}^2$ can be written as

$$Z = (x, y) = (1, 0)x + (0, 1)y = x + iy = x + yi$$

providing we equate x with $(x, 0) = (1, 0)x$.

We call such a point a *complex number*. The set of all complex numbers is written as $\mathbb{C}$. As a set it is the same as $\mathbb{R}^2$ but has the additional property of a multiplication. To multiply two complex numbers you multiply as if dealing with polynomials in i and replace i^2 by -1, i^3 by $(-1)i = i$ and so on. Addition of complex numbers is the same as for vectors.

Example 4.1

Let $Z = 3 - i,\ W = -2 + i$ (see Fig. 4.1) then

$$\begin{aligned} ZW &= (3 - i)(-2 + i) \\ &= -6 + 3i + (-i)(-2) - i^2 \\ &= -6 + 3i + 2i + 1 \\ &= -5 + 5i \end{aligned}$$

and

$$\begin{aligned} Z + W =& 3 - 2 - i + i \\ =& 1 + i0 \\ =& 1. \end{aligned}$$

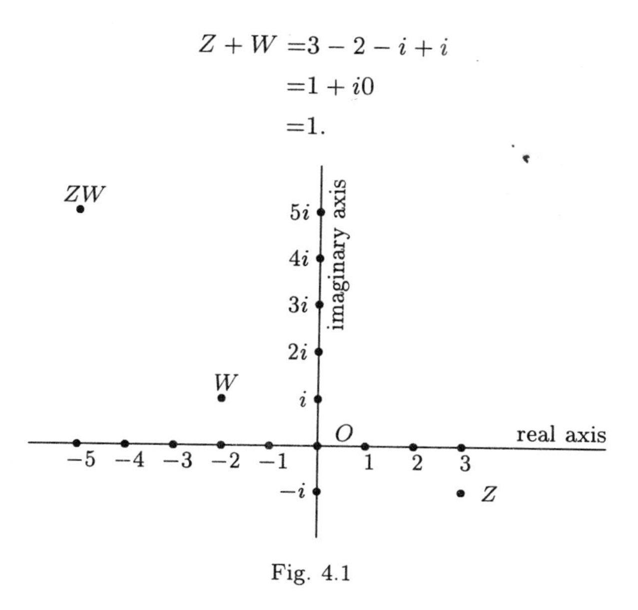

Fig. 4.1

The general formula for multiplication is

$$(x + iy)(x' + iy') = xx' + ixy' + iyx' + i^2yy' = xx' - yy' + i(xy' + yx').$$

In terms of pairs, $(x, y),\ (x', y') \in \mathbb{R}^2$, the formula is

$$(x, y)(x', y') = (xx' - yy', xy' + yx').$$

Exercise 4.1

Let $Z = 1 - i,\ W = 2 + 3i$. Find $Z + W,\ ZW,\ WZ$ and Z^2.

Exercise 4.2

Show that complex number multiplication is commutative, associative and is distributive with respect to addition (see Chapter 1 on real number multiplication).

A useful concept is *complex conjugation*. The complex conjugate of $Z = x + iy$ is given by

$$\overline{Z} = \overline{x + iy} = x - iy.$$

Conjugation means something different in the theory of groups so I will stick to the term complex conjugation.

Geometrically the map $Z \to \overline{Z}$ is reflection in the x-axis (Fig. 4.2). So we have

$$\overline{Z} = Z \text{ if and only if } Z \text{ is real.}$$

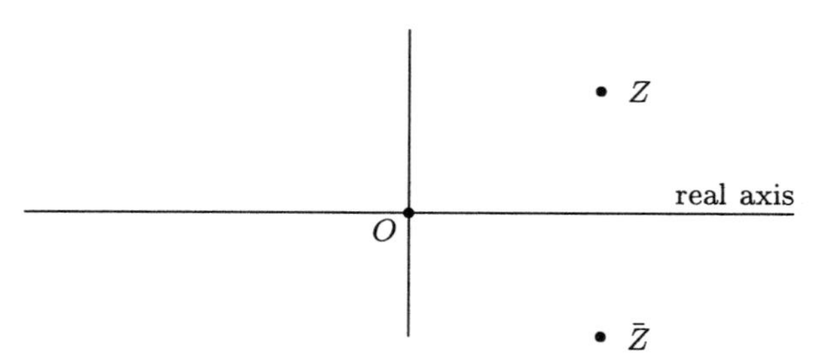

Fig. 4.2 Complex conjugation as reflection in the real axis

If $Z = x + iy$, let $x = \Re Z$, and $y = \Im Z$.

The numbers $\Re Z$ and $\Im Z$ are called the *real* and *imaginary* parts of $Z = x + iy$ respectively.

They are related to complex conjugation as follows

$$x = \Re Z = \frac{Z + \overline{Z}}{2}, \qquad y = \Im Z = \frac{Z - \overline{Z}}{2i}.$$

Numbers with $\Im Z = 0$ are of course real and numbers with $\Re Z = 0$ are called *purely imaginary.*

Exercise 4.3

Write $\sqrt{5 + 12i}$ in the form $x + iy$. Hint: square $x + iy$ and equate real and imaginary parts.

Exercise 4.4

Show that $\overline{Z + W} = \overline{Z} + \overline{W}$ and $\overline{ZW} = \overline{Z}\,\overline{W}$.

It is also true that $\overline{Z/W} = \overline{Z}/\overline{W}$ although we have not yet defined division. The above exercise shows that complex conjugation is a *field automorphism* of $\mathbb{C}$, that is, it respects all the field operations. In fact it is the only one, other than the identity, which is linear over the reals.

Exercise 4.5

Show that for all $Z,\ W \in \mathbb{C}$, the number $Z\overline{W} + \overline{Z}W$ is real.

4.2 Modulus and Division

The real number $x^2 + y^2$ is never negative and so we can take its square root. Then

$$|Z| = \sqrt{x^2 + y^2}$$

is called the *modulus* of the complex number Z.

Since $x^2 + y^2$ is zero if and only if Z is zero it follows that $|Z|$ is zero if and only if Z is zero. So if that is not the case then we can divide by Z as follows. Let $Z = x + iy$. Then

$$Z\overline{Z} = (x + iy)(x - iy) = x^2 + y^2.$$

This means that non-zero complex numbers have inverses. So

$$\frac{1}{Z} = \frac{1}{x+iy} = \frac{x - iy}{(x+iy)(x-iy)} = \frac{x-iy}{x^2+y^2} = \frac{\overline{Z}}{|Z|^2}.$$

In general, division is defined by

$$Z/W = ZW^{-1} = Z\overline{W}/|W|^2.$$

Example 4.2

Consider $Z = 1 - 4i$, then

$$\frac{1}{1-4i} = \frac{1+4i}{(1-4i)(1+4i)} = \frac{1+4i}{17}.$$

Taking modulus is well behaved with respect to multiplication and division. So

$$|ZW| = |Z||W| \text{ and } |Z/W| = |Z|/|W|.$$

Exercise 4.6

Prove $|ZW| = |Z||W|$ using the fact that $|Z|^2 = Z\overline{Z}$.

Exercise 4.7

Show that

$$|W + Z|^2 = |W|^2 + |Z|^2 + Z\overline{W} + \overline{Z}W.$$

Exercise 4.8

What exactly are the conditions on Z, W so that $|Z + W| = |Z| + |W|$?

Exercise 4.9

Show that
$$||Z| - |W|| \leq |Z \pm W| \leq |Z| + |W|.$$

In general $|Z + W| \neq |Z| + |W|$. But we do have
$$|Z + W| \leq |Z| + |W|$$
by the triangle inequality.

4.3 Unimodular Complex Numbers and the Unit Circle

Complex numbers U for which $|U| = 1$ are called *unimodular*. The set of unimodular numbers U is called the *unit circle* and its notation is S^1 (Fig. 4.3). So
$$S^1 = \{Z \in \mathbb{C} \mid |Z| = 1\} = \{(x, y) \in \mathbb{R}^2 \mid x^2 + y^2 = 1\}.$$

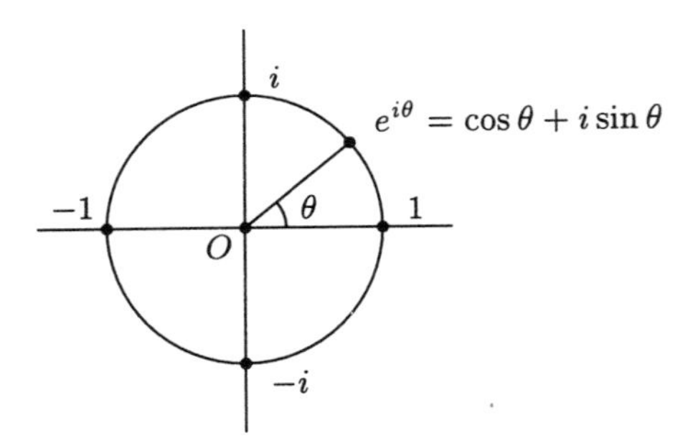

Fig. 4.3 The unit circle showing the points ± 1, $\pm i$ and $e^{i\theta}$

If U is unimodular then $U = \cos\theta + i\sin\theta$ for some unique angle θ.

If $Z \neq 0$ then $Z = rU = r(\cos\theta + i\sin\theta)$ where $r = |U|$. Then the angle θ is called the ***argument*** of Z and we write $\theta = \arg Z$.

Exercise 4.10

Find $\arg Z$ when $Z = 1,\ i,\ -1,\ -i,\ 1+i,\ -1-i$.

▮ The argument function satisfies
$$\mathbf{arg}\,(Z_1 Z_2) = \mathbf{arg}\, Z_1 + \mathbf{arg}\, Z_2.$$

Proof Consider $Z_1 = r_1(\cos\theta_1 + i\sin\theta_1)$ and $Z_2 = r_2(\cos\theta_2 + i\sin\theta_2)$. Then

$$\begin{aligned} Z_1Z_2 &= r_1(\cos\theta_1 + i\sin\theta_1)r_2(\cos\theta_2 + i\sin\theta_2) \\ &= r_1r_2(\cos\theta_1\cos\theta_2 - \sin\theta_1\sin\theta_2 + i(\cos\theta_1\sin\theta_2 + \sin\theta_1\cos\theta_2)) \\ &= r_1r_2(\cos(\theta_1+\theta_2) + i\sin(\theta_1+\theta_2)). \end{aligned}$$ □

A convenient notation for a general unimodular complex number is

$$U = \cos\theta + i\sin\theta = e^{i\theta}.$$

The number e here is Euler's constant $e = 2.7182818\cdots$. The reasons for this are by analogy with the Taylor expansions of cos, sin and exp. The details can be found in any book on complex variables.

For a general non-zero complex number $Z = r(\cos\theta + i\sin\theta)$ and we write $Z = re^{i\theta}$. This is the *polar form* of Z. The power notation is justified by the following simple form for multiplication of complex numbers in polar form,

$$(r_1e^{i\theta_1})(r_2e^{i\theta_2}) = r_1r_2e^{i\theta_1+\theta_2}.$$

Exercise 4.11

Write $Z = 2i, -1, 1+i, -1-i$ in polar form.

Multiplication of complex numbers can be pictured geometrically as follows. Consider the triangle with vertices O, E, Z, corresponding to 0, 1 on the real axis and a general complex number (Fig. 4.4). Now multiply by another complex number W to get a new triangle, similar to the old one, with vertices O, W, ZW. The new triangle is obtained from the old by a rotation about O through $\arg W$ and a scale change with factor $|W|$.

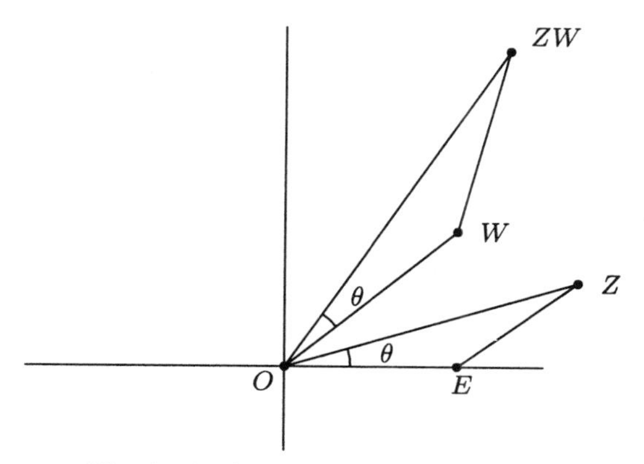

Fig. 4.4 Multiplying complex numbers

4.4 Lines and Circles in the Complex Plane

The Line

Let $A = a + ib$ and $Z = x + iy$. Then a simple calculation shows that

$$ax + by = \frac{\overline{A}Z + A\overline{Z}}{2}.$$

It follows that the equation of the line, $ax + by = c/2$, can be written in terms of complex numbers as

$$\overline{A}Z + A\overline{Z} = c$$

where c is real.

Exercise 4.12

Find the equation of the line through i and $1 + 2i$.

The vector A is perpendicular to the line $\overline{A}Z + A\overline{Z} = c$ and the vector iA is parallel. The nearest point to the origin, O, on the line is $cA/2|A|^2$.

The formulæ are considerably simplified if we assume, as we clearly can, that $|A| = 1$. For a general point B, the distance to the line is

$$\frac{|\overline{A}B + A\overline{B} - c|}{2}$$

and the nearest point is

$$\frac{B - A^2\overline{B} + cA}{2}.$$

It follows that reflection in the line is given by the map

$$S(Z) = cA - A^2\overline{Z}.$$

Exercise 4.13

Find the formula for reflection in the line $x + y = 1$.

Exercise 4.14

In which line is $S(Z) = \overline{Z} - i$ a reflection?

The Circle

Points on the circle centre C and radius r are given by the equation $|Z-C| = r$ or $(Z-C)(\overline{Z}-\overline{C}) = r^2$. When multiplied out this becomes

$$Z\overline{Z} - C\overline{Z} - \overline{C}Z + C\overline{C} = r^2.$$

Exercise 4.15

Find the equation of a circle centre i and radius 2.

Exercise 4.16

If B lies on the circle centre C and radius r, show that the equation of the tangent at B is

$$(\overline{B}-\overline{C})Z + (B-C)\overline{Z} = B\overline{B} - C\overline{C} + r^2.$$

4.5 Manipulating Complex Numbers

The operations on the complex numbers which we have considered so far can be tabulated as transformations on $\mathbb{C}$.

O:	Scale change by c	$Z \to cZ$	c a non-zero real number
I:	Translation by A	$Z \to Z + A$	A a fixed complex number
II:	Rotation about O through an angle θ	$Z \to AZ$	$A = e^{i\theta}$
III:	Reflection in x-axis	$Z \to \overline{Z}$	complex conjugation
IV:	Division	$Z \to 1/Z$	

Combinations of any number of these basic transformations are called *Möbius transformations.* We will write Möbius transformations as functions such as $T(Z)$ or $S(Z)$. They are divided into two types: those which involve an even number of reflections, called ***direct*** and those which involve an odd number of reflections, called ***indirect***. The direct Möbius transformations preserve orientation and the indirect Möbius transformations reverse orientation.

The scale change keeps O fixed and preserves any line through O set-wise. It is sometimes called a ***dilation*** with fixed point O. The effect of a scale change can be visualised as follows. Imagine an observer directly above O. A scale change either raises the observer if $c < 1$ or lowers the observer if $c > 1$. The fixed point can be made arbitrary by combining **O** and **I**. So $Z \to cZ + (1-c)A$ is a dilation with fixed point A. Dilations take lines into parallel lines, take triangles into similar triangles and generally change nothing but size.

A combination of **O**, **I** and **II** is called a (direct) *affine* transformation and is of the form

$$T(Z) = AZ + B, \qquad A \neq 0.$$

Taking the modulus of the image of two points gives $|T(Z) - T(Z')| = |A||Z - Z'|$. So if $|A| = 1$, distances are preserved and the transformation is an *isometry*. If $B = 0$ then the transformation is just multiplication by a complex number, that is a scale change combined with a rotation about O through an angle $\arg A$.

▮ The affine transformation $T(Z) = AZ + B$ is a translation if and only if $A = 1$. If $|A| = 1$ and $A \neq 1$ then T is a rotation about $B/(1 - A)$ through an angle $\arg A$.

Proof Clearly $T(Z) = Z + B$ is a translation. If $|A| = 1$ and $A \neq 1$, let F be a fixed point of T. Then $AF + B = F$ so $F = B/(1-A)$. Now $T(Z) - F = A(Z - F)$ and so T is the required rotation. □

Exercise 4.17

What affine map is represented by the equation $T(Z) = iZ + 1$?

Exercise 4.18

Write down in the form $Z \to AZ + B$ the following transformations of the complex plane:
(a) translation in the direction $(2, -3)$
(b) rotation about $(0, 1)$ through $\pi/4$.

Exercise 4.19

Find the formula for the rotation about $-i$ through $\pi/3$.

Exercise 4.20

Let ℓ_1, ℓ_2 be two lines in the plane meeting at O at an angle α. Let r_1, r_2 be the corresponding line reflections. Show that if $\alpha = \pi/2$ then $r_1 r_2 = r_2 r_1$ and if $\alpha = \pi/3$ then $r_1 r_2 r_1 = r_2 r_1 r_2$.

The general indirect affine transformation is given by

$$S(Z) = A\overline{Z} + B.$$

If $|A| = 1$ this may represent a reflection in a line but could represent a *glide reflection*, that is a reflection in a line (the axis) followed by a translation parallel to that axis. For example $S(Z) = -\overline{Z} + i$ is a glide reflection with axis the y-axis in the direction i. The conditions for when an indirect affine transformation is a reflection or glide reflection is contained in the following result.

■ Let $S(Z) = A\overline{Z} + B$ where $|A| = 1$ and let $\Delta_S = A\overline{B} + B$. Then S is a reflection if and only if $\Delta_S = 0$.
If $B = 0$ then S is the reflection in the line through O which makes an angle $(\arg A)/2$ with the x-axis.
If $B \neq 0$ and $\Delta_S = 0$ then S is the reflection in the line $\overline{B}Z + B\overline{Z} = |B|^2$.
If $\Delta_S \neq 0$ then S is the glide reflection with axis

$$(\overline{A}B - \overline{B})Z + (A\overline{B} - B)\overline{Z} = \frac{\overline{A}B + A\overline{B}}{2} - B\overline{B}$$

in the direction $\Delta_S/2$.

Proof Let $A = e^{i\theta}$. If $B = 0$ then $\arg(S(Z)) = \theta - \arg Z$ and $|S(Z)| = |Z|$. This represents a reflection in the line $\arg Z = \theta/2$.

Now assume that $B \neq 0$. Applying S twice gives, $S^2(Z) = Z + A\overline{B} + B = Z + \Delta_S$. This is equal to Z if and only if $\Delta_S = 0$, which is therefore a necessary condition for S to be a reflection. Conversely if $\Delta_S = 0$ then $A = -B/\overline{B}$ and $S(Z) = B(1 - \overline{Z}/\overline{B})$. This is the required reflection.

If $\Delta_S \neq 0$ then $S^2(Z)$ is a translation through Δ_S and it is easily seen that $S(Z)$ is the required glide reflection. □

A corollary to the above results is:

■ An affine transformation (direct or indirect) takes lines to lines and circles to circles. □

Example 4.3

Consider the transformation $S(Z) = i\overline{Z} + 1$. Then $\Delta_S = i + 1 \neq 0$ and so S represents a glide reflection. The axis passes through $B/2 = 1/2$ and is parallel to Δ_S. This is the line $y - x + 1/2 = 0$. Summing up, S is the glide reflection which is a reflection in the line $y - x + 1/2 = 0$ followed by a translation by $(i + 1)/2$.

Exercise 4.21

What transformations are represented by $S(Z) = i \pm \overline{Z}$?

Exercise 4.22

Let T, S be the reflections defined by $T(Z) = e^{i\theta}\overline{Z}$ and $S(Z) = e^{i\phi}\overline{Z}$. What are the compositions ST and TS? If R is the rotation $R(Z) = e^{i\psi}Z$ what is SR?

Exercise 4.23

Show that the composition of two direct or two indirect transformations is direct. Show that the composition of a direct with an indirect transformation is indirect.

Exercise 4.24

Show that the product of two reflections is either a translation or a rotation.

Exercise 4.25

What is the product of three reflections?

Exercise 4.26

What is the product of two rotations?

Exercise 4.27

If $T(Z) = AZ + B$ find the formula for the inverse transformation T^{-1}. Repeat for $S(Z) = A\overline{Z} + B$. In each case show that the inverse is of the same kind.

Exercise 4.28

Let $\mathbf{M}$ be a 2×2 matrix with real entries. By considering a complex number Z as a pair (x, y), show that the transformation $T(Z) = Z\mathbf{M}$ can be written in the form $T(Z) = AZ + B\overline{Z}$ for suitable complex numbers A, B.

Exercise 4.29

* Show that the equation $AX + B\overline{X} = C$, where X is an unknown and A, B, C are given complex numbers, has a unique solution if $|A| \neq |B|$. Show that if $|A| = |B|$ then $\overline{A}C = B\overline{C}$ is a necessary condition for a solution and if $A + \overline{A} \neq 0$ then $X = C/(A + \overline{A})$ is a solution.

Two affine transformations compose to form another affine transformation. An affine transformation has an inverse transformation which is also an affine transformation. The identity transformation is affine. So the set of affine transformations forms a group, the affine group.

4.6 Infinity and the Riemann Sphere

When we come to look at division of complex numbers in more detail then we will have to divide by 0 from time to time. So we need to add a number, ∞, to the complex plane corresponding to infinity.

This will have the following algebraic properties.

Firstly if Z is any complex number, we agree that

$$\boldsymbol{Z + \infty = \infty, \quad \frac{Z}{\infty} = 0.}$$

Secondly if W is any non-zero complex number then we agree that

$$\boldsymbol{W\infty = \infty, \quad \frac{W}{0} = \infty \quad (W \neq 0).}$$

On the other hand, combinations such as

$$\infty + \infty, \quad \infty\infty, \quad 0\infty, \quad \frac{\infty}{\infty}, \quad \frac{0}{0}$$

will have no meaning.

These simple rules will allow us to manipulate ∞ in the following as just another complex number (almost). The complex plane with ∞ added is called the *extended* complex plane and we write $\mathbb{C}^+ = \mathbb{C} \cup \{\infty\}$.

Another name for $\mathbb{C}^+$ is the *Riemann sphere* after B. Riemann (1826–1865). He was the first mathematician to identify the extended complex plane with a sphere under stereographic projection.

This identification allows us make geometric sense of the infinite. To define stereographic projection consider the 2-sphere, S^2, as the set of unit length points in 3-space, $\mathbb{R}^3$. Identify the complex plane as $z = 0$. So a complex number $Z = x + iy$ will be equated with the point $(x, y, 0)$. The intersection of the complex plane with S^2 may be considered as the equatorial circle on the sphere or the unit circle in the complex plane (Fig. 4.5). Let $N = (0, 0, 1)$ be the north pole on S^2. If P is a point on S^2 other than N let Z be the unique point in $\mathbb{C}$ such that Z, P, N lie on a line.

The correspondence $P \leftrightarrow Z$ is called stereographic projection. It defines a bijection between $S^2 - \{N\}$ and $\mathbb{C}$. Points in the northern hemisphere are

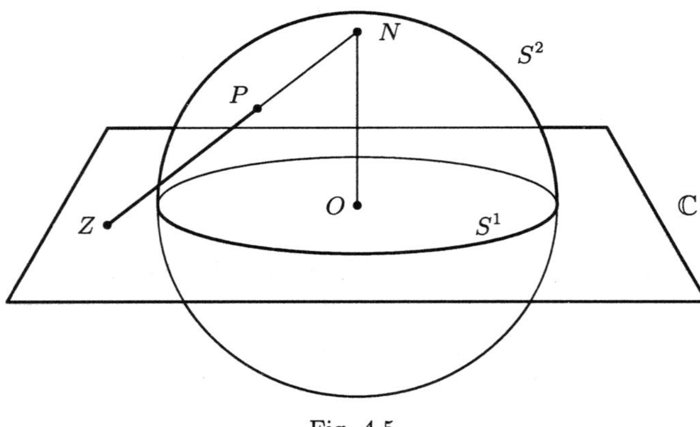

Fig. 4.5

projected to complex numbers with modulus greater than 1 and points in the southern hemisphere are projected to complex numbers with modulus less than 1. Moreover as P tends to N then the modulus of Z becomes arbitrarily large. It is reasonable therefore to equate N with ∞ and hence get a bijection between $\mathbb{C}^+$ and the whole sphere. There are many more interesting properties of stereographic projection as we will see.

Stereographic projection was used as far back as the second century AD by the astronomer Ptolemy to produce star maps.

Equating $\mathbb{C}^+$ with the sphere is an example of *one-point compactification.* That is a space which is non-compact such as the plane, $\mathbb{C} = \mathbb{R}^2$, has a point, ∞, added to make a compact space, the sphere S^2. There are other kinds of compactification, for example the projective plane which we will meet later is obtained from the standard plane by adding a line at infinity (actually a circle).

We now find a relationship between the coordinates of $P = (u, v, w)$ and $Z = x + iy$. Since P lies on the unit sphere, $u^2 + v^2 + w^2 = 1$. Let $\rho = \sqrt{u^2 + v^2}$ be the perpendicular distance from P to the line ON and let $r = \sqrt{x^2 + y^2}$ be the distance from Z to O. Then by similar triangles

$$\frac{\rho}{r} = \frac{1 - w}{1} = \frac{u}{x} = \frac{v}{y}.$$

So $x = u/(1 - w)$ and $y = v/(1 - w)$ giving

$$\boldsymbol{Z = \frac{u + iv}{1 - w}, \quad \overline{Z} = \frac{u - iv}{1 - w}}.$$

To find the inverse relationship, multiply Z and $\overline{Z}$ to get

$$Z\overline{Z} = \frac{u^2 + v^2}{(1 - w)^2} = \frac{1 + w}{1 - w}$$

and so

$$Z\overline{Z} - 1 = \frac{2w}{1-w}, \quad Z\overline{Z} + 1 = \frac{2}{1-w}.$$

The coordinates of P can now be expressed in terms of Z as follows,

$$u = \frac{2x}{Z\overline{Z}+1} = \frac{Z+\overline{Z}}{Z\overline{Z}+1}, \quad v = \frac{i(\overline{Z}-Z)}{Z\overline{Z}+1}, \quad w = \frac{Z\overline{Z}-1}{Z\overline{Z}+1}.$$

Exercise 4.30

Use the above formulæ to show that $NP \cdot NZ$ is constant.

Suppose now that P lies on a circle in S^2. Then P lies in the plane of the circle which is given by an equation of the form $au + bv + cw = d$. For the moment assume that $d \neq c$. Then using the equations above we see that Z satisfies

$$Z\overline{Z} - \overline{C}Z - C\overline{Z} + C\overline{C} = \frac{a^2+b^2+c^2-d^2}{(d-c)^2}$$

where $C = (a+ib)/(d-c)$.

Since this plane meets the unit sphere in a circle, the plane's nearest point to O must lie inside the unit sphere. But this is just the condition $a^2+b^2+c^2 > d^2$. It follows that Z lies on the circle centre C and radius $\sqrt{a^2+b^2+c^2-d^2}/|d-c|$.

On the other hand if $d = c$ then the plane passes through N and Z lies on the line

$$(a-ib)Z + (a+ib)\overline{Z} = 2d,$$

a fact which can easily be seen geometrically. Summing up,

■ Stereographic projection induces bijections between circles in the plane and circles on the sphere which do not pass through N and lines in the plane and circles on the sphere which do pass through N. □

It is reasonable then to look upon a line in the plane as a circle with infinite radius. Then the correspondence is between circles in the plane (with the augmented definition) and circles on the sphere.

Exercise 4.31

In terms of the above notation, what is the difference between $d > c$ and $d < c$?

Although we have not met inversion in a circle yet and will not consider inversions in a sphere at all, nevertheless it is interesting to point out that stereographic projection extends to the whole of space (except N) where it corresponds to inversion in a sphere centre N and radius $\sqrt{2}$. All the above properties now follow by the geometric properties of inversion. Moreover since these arguments are independent of the use of complex numbers it follows that corresponding results hold for stereographic projection in all dimensions.

Stereographic projection also has the interesting property of being *conformal*, that is angles are preserved. (Actually the angles are equal in magnitude but opposite in sign because the correspondence is indirect.)

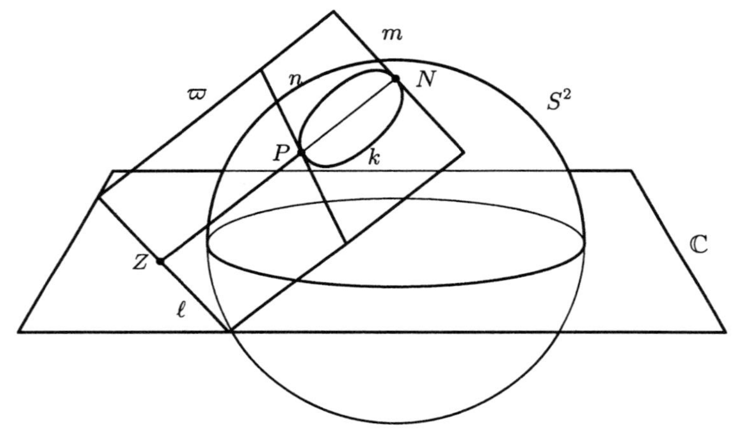

Fig. 4.6

To prove conformality consider a line ℓ in $\mathbb{C}$ through Z (as shown in Fig. 4.6). Let ϖ be the plane in $\mathbb{R}^3$ containing ℓ and N. Then ϖ meets S^2 in a circle, k say, passing through N and P. So ℓ is the image of $k-\{N\}$ under stereographic projection from N. Let m be the tangent to k at N and let n be the tangent to k at P. Then m is the intersection of the tangent plane to S^2 at N, the plane through N parallel to $\mathbb{C}$, with ϖ. It follows that m is parallel to ℓ.

Consider now another line ℓ' in $\mathbb{C}$ through Z making an angle θ with ℓ. Let ϖ', m', ... be the corresponding plane, tangent line, etc.

Because ℓ is parallel to m and ℓ' is parallel to m' it follows that m also makes an angle θ with m'. So the two circles on the sphere, k, k' have tangents, m, m', making an angle θ at one point of intersection, N. It follows that this also occurs at their other point of intersection, namely P, although the angle is equal and opposite. This proves that stereographic projection is conformal.

Exercise 4.32

Show that stereographic projection takes meridians of longitude θ on the sphere, given as $x = \cos\theta, y = \sin\theta$, into the lines $\arg Z = \theta$ in $\mathbb{C}$.

Exercise 4.33

Show that stereographic projection takes parallels of latitude ϕ on the sphere, given as $z = \sin\phi$, into the circles in $\mathbb{C}$ with centre O and radius $(1 + \tan\phi/2)/(1 - \tan\phi/2)$.

4.7 Division and Inversion

Having introduced ∞ in order to divide by 0 we can now return to our investigations of the elementary complex number operations by considering division. We compose with complex conjugation and consider the transformation S of the extended complex plane $\mathbb{C}^+$ given by $S(Z) = r^2/\overline{Z}$ where r is a positive real number.

It is easily seen that S has the following geometric properties,

(1) S takes points with modulus less than r into points with modulus greater than r and conversely;

(2) points with modulus r are fixed by S;

(3) Z and $S(Z)$ are on a line with O, the same side of O and satisfy $OZ \cdot OS(Z) = r^2$.

The last condition suffices to define S which is called *inversion* in the circle centre O and radius r. Inversion is like reflection except that the mirror is a circle instead of a line.

We can find the formula for inversion in a general circle, centre C and radius r by translating the origin to C. We get

$$S(Z) = \frac{C\overline{Z} - C\overline{C} + r^2}{\overline{Z} - \overline{C}}.$$

The point C is called the *centre of inversion*.

Example 4.4

Consider $S(Z) = \overline{Z}/(\overline{Z} + 1)$. Clearly $S(-1) = \infty$ so the circle of inversion has centre $C = -1$. Let r be the radius of the circle inversion. Since $-C\overline{C} + r^2 = -1 + r^2 = 0$ we see that the radius of the circle of inversion is 1.

Exercise 4.34

Find the formula for inversion in the circle centre i and radius 2.

Exercise 4.35

What inversion is represented by $S(Z) = (1-i)\overline{Z}/(\overline{Z}-1-i)$?

■ Under stereographic projection, spatial reflection in the plane $\mathbb{C}$ when restricted to the sphere corresponds to inversion in the unit circle.

Proof A calculation shows that under $Z \to 1/\overline{Z}$ the corresponding spatial coordinates transform by $(u, v, w) \to (u, v, -w)$ which is the required reflection. □

Exercise 4.36

What spatial transformations correspond to $S(Z) = \overline{Z}$ and $S'(Z) = -1/\overline{Z}$?

Consider the circle centre C and radius s given by the equation

$$Z\overline{Z} - Z\overline{C} - \overline{Z}C + C\overline{C} - s^2 = 0.$$

Under the inversion $W = r^2/\overline{Z}$ this is transformed into the equation

$$(C\overline{C} - s^2)W\overline{W} - r^2\overline{C}W - r^2C\overline{W} + r^4 = 0.$$

This is the equation of a circle unless $C\overline{C} = s^2$ when the equation is of a line (not through O). The condition $C\overline{C} = s^2$ means that the circle passes through O, the centre of inversion. Conversely lines correspond to circles unless the line passes through O when it corresponds to itself set-wise. This situation is unaltered by translating the origin and so we have the general result as follows.

■ Inversion takes circles into circles unless the circle passes through the centre of inversion when a circle inverts into a line not through the centre of inversion. Conversely a line not through the centre of inversion inverts into a circle passing through the centre of inversion. A line through the centre of inversion is invariant under inversion. □

Exercise 4.37

Show that under inversion in the unit circle a circle with centre C and radius r inverts into a circle centre $C/(C\overline{C}-s^2)$ and radius $s/(C\overline{C}-s^2)$.

Another way of seeing the results above is to relate them to the corresponding transformations of the sphere S^2. Since an inversion corresponds to a reflection in a plane and since reflection takes circles to circles we see that inversion also takes circles to circles if we think of lines as circles with infinite radius.

Since planar reflections are conformal like stereographic projection we easily see the following.

■ Inversion preserves the magnitude of angles and reverses their direction. □

Two circles k and k' are said to be *orthogonal* if they meet at two points P, Q where their tangents are orthogonal (Fig. 4.7). Suppose these tangents meet at O and O'. Then these will be the corresponding centres of the two circles.

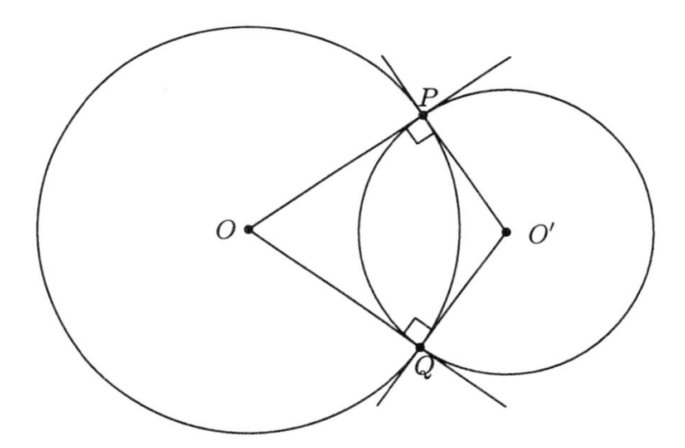

Fig. 4.7 Orthogonal circles

■ Under inversion in a circle k an orthogonal circle is invariant set-wise.

Proof With the notation above the points P, Q are fixed and the tangents to k' are invariant set-wise. So k' must also be invariant. Clearly the arc of k' outside of k is interchanged with the arc inside. □

Exercise 4.38

With the notation above, what is the image of the tangent line PO' under inversion in the circle k? What is the image of the circle with diameter OO'? Deduce the image of the point O'.

4.8 Möbius Transformations

The reciprocal map, $Z \to 1/Z$, is a combination of two maps, $Z \to \overline{Z}$ (reflection in the x-axis) and $Z \to 1/\overline{Z}$ (inversion in the unit circle). Since it is the combination of two indirect maps it is direct. If we combine this with all the direct maps we have met before we get the general (direct) *Möbius transformation* of the form

$$Z \to W = T(Z) = \frac{AZ+B}{CZ+D}, \qquad AD \neq BC.$$

The condition $AD \neq BC$ is necessary otherwise the top divides the bottom and then T is constant.

Exercise 4.39

Is $T(Z) = (Z+i)/(iZ-1)$ a Möbius transformation?

If $C \neq 0$ then the image of $Z = -D/C$ is ∞. Conversely writing $T(Z) = (A+B/Z)/(C+D/Z)$ we have $T(\infty) = A/C$. If $C = 0$ then $T(\infty) = \infty$.

■ A Möbius transformation defines a bijection between the extended plane $\mathbb{C}^+$ and itself. □

The general (indirect) Möbius transformation is of the form

$$\boldsymbol{Z \to W = T(Z) = \frac{A\overline{Z}+B}{C\overline{Z}+D} \quad AD \neq BC.}$$

Since a general Möbius transformation is a combination of simpler transformations which preserve circles etc. we have the following.

■ A Möbius transformation takes circles into circles, where we extend the definition of a circle so that a line is a circle of infinite radius. □

In a similar vein we have the following.

■ A Möbius transformation is conformal. If the Möbius transformation is indirect then the magnitude of the angle is preserved but the orientation is reversed. □

The image of a circle under a Möbius transformation can be determined by the image of three general points on the circle. If dealing with a line, find the image of two points on the line and infinity.

Example 4.5

To find the image of the unit circle under the Möbius transformation $Z \to \frac{Z}{1-Z}$ consider the image of three general points on the unit circle, say $1, -1, i$. These go to $\infty, -1/2, (i-1)/2$. Since these are collinear the image is the line $x = -1/2$.

Exercise 4.40

Find the image of the unit circle under the Möbius transformations $Z \to \frac{Z}{1+Z}$ and $Z \to \frac{Z}{1+2iZ}$.

Exercise 4.41

Show that for all $A, B \in \mathbb{C}$ with $A\overline{A} - B\overline{B} = 1$ the Möbius transformation

$$T(Z) = \frac{AZ + B}{\overline{B}Z + \overline{A}}$$

maps the unit circle to itself. Hint: consider $T(Z)\overline{T(Z)}$.

Exercise 4.42

* Show that for all $A \in \mathbb{C}$ with $|A| < 1$ and all unit complex numbers $e^{i\phi}$ the Möbius transformation $Z \to e^{i\phi}\frac{A-Z}{1-\overline{A}Z}$ preserves the unit disc set-wise. What happens if $|A| > 1$ or $|A| = 1$?

Exercise 4.43

What is the inverse image of the unit disc under the transformation $Z \to e^{i\phi}(Z - A)/(Z - \overline{A})$?

Exercise 4.44

Show that a Möbius transformation $Z \to \frac{AZ+B}{CZ+D}$ has either one or two fixed points. Show that it has just one, if and only if $(A+D)^2 = 4(AD - BC)$, $C \neq 0$. What happens if $C = 0$? (Remember that ∞ could be a fixed point!)

Let $GL_2(\mathbb{C})$ denote the group of invertible 2×2 matrices with complex entries,

$$GL_2(\mathbb{C}) = \left\{ \begin{pmatrix} A & B \\ C & D \end{pmatrix} \;\middle|\; AD - BC \neq 0 \right\}.$$

■ The set of Möbius transformations, **Möb**, forms a group under composition and the natural map

$$\begin{pmatrix} A & B \\ C & D \end{pmatrix} \to \left\{ Z \to \frac{AZ+B}{CZ+D} \right\}$$

is a surjective homomorphism $GL_2(\mathbb{C}) \to \mathbf{Möb}$ with kernel the diagonal matrices

$$\left\{ \begin{pmatrix} A & 0 \\ 0 & A \end{pmatrix} \middle| A \neq 0 \right\}.$$

Proof We must check that multiplication of matrices corresponds to composition of Möbius transformations. The kernel corresponds to cancellation in the formula for Möbius transformations of common factors in the top and bottom of the fractions. □

Exercise 4.45

Show that two Möbius transformations of the form

$$T(Z) = \frac{AZ+B}{-\overline{B}Z+\overline{A}}$$

compose to make a third of the same kind. That is they form a subgroup of **Möb**.

4.9 Cross Ratios

Let A, B, C be three complex numbers. Then the complex number, in polar form $re^{i\theta}$ and defined as the quotient

$$re^{i\theta} = \frac{A-C}{B-C}$$

has modulus $r = |A-C|/|B-C|$ and argument $\theta = \arg(A-C) - \arg(B-C) = \angle ACB$.

So $re^{i\theta}$ measures the difference between the vectors $A-C$ and $B-C$ or alternatively the difference between A and B with the origin at C.

Let A, B, C, D be four complex numbers, no three of which are equal. Then their *cross ratio* is defined to be the quotient

$$(\boldsymbol{AB}, \boldsymbol{CD}) = \frac{\boldsymbol{A}-\boldsymbol{C}}{\boldsymbol{B}-\boldsymbol{C}} \bigg/ \frac{\boldsymbol{A}-\boldsymbol{D}}{\boldsymbol{B}-\boldsymbol{D}} = \frac{(\boldsymbol{A}-\boldsymbol{C})(\boldsymbol{B}-\boldsymbol{D})}{(\boldsymbol{A}-\boldsymbol{D})(\boldsymbol{B}-\boldsymbol{C})}.$$

The cross ratio is defined even if one of the numbers, D say, is ∞. The value is found by replacing D by a fraction and letting the denominator tend to zero. It is easily seen that

$$(AB, C\infty) = \frac{A - C}{B - C} \text{ and in particular } (A1, 0\infty) = A.$$

The argument of the cross ratio (AB, CD) is $\angle ACB - \angle ADB$. We can use this fact to deduce the following important property of the cross ratio.

■ **The real cross ratio theorem** The cross ratio (AB, CD) is a real number if and only if the four points A, B, C, D are concyclic or collinear.

Proof The cross ratio (AB, CD) is real if and only if $\arg(AB, CD)$ is zero or π. That is either $\angle ACB = \angle ADB$ or $\angle ACB - \angle ADB = \pi$. But if the points are not collinear this is precisely the condition that they lie on a circle (Fig. 4.8). □

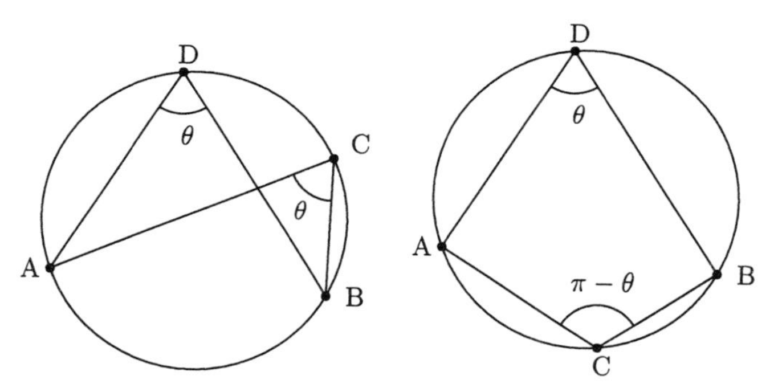

Fig. 4.8 Real cross ratios

Exercise 4.46

Find the cross ratio of the points 1, $i + 2$, $\frac{7+4i}{5}$, $\frac{14+3i}{5}$ and verify that they all lie on a circle. What is the equation of the circle?

Exercise 4.47

If $Z, S(Z)$ and $W, S(W)$ are inverse pairs of points, show that $Z, S(Z), W, S(W)$ are concyclic (or collinear).

The following is an elegant application of the real cross ratio theorem.

■ Consider four circles, s_1, s_2, s_3, s_4. Let s_1, s_2 meet in the points Z_1, W_1; let s_2, s_3 meet in Z_2, W_2; let s_3, s_4 meet in Z_3, W_3; and let s_4, s_1 meet in Z_4, W_4

(Fig. 4.9). Then the points $Z_1Z_2Z_3Z_4$ are concyclic or collinear if and only if the points $W_1W_2W_3W_4$ are concyclic or collinear.

Proof

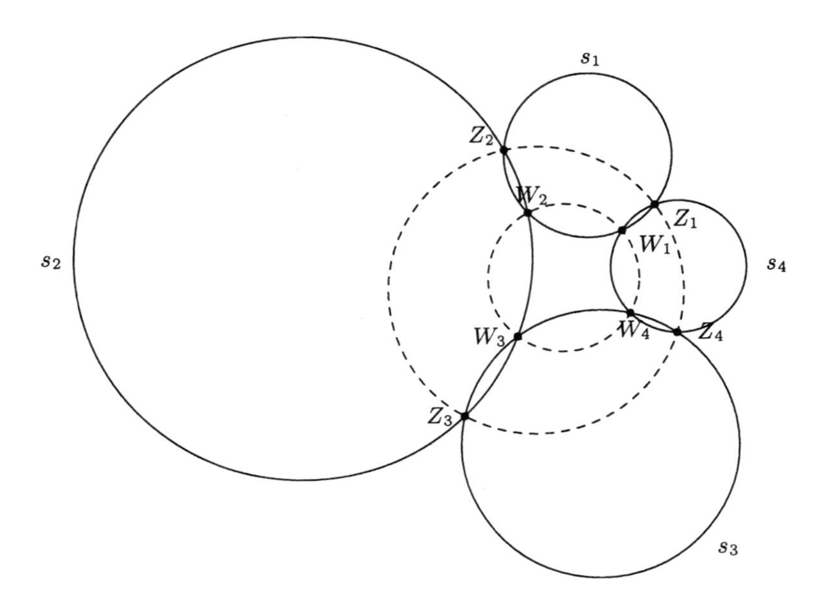

Fig. 4.9

Consider the cross ratios

$$A_1 = (Z_1W_2, Z_2W_1), \quad A_2 = (Z_2W_3, Z_3W_2),$$
$$A_3 = (Z_3W_4, Z_4W_3), \quad A_4 = (Z_4W_1, Z_1W_4).$$

All four are real because the corresponding points all lie on circles. For example $Z_1W_2Z_2W_1$ lie on s_2. The product

$$\frac{A_1A_3}{A_2A_4} = ZW$$

where $Z = (Z_1Z_3, Z_2Z_4)$ and $W = (W_1W_3, W_2W_4)$ is therefore real. It follows that Z is real if and only if W is real. The result now follows by the real cross ratio theorem. □

Values of the Cross Ratio

If $X = (AB, CD)$ is a cross ratio then in general it can take any value in $\mathbb{C}$. However there are three singular values, $0, 1, \infty$, which are attained when two of the points coincide. We shall always assume that at least three of the points

A, B, C, D are distinct in which case the singular values occur in the following situations.

$$(AB, CD) = 0 \text{ if } A = C \text{ or } B = D,$$
$$(AB, CD) = 1 \text{ if } A = B \text{ or } C = D,$$
$$(AB, CD) = \infty \text{ if } A = D \text{ or } B = C.$$

If we now permute A, B, C, D then we find that the cross ratio is unaltered by the following four permutations

$$(AB, CD) = (BA, DC) = (CD, AB) = (DC, BA).$$

Otherwise the cross ratio takes one of six possible values

$$\begin{aligned} (AB, CD) &= X, & (AB, DC) &= 1/X, \\ (AC, BD) &= 1 - X, & (AC, DB) &= 1/(1 - X), \\ (AD, BC) &= (X - 1)/X, & (AD, CB) &= X/(X - 1). \end{aligned}$$

In general these will be six distinct values. A *special value* is defined to be a non-singular value which has less than six values under permutation of A, B, C, D. The special values can be found by equating two of the above six values.

One possibility is that $X = -1$ or 2 or $1/2$. In that case the points A, B, C, D are said to be a *harmonic set* of points. Elements of a harmonic set lie on a circle or line and are related by a geometric construction as illustrated for example in Fig. 4.10.

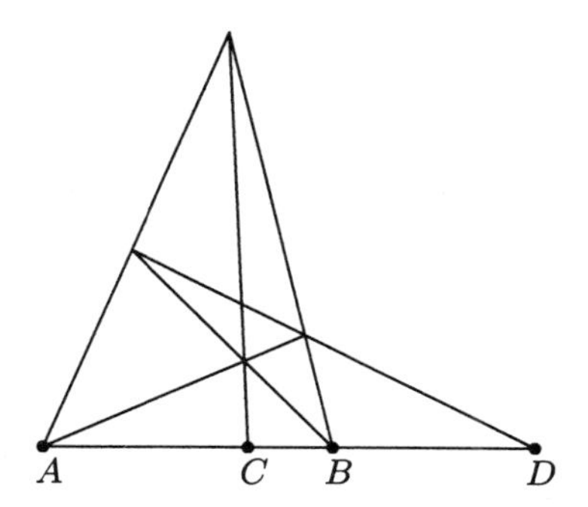

Fig. 4.10 Harmonic points

Exercise 4.48

Show that if $(AB, CD) = -1$ and D tends to infinity then C tends to the midpoint of AB.

The final possibility is that

$$X = \frac{1 \pm i\sqrt{3}}{2}.$$

In that case the points are said to form an *equianharmonic set* of points. Again they are related by a geometric construction. For example if A, B, C are the vertices of an equilateral triangle then D is either the centre or ∞.

The permutations which leave the cross ratio unaltered form a normal subgroup of the permutation group S_4 of order 4. The quotient group is S_3.

Exercise 4.49

If $2(AD + BC) = (A + D)(B + C)$, show that $\{A, B, C, D\}$ is harmonic.

Cross Ratios and Möbius Transformations

Let Z, Z' be transformed into W, W' by the Möbius transformation $Z \to (AZ + B)/(CZ + D)$. Then a calculation gives

$$W - W' = (Z - Z')\frac{AD - BC}{(CZ + D)(CZ' + D)}.$$

It follows that if Z_1, Z_2, Z_3, Z_4 goes to W_1, W_2, W_3, W_4 then $(Z_1Z_2, Z_3Z_4) = (W_1W_2, W_3W_4)$. Hence

■ Under direct Möbius transformations the cross ratio is invariant. □

Exercise 4.50

Show that under an indirect Möbius transformation the cross ratio of four points goes to its complex conjugate.

The invariance of the cross ratio under Möbius transformations combined with the real cross ratio theorem gives another proof that Möbius transformations take lines or circles into lines or circles.

Exercise 4.51

Show that under Möbius transformations, a harmonic (equianharmonic) set is taken to another harmonic (equianharmonic) set.

Assume now that A, B, C are distinct complex numbers. Then the cross ratio provides a transformation of the form

$$T(Z) = (ZA, BC) = \frac{(Z-B)(A-C)}{(Z-C)(A-B)}.$$

Expanding the brackets we see that this is the Möbius transformation $T(Z) = (A'Z + B')/(C'Z + D')$ where

$$A' = A - C, \quad B' = B(C-D), \quad C' = A - B, \quad D' = C(B-A).$$

This really is a Möbius transformation because

$$A'D' - B'C' = -(A-B)(B-C)(C-A)$$

and this is non-zero if A, B, C are distinct. Under this Möbius transformation

$$T(A) = 1, \quad T(B) = 0, \quad T(C) = \infty.$$

In fact there is a unique Möbius transformation taking any three distinct points A, B, C to any other three distinct points A', B', C'. If $Z \to W$, its formula is

$$(ZA, BC) = (WA', B'C').$$

Because the cross ratio is invariant we have

■ There is a Möbius transformation taking the four points A, B, C, D to the four points A', B', C', D' if and only if the cross ratios (AB, CD) and $(A'B', C'D')$ are equal. □

4.10 A Formula for the Cross Ratio

Let A, B, C, D be four distinct complex numbers lying on a line ℓ. Let P be a point not on ℓ and let A, B, C, D make angles θ, ϕ, ψ at P as shown in Fig. 4.11.

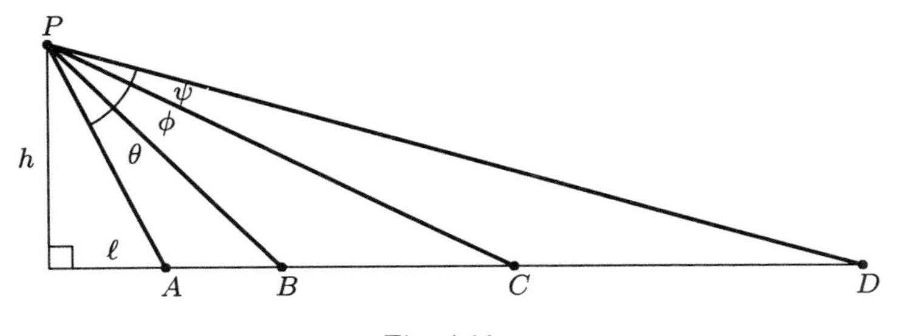

Fig. 4.11

■ With the notation above the cross ratio of the points A, B, C, D is given by the formula

$$\boldsymbol{(AB, CD) = \frac{\sin(\theta + \phi)\sin(\phi + \psi)}{\sin(\theta + \phi + \psi)\sin(\phi)}}.$$

Proof Let h be the perpendicular distance from P to the line ℓ. Then the area of the triangle ACP is given by two identities

$$\triangle ACP = \frac{1}{2}AC\ h = \frac{1}{2}PA\ PC\ \sin(\theta + \phi).$$

So we deduce that

$$AC = \frac{PA\ PC\ \sin(\theta + \phi)}{h}.$$

Similarly

$$AD = \frac{PA\ PD\ \sin(\theta + \phi + \psi)}{h},$$

$$BC = \frac{PB\ PC\ \sin(\phi)}{h},$$

$$BD = \frac{PB\ PD\ \sin(\phi + \psi)}{h}.$$

It follows that the cross ratio is

$$\begin{aligned}(AB, CD) &= \frac{AC.BD}{AD.BC} \\ &= \frac{PA\ PC\ \sin(\theta + \phi)/h\ PB\ PD\ \sin(\phi + \psi)/h}{PA\ PD\ \sin(\theta + \phi + \psi)/h\ PB\ PC\ \sin(\phi)/h} \\ &= \frac{\sin(\theta + \phi)\sin(\phi + \psi)}{\sin(\theta + \phi + \psi)\sin(\phi)}.\square\end{aligned}$$

Since the formula only involves the angles subtended at P it follows that the cross ratio is independent of the line ℓ. That is, if another line, ℓ', met the lines PA, $PB, \ldots$ in points A', $B', \ldots$ then they would have the same cross ratio, $(AB, CD) = (A'B', C'D')$.

This result can be summed up as follows.

■ The cross ratio of four points on a line is an invariant of perspective. □

4.11 Roots of Unity

Any solution to the equation $Z^n = 1$ is called an nth *root of unity*. For example, the fourth roots of unity are $\{1, -1, i, -i\}$. Of course 1 will always be an nth root of unity but the fundamental theorem of algebra says there must be $n-1$ more. Taking moduli we see that $|Z|^n = 1$ so $|Z| = 1$ and therefore Z is unimodular. Let $Z = e^{i\theta}$ where $\theta = \arg Z$. Then $Z^n = e^{in\theta}$ and so $n\theta$ is some integer multiple of 2π. It follows that the nth roots of unity are $\{1, e^{\frac{2\pi i}{n}}, e^{\frac{4\pi i}{n}}, \ldots, e^{\frac{2(n-1)\pi i}{n}}\}$. So the nth roots of unity are distributed evenly around the unit circle and form the vertices of a regular n-gon (see, for example, Fig. 4.12). If $W = e^{\frac{2\pi i}{n}}$ then the roots are

$$\{1, W, W^2, \ldots, W^{n-1}\}.$$

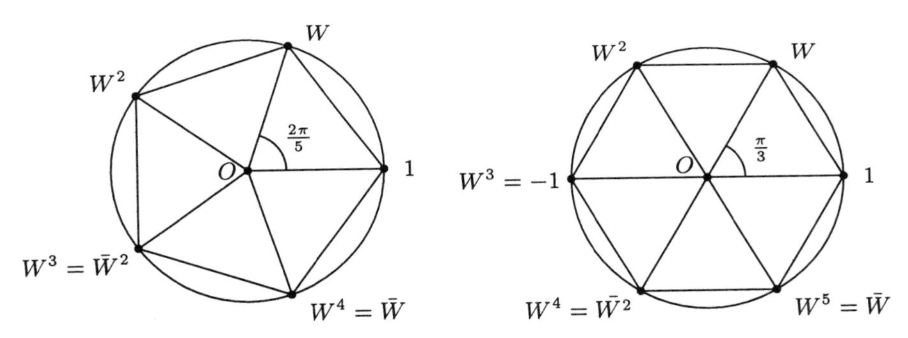

Fig. 4.12 Fifth and sixth roots of unity

Exercise 4.52

Show that the nth roots of unity apart from 1 or -1 form conjugate pairs. Show that if n is even then W is an nth root of unity if and only if $-W$ is an nth root of unity. If n is odd and the nth roots of unity are $\{1, W, W^2, \ldots, W^{n-1}\}$ show that the $2n$th roots of unity are $\{\pm 1, \pm W, \pm W^2, \ldots, \pm W^{n-1}\}$.

4.12 Formulæ for the nth Roots of Unity

We know that the nth roots of unity are $e^{2k\pi i/n}$ where $k = 0, 1, 2, \ldots, n-1$. But with a bit of manipulation other forms for these can be given. For example, suppose that the cube roots of unity are $\{1, W, W^2\}$ where $W = e^{2\pi i/3}$. Then W, W^2 are roots of

$$\frac{x^3 - 1}{x - 1} = x^2 + x + 1.$$

So solving this quadratic equation gives

$$W = \frac{-1 + i\sqrt{3}}{2}, \quad W^2 = \frac{-1 - i\sqrt{3}}{2}.$$

Since $W = \cos 2\pi/3 + i \sin 2\pi/3$, comparing real and imaginary parts verifies the well-known formulæ $\cos 2\pi/3 = -1/2$ and $\sin 2\pi/3 = \sqrt{3}/2$.

The unimodular numbers $U = e^{\pm i\theta} = \cos\theta \pm i \sin\theta$ are roots of the quadratic equation

$$x^2 - 2x\cos\theta + 1 = 0$$

as may be easily verified. Hence if n is odd there is a factorisation

$$x^{n-1} + x^{n-2} + \cdots + x + 1 = (x^2 - 2\cos 2\pi/n + 1)(x^2 - 2\cos 4\pi/n + 1) \\ \cdots (x^2 - 2\cos 2(n-1)\pi/n + 1).$$

If n is even there are similar factors with an extra factor of $x + 1$.

This factorisation can be used to find explicit formulæ for the values of $\cos 2\pi/n$. For example if $n = 5$ let us try the following factorisation

$$x^4 + x^3 + x^2 + x + 1 = (x^2 + ax + 1)(x^2 + bx + 1)$$

where a, b are unknown. Comparing coefficients of x^3 and x^2 yields the two equations

$$a + b = 1, \quad ab = -1.$$

So a, b are the roots of the quadratic in t given by $t^2 - t - 1$. Solving gives

$$a = \frac{1 + \sqrt{5}}{2}, \quad b = \frac{1 - \sqrt{5}}{2}.$$

So

$$\cos(\frac{2\pi}{5}) = \frac{\sqrt{5} - 1}{4}, \quad \cos(\frac{4\pi}{5}) = \frac{-\sqrt{5} - 1}{4}.$$

Exercise 4.53

If $U = e^{i\theta}$ show that

$$2\cos n\theta = U^n + \frac{1}{U^n}.$$

Deduce that

$$\cos^3\theta = \frac{1}{4}(\cos 3\theta + 3\cos\theta)$$

and

$$\cos^5\theta = \frac{1}{16}(\cos 5\theta + 5\cos 3\theta + 10\cos\theta).$$

Exercise 4.54

Factorise $x^5 + x^4 + x^3 + x^2 + x + 1$ into linear and quadratic factors with real coefficients.

Exercise 4.55

Solve $x^5 = i$.

Exercise 4.56

Solve $x^9 + x^6 - x^4 = 1$.

Exercise 4.57

Factorise $x^6 + x^5 + x^4 + x^3 + x^2 + x + 1$ into three quadratic factors with real coefficients.

Ruler and Compass Constructions

Plato considered the straight line and the circle as the only perfect geometric figures. Accordingly the ancient Greeks tended to restrict their considerations to objects which could be drawn with a ruler (actually a straight edge) and a compass. With this restriction many constructions were possible. The reader will have no difficulty in drawing the perpendicular bisector of a line segment or the bisector of an angle. Equilateral triangles and regular hexagons are also easy to inscribe in a given circle. Figure 4.13 illustrates a ruler and compass construction for an inscribed regular pentagon.

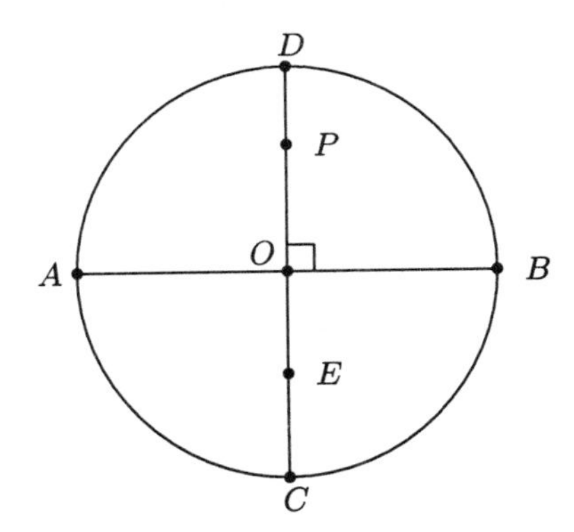

Fig. 4.13 Constructing a pentagon: $OE = OC/2$ and $EP = EA$

(1) In a circle centre O construct two orthogonal diameters AB and CD;

(2) bisect OC at E;

(3) with centre E, radius EA, draw a circle to cut OD at P.

Then AP is the side of a regular pentagon inscribed in the circle.

Exercise 4.58

Can you prove it is a pentagon?

Nevertheless there remain constructions which are impossible with ruler and compass alone. Most notorious is the trisection of a general angle. Gauss, using algebraic properties of roots of unity, decided which regular polygons could be constructed with ruler and compass alone. For example the 17-gon (his age at the time!) could be constructed but the heptagon and nonagon could not. It was this discovery which prompted his career in mathematics.

4.13 Solving Cubic and Biquadratic Polynomials

Everyone is familiar with the formula

$$x = \frac{-b \pm \sqrt{b^2 - 4ac}}{2a}$$

for the solution of the quadratic polynomial equation $ax^2 + bx + c = 0$. But what about solving the cubic, $ax^3 + bx^2 + cx + d = 0$?

A formula for the solution of the general cubic equation is due to the 16th century Italian algebraists Ferro, Tartaglia and Cardano. We first simplify the equation as follows. Dividing the equation by a we can assume that $a = 1$. Replacing x by $x - b/3$ reduces the general cubic to the simpler form

$$x^3 + px + q = 0$$

where $p = c - b^2/3$ and $q = 2b^3/27 - bc/3 + d$. We will now solve the cubic in this form. Suppose the unknown x can be written as a sum $x = u + v$. Then

$$x^3 = u^3 + 3uv(u + v) + v^3 = 3uvx + u^3 + v^3.$$

Comparison with the original cubic equation yields

$$3uv = -p \text{ and } u^3 + v^3 = -q.$$

It follows that u^3 and v^3 are the roots of the quadratic in t given by

$$t^2 + qt - \frac{p^3}{27} = 0.$$

Using the formula for the solution of the quadratic we see that u^3 and v^3 are equal to

$$\frac{-q}{2} \pm \sqrt{\frac{q^2}{4} + \frac{p^3}{27}}.$$

So the solution to the simplified cubic is

$$x = u + v = \sqrt[3]{-\frac{q}{2} + \sqrt{\left(\frac{q}{2}\right)^2 + \left(\frac{p}{3}\right)^3}} + \sqrt[3]{-\frac{q}{2} - \sqrt{\left(\frac{q}{2}\right)^2 + \left(\frac{p}{3}\right)^3}}.$$

However there are three cube roots of any non-zero complex number so picking one cube root in both cases and bearing in mind that uv is a constant, the other two roots, corresponding to the other cube roots of unity, are

$$W\sqrt[3]{-\frac{q}{2} + \sqrt{\left(\frac{q}{2}\right)^2 + \left(\frac{p}{3}\right)^3}} + W^2\sqrt[3]{-\frac{q}{2} - \sqrt{\left(\frac{q}{2}\right)^2 + \left(\frac{p}{3}\right)^3}}$$

and

$$W^2\sqrt[3]{-\frac{q}{2} + \sqrt{\left(\frac{q}{2}\right)^2 + \left(\frac{p}{3}\right)^3}} + W\sqrt[3]{-\frac{q}{2} - \sqrt{\left(\frac{q}{2}\right)^2 + \left(\frac{p}{3}\right)^3}}$$

where $W = e^{2\pi/3}$.

Example 4.6

Consider the equation

$$x^3 - 15x - 126 = 0.$$

Using the notation above, u^3, v^3 are the roots of the quadratic

$$t^2 - 126t + 125 = 0,$$

that is 125 and 1. So $u = 5$ and $v = 1$ say. (For convenience we have chosen the real roots but it would have made no difference if we had chosen imaginary roots.) Hence one root is $5 + 1 = 6$ and the others are

$$5\frac{-1+\sqrt{3}i}{2} + \frac{-1-\sqrt{3}i}{2} = -3 + 2\sqrt{3}i$$

and

$$5\frac{-1-\sqrt{3}i}{2} + \frac{-1+\sqrt{3}i}{2} = -3 - 2\sqrt{3}i.$$

It was the discovery of this formula which led inevitably to the discovery of complex numbers. Assume that the coefficients of the cubic equation are real.

If $(\frac{q}{2})^2 + (\frac{p}{3})^3$ is negative, the so-called casus irreducibilis, then we are obliged to take the square root of a negative number. However the resulting complex numbers are complex conjugates and so the imaginary parts cancel out in the sum and we are left with a real root! This is best illustrated by an example. Consider

$$x^3 - 7x + 6 = 0.$$

Then $(\frac{q}{2})^2 + (\frac{p}{3})^3 = -\frac{100}{27}$ so we are looking at cube roots of $-3 \pm 10i/3\sqrt{3}$, for example $1 \pm 2i/\sqrt{3}$. Their sum is 2, the imaginary terms $\pm 2i/\sqrt{3}$ having cancelled. So 2 is one root, the others being 1 and -3.

Exercise 4.59

Solve the equation $x^3 - 2x^2 + 2x - 1 = 0$. Hint: make a suitable change of unknown x to get it in the form $x^3 + px + q = 0$.

Exercise 4.60

If p, q are real and $(\frac{q}{2})^2 + (\frac{p}{3})^3 \leq 0$, show that all three roots of $x^3 + px + q = 0$ are real.

Exercise 4.61

By making the substitution $x = k\cos\theta$ in $x^3 + px + q = 0$, where $k^2 = -4p/3$, show that the casus irreducibilis can be reduced to the trigonometric equation

$$\cos 3\theta = \frac{3q}{2}\sqrt{-\frac{3}{p^3}}.$$

Hint: use the identity $\cos 3\theta = 4\cos^3\theta - 3\cos\theta$.

The biquadratic or quartic equation can be solved by writing it as the product of two quadratic factors, for example

$$x^4 + px^2 + qx + r = (x^2 + ax + b)(x^2 - ax + c)$$

where a, b, c are to be determined. Equating coefficients gives

$$-a^2 + b + c = p, \quad a(c - b) = q, \quad bc = r$$

from which b, c can be eliminated to give a cubic equation for a^2, namely

$$a^6 + 2pa^4 + (p^2 - 4r)a^2 - q^2 = 0.$$

It turns out that if p, q, r are real then this cubic always has a real positive solution in a^2 and so a, b, c can be found and the roots of the original quartic are the roots of the two quadratics

$$x^2 + ax + b = 0 \text{ and } x^2 - ax + c = 0.$$

Exercise 4.62

Show that the cubic equation associated to $x^4 - 2x^2 + 8x - 3 = 0$ is

$$a^6 - 4a^4 + 16a^2 - 64 = 0$$

which has a root $a^2 = 4$. Deduce that the roots of the original quartic are

$$-1 \pm \sqrt{2}, \quad 1 \pm i\sqrt{2}.$$

The above formulæ for the cubic and the quartic equations, are known as solutions by *radicals*. That is they just involve the usual field operations and the extractions of roots in the coefficients. It was shown by Abel and Galois that this is not the case for higher degree equations. For example the quintic equation $x^5 - 6x + 3 = 0$ does not have a solution by radicals.

Answers to Selected Questions in Chapter 4

4.1 $Z + W = 3 + 2i,\ ZW = WZ = 5 + i,\ Z^2 = -2i.$

4.3 $3 + 2i$ or $-3 - 2i$.

4.5 $\overline{Z\overline{W} + \overline{Z}W} = \overline{Z\overline{W}} + \overline{\overline{Z}W} = \overline{Z}W + Z\overline{W}$ so $Z\overline{W} + \overline{Z}W$ is unaltered by complex conjugation and is therefore real.

4.8 $Z = aW$ or $W = aZ$ where a is real and $a \geq 0$.

4.10 $\arg = 0,\ \pi/2,\ \pi,\ 3\pi/2,\ \pi/4,\ 5\pi/4.$

4.11 $2e^{i\pi/2},\ e^{i\pi},\ \frac{1}{\sqrt{2}}e^{i\pi/4},\ \frac{1}{\sqrt{2}}e^{i5\pi/4}.$

4.12 $-x + y = 1$ or $(-1 - i)Z + (-1 + i)\overline{Z} = 2.$

4.13 $S(Z) = 1 + i - i\overline{Z}.$

4.14 $y = -1/2.$

4.15 $Z\overline{Z} - i\overline{Z} + iZ = 3.$

4.17 The fixed point F satisfies $F = iF + 1$, so $F = (1 + i)/2$. Since $|i| = 1$ this map is an isometry and is a rotation about $(1 + i)/2$ through $\arg i = \pi/2$.

4.18 (a) $Z \to Z + 2 - 3i$, (b) $Z \to AZ - iA + i$ where $A = \frac{1+i}{\sqrt{2}}$.

4.19 Let the image of Z be $T(Z)$. Then $T(Z) + i = e^{i\pi/3}(Z + i)$. Since $e^{i\pi/3} = 1/2 + i\sqrt{3}/2$ it follows that $T(Z) = (1/2 + i\sqrt{3}/2)Z - (\sqrt{3}/2 + i/2)$.

4.20 In the first case we might as well take the lines to be the coordinate axes and then the product of the two reflections is the half turn $Z \to -Z$ in either order. You could do something similar in the next case or argue that r_1r_2 has order 3 so $r_1r_2r_1r_2r_1r_2 = 1$. Now using the fact that a reflection has order 2 gives the result.

4.21 Reflection in the line $y = 1/2$ and glide reflection by i with axis $x = 0$.

4.22 $ST(Z) = e^{i(\phi-\theta)}Z$ which is a rotation about the origin through $\phi - \theta$. TS is a similar rotation through $\theta - \phi$. Since T is the reflection in a line making an angle $\theta/2$ with the x-axis and S is the reflection in a line making an angle $\phi/2$ with the x-axis it follows that the composition is a rotation about the common point of the lines through an angle twice that of the angle between them. $SR(Z) = e^{i(\phi-\psi)}\overline{Z}$ a reflection in the line making an angle $(\phi - \psi)/2$ with the x-axis.

4.24 A reflection is indirect so their product is direct. Hence the product is either a rotation or a translation. Either can occur.

4.25 The product is indirect so is either a reflection or a glide reflection. Either can occur.

4.26 Their product is direct. Hence the product is either a rotation or a translation. Either can occur. The product is a translation if the angles of rotation are equal and opposite.

4.27 $T^{-1}(Z) = A^{-1}Z - A^{-1}B$ and $S^{-1}(Z) = \overline{A}^{-1}\overline{Z} - \overline{A}^{-1}\overline{B}$.

4.28 Suppose $\mathbf{M} = \begin{pmatrix} a & b \\ c & d \end{pmatrix}$ then $A = (a + d + i(b - c))/2$ and $B = (a - d + i(b + c))/2$.

4.29 Taking complex conjugate of the equation gives $\overline{B}X + \overline{AX} = \overline{C}$. Eliminating $\overline{X}$ yields $(A\overline{A} - B\overline{B})X = \overline{A}C - B\overline{C}$. If $|A| \neq |B|$ there is a unique solution $(\overline{A}C - B\overline{C})/(|A|^2 - |B|^2)$. If not, and $\overline{A}C = B\overline{C}$ then it is easy to check that $X = C/(A + \overline{A})$ is a solution.

4.30 $(NP \cdot NZ)^2 = (u^2 + v^2 + (1 - w)^2)(x^2 + y^2 + 1) = (2 - 2w)(2/(1 - w)) = 4$.

4.34 $Z \to (i\overline{Z} + 3)/(\overline{Z} + i)$.

4.35 Centre $1 - i$, radius $\sqrt{2}$.

4.36 Reflection in the plane $v = 0$ and reflection in the origin O.

4.38 The tangent line becomes a circle with diameter OP. The circle becomes the line PQ. The image of the point O' is the point where PQ and OO' meet.

4.39 Sadly no, as $AD - BC = 0$.

4.40 The images are the line $x = 1/2$ and the circle centre $-2i/3$, radius $1/3$ respectively.

4.42 Let $W = e^{i\phi}\frac{A-Z}{1-\overline{A}Z}$, then

$$\begin{aligned} 1 - W\overline{W} &= 1 - \frac{(A-Z)(\overline{A}-\overline{Z})}{(1-\overline{A}Z)(1-A\overline{Z})} \\ &= \frac{(1-A\overline{A})(1-Z\overline{Z})}{|1-\overline{A}Z|^2} \\ &> 0 \text{ if } |A| < 1 \text{ and } |Z| < 1. \end{aligned}$$

If $|A| > 1$ the unit disc is turned inside out but the unit circle is preserved set-wise. If $|A| = 1$ then the image of the transformation is the single point $e^{i\phi}A$.

4.46 The cross ratio is

$$\frac{(1-\frac{7+4i}{5})(i+2-\frac{14+3i}{5})}{(1-\frac{14+3i}{5})(i+2-\frac{7+4i}{5})} = -\frac{2}{3}.$$

Since this is real all four points are concyclic or collinear. However the latter case is clearly not true. The equation of the circle is $|Z-2| = 1$ or $(x-2)^2 + y^2 = 1$ (centre $(2,0)$ and radius 1).

4.47 We may as well assume that inversion is with respect to the unit circle. So $S(Z) = 1/\overline{Z}$. Then the cross ratio $(ZS(Z), WS(W))$ is equal to $(Z - W)(\overline{Z} - \overline{W})/(Z\overline{W} - 1)(\overline{Z}W - 1)$ which is real as it is unchanged under conjugation. The result now follows by the real cross ratio theorem.

4.49 A calculation shows that $(AB, CD) = -1$.

4.50 Let $W = \frac{A\overline{Z}+B}{C\overline{Z}+D}$ etc. Then $W - W' = (\overline{Z} - \overline{Z'})\frac{AD-BC}{(C\overline{Z}+D)(C\overline{Z'}+D)}$ and the result follows as in the direct case.

4.51 The cross ratio (AB, CD) is taken to (AB, CD) or $\overline{(AB, CD)}$. If the transformation is direct then (AB, CD) is unaltered. Otherwise $-1, 2, 1/2$ is unaltered and $(1 \pm i\sqrt{3})/2$ is taken to $(1 \mp i\sqrt{3})/2$.

4.54 $(x+1)(x^2-x+1)(x^2+x+1)$. It might be helpful to see this by noting that the roots are all the sixth roots of unity other than 1.

4.55 Let $w = e^{i\pi/10}$. Then the roots are $\{w,\ -\overline{w},\ -w^3,\ \overline{w}^3\}$.

4.56 $e^{(2r-1)\pi i/4}, r = 1, \ldots, 4$ and $e^{2s\pi i/5}, s = 0, \ldots, 4$.

4.57 $(x^2 - 2\cos(2\pi/7)x + 1)(x^2 - 2\cos(4\pi/7)x + 1)(x^2 - 2\cos(6\pi/7)x + 1)$.

4.59 $1, 1/2 \pm \sqrt{3}i/2$.

4.60 One way of seeing this is to show that the graph of $f(x) = x^3 + px + q$ crosses the x-axis three times. The maxima and minima of f occur at $x^2 = -p/3$. Since p is clearly negative by the condition there are two solutions. If we substitute in the equation we see that $f(-\sqrt{-p/3}) \geq 0$ and $f(\sqrt{-p/3}) \leq 0$ so there are three real roots.

5
Solid Geometry

O, that this too too solid flesh would melt
Thaw and resolve itself into a dew!

Hamlet

The geometry of the three-dimensional world in which we live must be one of the most interesting and useful topics to study. In this chapter we will only scratch the foreshore of possibilities. The undiscovered hinterland teems with unknown polyhedra, strange non-measurable sets, topological knots and links etc. We will only consider the simplest objects; points, lines, planes and a few polyhedra including the platonic solids.

The space $\mathbb{R}^3$ has a vector product which simplifies many formulæ. In this respect it is almost unique. The only other dimension which has a vector product is seven.

I have tried to make this chapter as self-contained as possible so there are overlaps with other chapters. Consequently some results are given without proof. More details on coordinates and the basics of linear algebra may be found in the chapter on coordinate geometry.

5.1 Points and Coordinates

Points in space, $\mathbb{R}^3$, are determined by three real *cartesian coordinates*. We write $P = (x, y, z)$ for a general point in space. Sometimes we use the suffix notation

$A = (a_1, a_2, a_3)$. The numbers x, y, z or a_1, a_2, a_3 are, naturally enough, called the x, y and z-coordinates. To make explanations simpler we will think of the z-coordinate as the height (or depth) above a horizontal plane (think of a table top) consisting of those points for which the z-coordinate is zero. In fact we can think of this horizontal plane as a copy of the standard two-dimensional space $\mathbb{R}^2$. The point (x, y) in $\mathbb{R}^2$ will be identified with the point $(x, y, 0)$ in space $\mathbb{R}^3$.

The x-axis is the set of points $(x, 0, 0)$. The y and z-axis are similarly defined. These three coordinate axes, determined by two zero coordinates, meet at the origin $O = (0, 0, 0)$.

We will further assume that the axes meet orthogonally. That is the system is *rectangular*. If the plane containing the x and y-axes is called the xy-coordinate plane with similar definitions for the yx and zx-coordinate planes then all three planes meet at right angles.

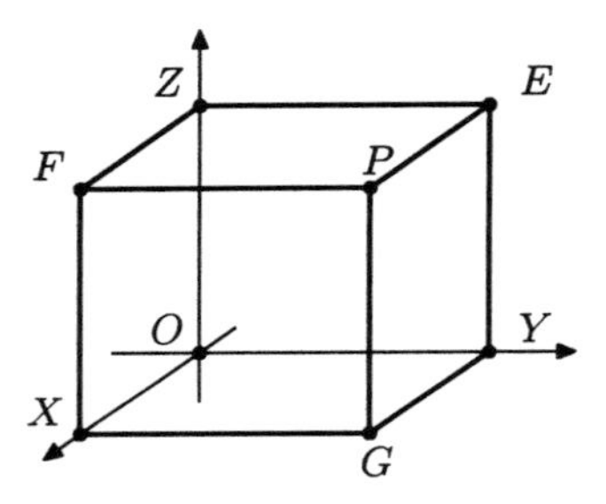

Fig. 5.1 Coordinate axes and planes

Suppose $P = (x, y, z)$ is a general point. Let

$$E = (0, y, z), \quad F = (x, 0, z), \quad G = (x, y, 0)$$

be the nearest points of the coordinate planes to P and let

$$X = (x, 0, 0), \quad Y = (0, y, 0), \quad Z = (0, 0, z)$$

be the nearest points of the coordinate axes.

Then O, X, G, Y, Z, F, P, E form the vertices of a rectangular box (Fig. 5.1).

Points may be added

$$A + B = (a_1, a_2, a_3) + (b_1, b_2, b_3) = (a_1 + b_1, a_2 + b_2, a_3 + b_3)$$

or multiplied by a real number

$$cA = c(a_1, a_2, a_3) = (ca_1, ca_2, ca_3).$$

We will denote by $I = (1, 0, 0)$, $J = (0, 1, 0)$ and $K = (0, 0, 1)$ the unit points on the coordinate axes. Since every point $P = (x, y, z)$ can be written uniquely as

$$\boldsymbol{P = xI + yJ + zK}$$

it follows that $\{I, J, K\}$ is a *basis* for $\mathbb{R}^3$. Any weighted sum such as $P = xI + yJ + zK$ is called a *linear combination* of $\{I, J, K\}$.

Exercise 5.1

Let $I' = (1, 1, 1), J' = (1, 1, 0), K' = (1, 0, 0)$. Show that any point can be written uniquely as a linear combination of $\{I', J', K'\}$. Deduce that $\{I', J', K'\}$ is a basis.

An ordered basis $\{A, B, C\}$ is called *right handed* if a rotation from A to B causes a right-handed screw to move in the positive direction along C. Otherwise the system is called *left handed.* We will always assume that our basis $\{I, J, K\}$ is right-handed.

If $\{A, B, C\}$ is a right-handed basis then so are $\{B, C, A\}$ and $\{C, A, B\}$. On the other hand $\{A, C, B\}$, $\{C, B, A\}$ and $\{B, A, C\}$ are left-handed.

It is very difficult to explain right-handedness except by using our common cultural knowledge of what a right-handed screw is. However if you wish to communicate the difference between left and right to an intelligent alien via a transgalactic telephone then you could argue that particles produced by beta decay of a neutron are preferentially displaced by an electric current. However this would all break down if the alien were made of antimatter.

The distance $|P|$ of the point $P = (x, y, z)$ from the origin O is given by

$$|\boldsymbol{P}| = \sqrt{\boldsymbol{x^2 + y^2 + z^2}}$$

which is an extension of Pythagoras' theorem. The number $|P|$ is called the *modulus* of P.

The distance of the point $A = (a_1, a_2, a_3)$ from the point $B = (b_1, b_2, b_3)$ is the modulus of the difference $A - B$, that is

$$|\boldsymbol{A} - \boldsymbol{B}| = \sqrt{(\boldsymbol{a_1} - \boldsymbol{b_1})^2 + (\boldsymbol{a_2} - \boldsymbol{b_2})^2 + (\boldsymbol{a_3} - \boldsymbol{b_3})^2}.$$

The points with unit modulus all lie on the unit sphere S^2. Any non-zero point P can be converted into one with unit modulus by

$$\hat{\boldsymbol{P}} = \frac{\boldsymbol{P}}{|\boldsymbol{P}|}, \qquad \boldsymbol{P} \neq \boldsymbol{O}.$$

If P is non-zero then the coordinates of $\hat{P}$ are called the *direction cosines* of P. If $\hat{P} = (\cos\alpha, \cos\beta, \cos\gamma)$ then $\cos\alpha^2 + \cos\beta^2 + \cos\gamma^2 = 1$ and the angles α, β, γ are the angles which the line OP makes with the coordinate axes (Fig. 5.2).

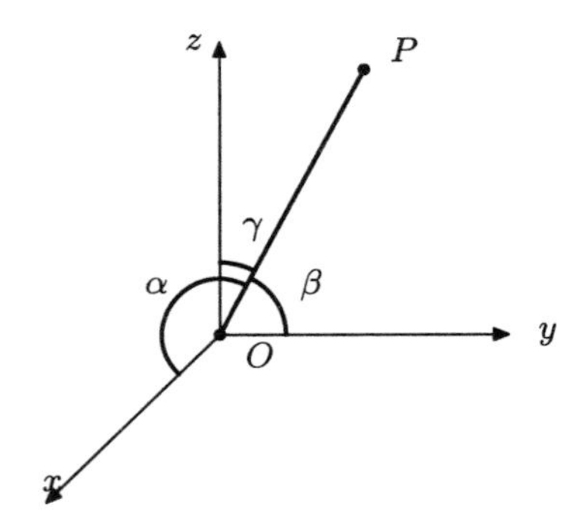

Fig. 5.2 Direction cosines

Exercise 5.2

If $P = (1, 2 - 2)$ find $\hat{P}$ and the angles which the line OP makes with the coordinate axes.

Let $G = (x, y, 0)$ be a point in the plane $z = 0$. This is the plane which we have identified with $\mathbb{R}^2$. Then we can consider its polar form $G = (r\cos\theta, r\sin\theta, 0)$ where $r = \sqrt{x^2 + y^2}$ and θ is the angle which OG makes with the x-axis. For a general point $P = (x, y, z)$ let $\rho = |P| = \sqrt{x^2 + y^2 + z^2}$ and let ϕ be the angle which OP makes with the z-axis, that is the third direction cosine of P (Fig. 5.3). Then $z = \rho\cos\phi$ and

$$r^2 = \rho^2 - z^2 = \rho^2 - \rho^2\cos^2\phi = \rho^2\sin^2\phi.$$

So the coordinates of P may be written $\rho(\cos\theta\sin\phi, \sin\theta\sin\phi, \cos\phi)$.

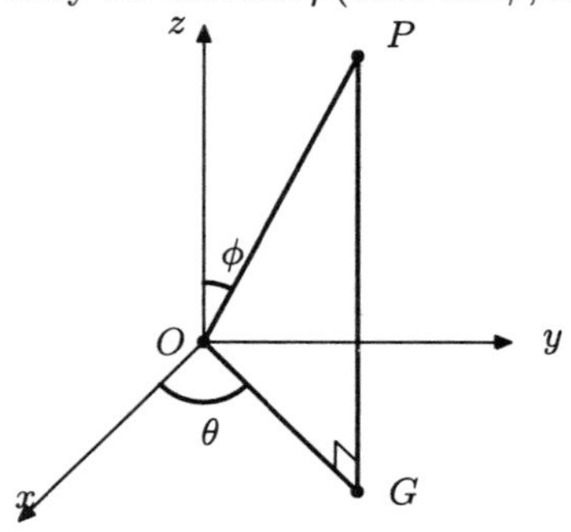

Fig. 5.3 Spherical polar coordinates

The numbers (ρ, θ, ϕ) are called the *spherical polar coordinates* of P. Beware: some authors interchange ϕ and θ. However I think it is more logical to keep θ as for polar plane coordinates.

Exercise 5.3

Find the spherical polar coordinates of the point with rectangular cartesian coordinates $(2, 1, -2)$.

5.2 Scalar Product

The *scalar* or *dot product* takes two points and generates a real number by the rule

$$\boldsymbol{A} \cdot \boldsymbol{B} = \boldsymbol{a_1 b_1 + a_2 b_2 + a_3 b_3}.$$

We can easily check that the dot product satisfies

$$\begin{aligned} A \cdot A &= |A|^2 \\ A \cdot B &= B \cdot A \\ (cA) \cdot B &= A \cdot (cB) = c(A \cdot B) \\ (A + B) \cdot C &= A \cdot C + B \cdot C \end{aligned}$$

A more intrinsic definition (coordinate free) is

$$\boldsymbol{A} \cdot \boldsymbol{B} = |\boldsymbol{A}||\boldsymbol{B}| \cos \boldsymbol{\theta}$$

where θ is the angle $\angle AOB$. So A, B are orthogonal if and only if $A \cdot B = 0$.

Exercise 5.4

In the coordinate axes box illustrated in Fig. 5.1 find the distance PX and the angle $\angle GOP$.

Exercise 5.5

Find the angle between any two diagonals of a cube.

Exercise 5.6

Show that if two pairs of edges of a tetrahedron are orthogonal then so is the third.

The dot product multiplication table for the basis $\{I, J, K\}$ is

$$\begin{array}{c|ccc} \cdot & \boldsymbol{I} & \boldsymbol{J} & \boldsymbol{K} \\ \boldsymbol{I} & 1 & 0 & 0 \\ \boldsymbol{J} & 0 & 1 & 0 \\ \boldsymbol{K} & 0 & 0 & 1 \end{array}$$

Any basis having this property is called *orthonormal.* This is a useful property because if $\{A, B, C\}$ is orthonormal then the coefficients in the equation

$$\boldsymbol{P} = x\boldsymbol{A} + y\boldsymbol{B} + z\boldsymbol{C}$$

can be found by the rule

$$x = \boldsymbol{A} \cdot \boldsymbol{P}, \quad y = \boldsymbol{B} \cdot \boldsymbol{P}, \quad z = \boldsymbol{C} \cdot \boldsymbol{P}.$$

Exercise 5.7

Let $A = (1/\sqrt{2}, 1/\sqrt{2}, 0), B = (1/\sqrt{2}, -1/\sqrt{2}, 0)$. Find the point C so that $\{A, B, C\}$ is a right-handed orthonormal basis.

5.3 Cross Product

The *vector product* or *cross product* takes two points or vectors $A = (a_1, a_2, a_3)$, $B = (b_1, b_2, b_3)$ and supplies a third $A \times B$ by the rule

$$\boldsymbol{A} \times \boldsymbol{B} = (\boldsymbol{a_2 b_3} - \boldsymbol{a_3 b_2}, \boldsymbol{a_3 b_1} - \boldsymbol{a_1 b_3}, \boldsymbol{a_1 b_2} - \boldsymbol{a_2 b_1}).$$

The cross product table for the standard right-handed basis $\{I, J, K\}$ is

$$\begin{array}{c|ccc} \times & \boldsymbol{I} & \boldsymbol{J} & \boldsymbol{K} \\ \boldsymbol{I} & \boldsymbol{O} & \boldsymbol{K} & -\boldsymbol{J} \\ \boldsymbol{J} & -\boldsymbol{K} & \boldsymbol{O} & \boldsymbol{I} \\ \boldsymbol{K} & \boldsymbol{J} & -\boldsymbol{I} & \boldsymbol{O} \end{array}.$$

It is easy to check that

$$\begin{aligned} (cA) \times B &= A \times (cB) = c(A \times B) \\ (A + B) \times C &= A \times C + B \times C \\ A \times (B + C) &= A \times B + A \times C. \end{aligned}$$

Unlike the dot product, which is symmetric, the cross product is *anti-symmetric*, that is

$$A \times B = -B \times A.$$

In particular $A \times A = O$. So A is parallel to B if and only if $A \times B = O$.

Exercise 5.8

Show that the anti-symmetry of the cross product follows from the property $A \times A = O$ for all A.

We now give an intrinsic definition of $A \times B$.

■ Let A, B be non-parallel and let N have unit modulus and be orthogonal to both A and B so that $\{A, B, N\}$ is a right-handed basis. Then $A \times B = (|A||B| \sin\theta)N$ where θ is the angle between A and B.

Proof It is an easy manipulation to prove that $A \cdot (A \times B) = B \cdot (A \times B) = 0$. So $A \times B$ is orthogonal to to both A and B. Hence $A \times B = \pm|A \times B|N$. Now

$$\begin{aligned}|A \times B|^2 &= (a_2b_3 - a_3b_2)^2 + (a_3b_1 - a_1b_3)^2 + (a_1b_2 - a_2b_1)^2 \\ &= a_1^2(b_2^2 + b_3^2) + a_2^2(b_1^2 + b_3^2) + a_3^2(b_1^2 + b_2^2) \\ &\quad - 2(a_1a_2b_1b_2 + a_2a_3b_2b_3 + a_3a_1b_3b_1) \\ &= |A|^2|B|^2 - a_1^2b_1^2 - a_2^2b_2^2 - a_3^2b_3^2 \\ &\quad - 2(a_1a_2b_1b_2 + a_2a_3b_2b_3 + a_3a_1b_3b_1) \\ &= |A|^2|B|^2 - (a_1b_1 + a_2b_2 + a_3b_3)^2 \\ &= |A|^2|B|^2 - |A|^2|B|^2\cos^2\theta \\ &= |A|^2|B|^2\sin^2\theta.\end{aligned}$$

It remains to chose the right sign when taking the square root. The choice is clearly continuous and does not change provided the cross product never vanishes. Rotate A, B so that they lie in the xy-plane. Say $A = (|A|, 0, 0)$ and $B = (|B|\cos\theta, |B|\sin\theta, 0)$. Then $A \times B = (0, 0, |A||B|\sin\theta)$ and since our original choice of axes was right-handed so is $A, B, A \times B$. □

■ If A, B are non-parallel then $\{A, B, A \times B\}$ is a right-handed basis for $\mathbb{R}^3$. If in addition A, B are unit length and orthogonal then $\{A, B, A \times B\}$ is a right-handed orthonormal basis for $\mathbb{R}^3$. □

Exercise 5.9

Let $A = (1, -1, 1), B = (1, 1, 1)$. Find a point C so that $\{A, B, C\}$ is a left-handed basis for $\mathbb{R}^3$.

■ The area of the parallelogram with vertices $O, A, B, A+B$ is $|A \times B|$.

Proof Using the "base times height" formula, take the base to be $|A|$ and the height to be $|B| \sin\theta$ with θ the angle $\angle AOB$. This gives the result. □

Exercise 5.10

Find the area of the parallelogram with vertices $(-1,0,0), (0,1,1)$, $(0,-1,1)$ and $(1,0,2)$.

Exercise 5.11

Find the area of the triangle with vertices $A = (1,0,1)$, $B = (0,2,3)$ and $C = (2,1,0)$.

5.4 The Scalar Triple Product

If we combine the cross and the dot products we get the *scalar triple product*,

$$(\boldsymbol{A} \times \boldsymbol{B}) \cdot \boldsymbol{C} = \boldsymbol{a_2 b_3 c_1 - a_3 b_2 c_1 + a_3 b_1 c_2 - a_1 b_3 c_2 + a_1 b_2 c_3 - a_2 b_1 c_3}.$$

The notation $[ABC]$ is used for $(A \times B) \cdot C$. The scalar triple product can be written as the 3×3 determinant

$$[\boldsymbol{ABC}] = (\boldsymbol{A} \times \boldsymbol{B}) \cdot \boldsymbol{C} = \begin{vmatrix} \boldsymbol{a_1} & \boldsymbol{a_2} & \boldsymbol{a_3} \\ \boldsymbol{b_1} & \boldsymbol{b_2} & \boldsymbol{b_3} \\ \boldsymbol{c_1} & \boldsymbol{c_2} & \boldsymbol{c_3} \end{vmatrix}.$$

It is easy to see that $[ABC]$ is linear in each variable. For example

$$[(A+B)CD] = [ACD] + [BCD] \text{ and } [(cA)BC] = c[ABC].$$

An elementary property of determinants is that interchanging two rows multiplies it by -1. Hence

$$[ABC] = [BCA] = [CAB] = -[ACB] = -[CBA] = -[BAC].$$

We also have $[ABA] = [ABB] = 0$ and so we have another proof that $A \times B$ is orthogonal to A and B.

■ The points $\{A, B, C\}$ form a basis if and only if $[ABC] \neq 0$. It is a right-handed basis if $[ABC] > 0$ and a left-handed one if $[ABC] < 0$.

Proof If $[ABC] \neq 0$ then $A \times B$ cannot be zero and so A, B are not parallel. Also C cannot lie in the plane determined by A, B and so $\{A, B, C\}$ form a basis. Conversely if $[ABC] = 0$ then C is orthogonal to $A \times B$ and so lies in the plane containing A, B. Hence $\{A, B, C\}$ cannot form a basis. If $[ABC] = A \times B \cdot C > 0$ then C lies on the same side of the plane containing A, B as $A \times B$ and so $\{A, B, C\}$ form a right-handed basis. If $[ABC] < 0$ then C lies on the other side of the plane and the basis is a left-handed one. □

A *parallelepiped* is a three-dimensional generalisation of a parallelogram. So if O, A, B, C are four of the vertices then the other four are $A+B, B+C, C+A$ and $A+B+C$ (Fig. 5.4).

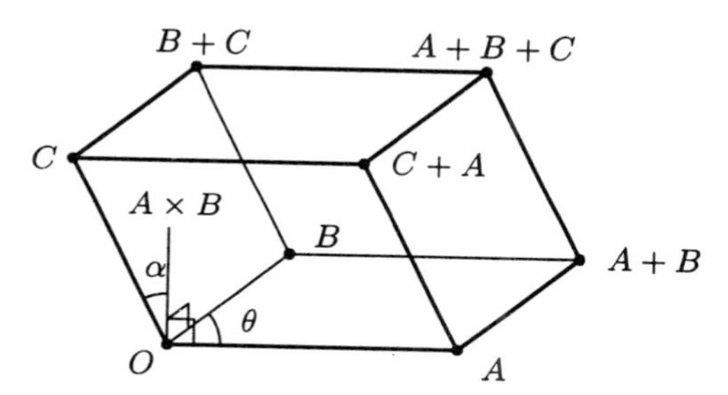

Fig. 5.4 A parallelepiped

■ With the notation above the volume of the parallelepiped is $[\boldsymbol{ABC}]$.

Proof The volume is "area of the base times height". The area of the base is the area of the parallelogram, that is $|A \times B|$. The height is $|C| \cos \alpha = C \cdot A \times B / |A \times B|$. The product is $[ABC]$. □

Exercise 5.12

What is the significance of negative volume?

■ The volume of a tetrahedron with vertices O, A, B, C is

$$[\boldsymbol{ABC}]/6 = \frac{1}{6} \begin{vmatrix} a_1 & a_2 & a_3 \\ b_1 & b_2 & b_3 \\ c_1 & c_2 & c_3 \end{vmatrix}.$$

Proof An elementary property of volumes is that the volume of a pyramid is "one third area of the base times height". Since the area of the base is $|A \times B|/2$ the result follows as for the volume of a parallelepiped. □

Exercise 5.13

Find the volume of the tetrahedron with vertices $(1,-2,0),(0,1,1)$, $(1,0,1)$ and $(-2,1,1)$.

5.5 The Vector Triple Product

Unlike ordinary multiplication, the cross product is not in general associative. That is $A \times (B \times C) \neq (A \times B) \times C$. However there is a formula which allows us to identify both sides.

The triple vector product formula:

$$(\boldsymbol{A} \times \boldsymbol{B}) \times \boldsymbol{C} = (\boldsymbol{A} \cdot \boldsymbol{C})\boldsymbol{B} - (\boldsymbol{B} \cdot \boldsymbol{C})\boldsymbol{A}.$$

Proof One method of proof would be to expand both sides using our various formulæ and check that the results were equal. But this would not give a great deal of insight into the situation.

Here is another way. Assume that A, B are not parallel. Now $X = (A \times B) \times A$ is orthogonal to A and lies in the plane of A, B. Also by right-handedness B and X lie on the same side of the plane containing A and $A \times B$. Moreover $|X|$ is $|A|^2|B|\sin\theta$ where θ is the angle $\angle AOB$.

Then $(A \cdot A)B - (A \cdot B)A$ satisfies these requirements which uniquely define X and so

$$X = (A \times B) \times A = (A \cdot A)B - (A \cdot B)A.$$

Similarly

$$(A \times B) \times B = (A \cdot B)B - (B \cdot B)A.$$

Now $\{A, B, A \times B\}$ is a basis for $\mathbb{R}^3$ and so for some x, y, z we have

$$C = xA + yB + zA \times B.$$

Expanding $(A \times B) \times (xA + yB + zA \times B)$ and using the fact that $A \times B$ is orthogonal to A and B we get

$$\begin{aligned}(A \times B) \times C &= x(A \cdot A)B - x(A \cdot B)A + y(A \cdot B)B - y(B \cdot B)A \\ &= (-xA - yB - z(A \times B)) \cdot B\ A \\ &\qquad + (xA + yB + z(A \times B)) \cdot A\ B \\ &= (A \cdot C)B - (B \cdot C)A.\end{aligned}$$

□

The three-dimensional nature of the above proof is essential. The vector product in $\mathbb{R}^7$ does not have such an expansion.

Exercise 5.14

Show that $A \times (B \times C) = (A \cdot C)B - (A \cdot B)C$.

Exercise 5.15

Find conditions on A, B, C for which $A \times (B \times C) = (A \times B) \times C$.

Exercise 5.16

Prove the Jacobi identity,

$$A \times (B \times C) + B \times (C \times A) + C \times (A \times B) = O.$$

Exercise 5.17

Prove the Lagrange identity,

$$(A \times B) \cdot (C \times D) = (A \cdot C)(B \cdot D) - (B \cdot C)(A \cdot D).$$

Exercise 5.18

By expanding $(A \times B) \times (C \times D)$ in two ways prove that

$$[ACD]B - [BCD]A = [ABD]C - [ABC]D.$$

5.6 Planes

A plane is the set of points satisfying an equation of the form

$$ax + by + cz = d.$$

If $A = (a, b, c)$ and $P = (x, y, z)$ then this can be written as

$$\boldsymbol{A \cdot P = d.}$$

The line OA is perpendicular to the plane $A \cdot P = d$ because if P_1 and P_2 lie in the plane then $A \cdot (P_1 - P_2) = d - d = 0$. So the angle between two planes $A_1 \cdot P = d_1$ and $A_2 \cdot P = d_2$ can be defined to be the angle between the lines OA_1 and OA_2.

Example 5.1

To find the angle between the planes given by $x+y+z=2$ and $y-z=4$ note that $A_1=(1,1,1)$ and $A_2=(0,1,-1)$. Now $A_1 \cdot A_2 = 0$ so the planes are at right angles.

The proof of the following can be found in Chapter 2.

■ The nearest point on the plane $A \cdot P = d$ to a point B is

$$\frac{d - A \cdot B}{|A|^2} A + B$$

and its distance from B is

$$\frac{|d - A \cdot B|}{|A|}.$$

□

Exercise 5.19

Consider the tetrahedron with vertices $(1,-2,0),(0,1,1),(1,0,1),$ $(-2,1,1)$. Find the distance from the vertex $(-2,1,1)$ to the opposite face.

The equation of a plane is determined by three non-collinear points on the plane. We can substitute these in the equation and solve the resulting linear equations. An alternative method is to note that if A, B, C lie in the plane then $(B-A) \times (C-A)$ is perpendicular to the plane and so its equation is $(B-A) \times (C-A) \cdot P = d$ for suitable d. The value $d = [ABC]$ can be found by substituting P for any point on the plane, A say. Then after simplification, the equation of the plane is

$$(\boldsymbol{A} \times \boldsymbol{B} + \boldsymbol{B} \times \boldsymbol{C} + \boldsymbol{C} \times \boldsymbol{A}) \cdot \boldsymbol{P} = [\boldsymbol{ABC}].$$

Example 5.2

To find the equation of the plane through the points $(0,1,1),(1,0,1)$ and $(-2,1,2)$ we find $((1,0,1)-(0,1,1)) \times ((-2,1,2)-(0,1,1)) = (1,-1,0) \times (-2,0,1) = (-1,-1,-2)$ so the equation is $-x-y-2z=d$ for some d. Substituting $(0,1,1)$ we see that $d=-3$ so the equation is $x+y+2z=3$.

5.7 Lines in Space

A line is determined by two points on it, say A and B. Then a general point on the line can be given in *parametric form* as $P = tA + (1 - t)B$ where the parameter t may be any real number. In particular if $t = 1/2$ then

$$\boldsymbol{M} = \frac{\boldsymbol{A} + \boldsymbol{B}}{\mathbf{2}}$$

is the midpoint of AB.

Any non-zero multiple of $A - B$ defines the direction of the line. The angle between two lines is the angle between their directions.

A line can also be considered as the intersection of two planes. If $A = (a, b, c)$ is a point on a line with direction (l, m, n) then a general point, $P = (x, y, z)$ on the line satisfies,

$$\frac{x-a}{l} = \frac{y-b}{m} = \frac{z-c}{n}.$$

This is called the *symmetric form* of the line equation. The line is the intersection of the planes given by equations

$$x/l - y/m = a/l - b/m \text{ and } y/m - z/n = b/m - c/n.$$

If the line passes through the two points (a_1, a_2, a_3) and (b_1, b_2, b_3), the symmetric form of the line is

$$\frac{x-a_1}{b_1-a_1} = \frac{y-a_2}{b_2-a_2} = \frac{z-a_3}{b_3-a_3}.$$

Exercise 5.20

Find the symmetric form of the line through $(11, 2, -1)$ and $(-9, 4, 5)$.

Exercise 5.21

Show that lines joining the midpoints of opposite edges of a tetrahedron have a common middle point.

If a line ℓ is the intersection of two planes $A \cdot P = d$ and $B \cdot P = d'$ then ℓ lies in both and so is orthogonal to the normals to both planes. So the direction of ℓ is $A \times B$.

There is a unique nearest point P on a line ℓ to a point Q not on the line. This is the intersection of the line through Q in the plane determined by the point Q and the line ℓ which meets ℓ at right angles.

Example 5.3

To find the nearest point on the line through $(11, 0, -1)$ and $(-9, 4, 5)$ to the origin consider a general point on the line, denoted by $P_t = t(11, 0, -1) + (1 - t)(-9, 4, 5) = (20t - 9, -4t + 4, -6t + 5)$. Then OP_t is orthogonal to the direction of the line when

$$(20t - 9, -4t + 4, -6t + 5) \cdot (20, -4, -6) = 0.$$

That is $t = 1/2$. So $P_{1/2} = (1, 2, 2)$ is the nearest point.

Vector Equation of Lines

The vector product can be used to give an equation for a line in space.

■ Let A, B be orthogonal vectors with A of unit length. Then the set of points P satisfying the equation

$$\boldsymbol{P} \times \boldsymbol{A} = \boldsymbol{B}$$

is a line parallel to A passing through $A \times B$. Conversely any line in space has an equation of this form which is unique up to sign.

Proof Substituting $P = A \times B$ in $P \times A$ gives

$$(A \times B) \times A = (A \cdot A)B - (A \cdot B)A = B$$

using the usual expansion rules for the triple vector product. So $A \times B$ is a point on the line.

If P_1 and P_2 are two points satisfying the equation then $(P_1 - P_2) \times A = B - B = O$ and so $P_1 - P_2$ is parallel to A. So all points satisfying the equation above lie on a line parallel to A passing through $A \times B$.

Conversely consider a line ℓ in space. Suppose that ℓ is parallel to the unit vector A. Then A is well defined up to sign. Let C be the point of ℓ nearest the origin O. Put $B = C \times A$. □

Exercise 5.22

Find the vector equation $P \times A = B$ for the line through the two points $(1, 1, 1)$ and $(-1, -1, 1)$.

Exercise 5.23

Show that the vector equation $P \times A = B$ for the line through two points P_0 and P_1 has

$$A = \frac{P_1 - P_0}{|P_1 - P_0|}, \qquad B = \frac{P_0 \times P_1}{|P_1 - P_0|}.$$

Exercise 5.24

Show that the line $P \times A = B$ passes through the origin if and only if $B = O$. If $B \neq O$ show that the line lies in the plane $P \cdot B = 0$.

Exercise 5.25

Let

$$P \cdot C = p, \qquad P \cdot C' = p'$$

be the equation of two planes which meet in a line. Show that the equation of the line is $P \times A = B$ where

$$A = \frac{C \times C'}{|C \times C'|}, \qquad B = \frac{p'C - pC'}{|C \times C'|}.$$

Configurations of Lines

Two lines in space either meet in a point, are parallel or do not lie in a common plane. In the latter case the lines are called *skew*.

Exercise 5.26

Show that the lines

$$\begin{aligned}(x-5)/4 = (y-7)/4 = -(z+3)/5,\\ (x-8)/7 = y-4 = (z-5)/3\end{aligned}$$

are coplanar. Find their common point and the equation of the plane in which they lie.

Exercise 5.27

* Show that the lines $P \times A = B$ and $P \times A' = B'$ are skew if and only if $A \times A' \neq O$ and $A \cdot B' + A' \cdot B \neq 0$.

A *transversal* of a set of lines is a line passing through all of them. There is a unique transversal to two skew lines which is perpendicular to both lines. To see this let the lines be ℓ_1 and ℓ_2. Let ϖ be a plane containing ℓ_1 and parallel to ℓ_2. To each point on ℓ_2 there is a unique nearest point on ϖ. The set of these points defines a line, ℓ_3, in ϖ. Let O_1 be the intersection of ℓ_1 and ℓ_3. Let O_2 be the point on ℓ_2 corresponding to O_1 on ℓ_3. Then O_1O_2 is the desired transversal.

Exercise 5.28

Find the shortest distance between the lines

$$x - 10 = \frac{y-9}{3} = -\frac{z+2}{2} \text{ and } \frac{x+1}{2} = \frac{y-12}{4} = z - 5.$$

Exercise 5.29

Show that the shortest distance between two skew lines $P \times A = B$ and $P \times A' = B'$ is

$$\frac{A \cdot B' + A' \cdot B}{|A \times A'|}.$$

If ℓ_1 and ℓ_2 are two skew lines and O is a point on neither, then the plane containing O and ℓ_1 meets ℓ_2 in a point O'. The line OO' is the unique transversal of ℓ_1 and ℓ_2 passing through O.

If O above moves along a further skew line ℓ_3 then the set of transversals of ℓ_1, ℓ_2 and ℓ_3 forms a quadric surface called a regulus. We will have more to say about these in Chapter 7.

Exercise 5.30

Let ℓ_1 be the line $z = 1,\ x+y+1 = 0$ and let ℓ_2 be the line $y = 1,\ x-z = 1$. Find the transversal of ℓ_1 and ℓ_2 passing through the origin O.

Configurations of Planes

Two planes are either parallel or meet in a line. If the planes $A \cdot P = d$ and $A' \cdot P = d'$ are parallel then A and A' are parallel and so $A \times A' = O$.

There are five possible configurations of three distinct planes. They are

Triple intersection: The planes meet in pairs in three lines all passing through a common point. The coordinate planes are a good model for this.

Triangular prism: The planes meet in pairs in three parallel but distinct lines.

Coaxial: The planes meet in a common line.

Two parallel planes:

Three parallel planes:

Illustrations for these cases can be found in Chapter 2, Fig. 2.16.

■ Three planes $A \cdot P = d, B \cdot P = d', C \cdot P = d''$ intersect in a triple point if and only if $[ABC] \neq 0$.

Proof To find a common point of the three planes we must solve the simultaneous linear equations

$$\begin{aligned} a_1x + a_2y + a_3z &= d \\ b_1x + b_2y + b_3z &= d' \\ c_1x + c_2y + c_3z &= d''. \end{aligned}$$

This has a unique solution if and only if

$$\begin{vmatrix} a_1 & a_2 & a_3 \\ b_1 & b_2 & b_3 \\ c_1 & c_2 & c_3 \end{vmatrix} = [ABC] \neq 0.$$

□

Exercise 5.31

Show that the planes

$$x + y + z = 0, \quad x - 2y + z = 6, \quad y - z = -3$$

have a triple point intersection and find this point.

Exercise 5.32

Show that the planes

$$\varpi_1 : x + y + z = 0, \quad \varpi_2 : 5x - y - 4z = 4, \quad \varpi_3 : 4x + 2y + z = 8$$

form an example of the second kind of configuration and find the area of the triangular cross-section.

Exercise 5.33

Show that the planes

$$8x + 3y + 5z = 4, \quad 18x - 3y + 13z = 6, \quad 5x - 3y + 4z = 1$$

contain a common line.

5.8 Isometries of Space

Any transformation of space which preserves distance is called an *isometry*. By this we mean that if $X \to T(X)$ is a transformation of space then T is an isometry if $|T(X) - T(Y)| = |X - Y|$ for all X, Y. It quickly follows that an isometry also preserves angle. The simplest isometry is the translation by A which adds A to each point. This sends the origin to A and transports the coordinate axes in a parallel fashion so that they are then based at A.

The isometries which keep the origin O fixed are called *orthogonal* transformations. They preserve the scalar product and the vector product is preserved up to sign.

Orthogonal transformations consist of three types.

Rotations

Let ℓ be the line OA. Any point X lies in a unique plane ϖ orthogonal to ℓ with equation in P given by $A \cdot P = A \cdot X$. The point

$$O' = \frac{A \cdot X}{A \cdot A} A$$

is the intersection of the line ℓ with the plane ϖ.

Let X' be the point of ϖ such that $|X - O'| = |X' - O'|$ and $\angle XO'X' = \theta$, the angle being chosen so that rotation from X to X' causes a right-handed screw to move forward along OA (Fig. 5.5).

The transformation $X \to X'$ is called a *rotation* about ℓ through an angle θ.

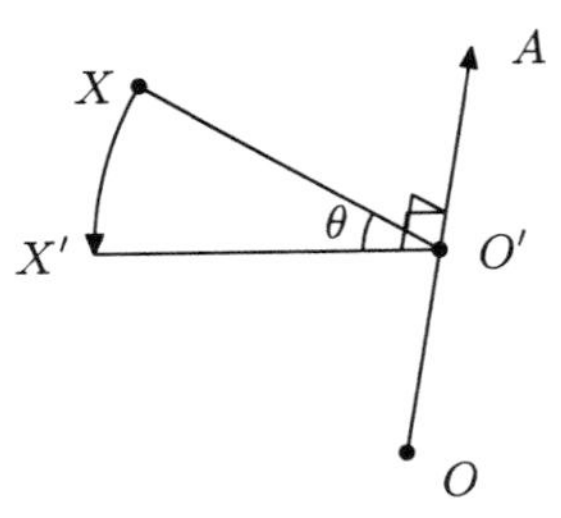

Fig. 5.5 A spatial rotation

It is important that the line ℓ be oriented by the choice of A in order to get the sense of rotation for θ. The only angles for which this is unimportant are 0 and π. In the latter case the rotation is a reflection in the line ℓ or a *half turn* about ℓ.

We can associate to every point P in the ball $|P| \leq \pi$ a rotation about OP through an angle $|P|$. Points on the boundary sphere $|P| = \pi$ correspond to half turns. Since a point P on the boundary gives rise to the same half turn as the antipodal point $-P$, the space of rotations may be identified with the corresponding quotient space, the real projective space.

The rotation is a *direct* isometry because handedness is preserved.

Exercise 5.34

Show that $X \to X\mathbf{M}$ where M is the matrix

$$\mathbf{M} = \begin{pmatrix} \cos\theta & \sin\theta & 0 \\ -\sin\theta & \cos\theta & 0 \\ 0 & 0 & 1 \end{pmatrix}$$

is a rotation about the z-axis through an angle θ.

Exercise 5.35

Let $X \to X\mathbf{M}$ be a rotation in space. Show that the axis of the rotation is the line in the direction A where A is an eigenvector corresponding to the eigenvalue 1. Show that the rotation is through an angle θ where $1 + 2\cos\theta$ is the trace of $\mathbf{M}$ (the sum of the diagonal elements).

Exercise 5.36

Show that the rotation $X \to X\mathbf{M}$ where M is the matrix

$$\mathbf{M} = \begin{pmatrix} 0 & 1 & 0 \\ 0 & 0 & 1 \\ 1 & 0 & 0 \end{pmatrix}$$

is about an axis along $(1, 1, 1)$ through an angle $2\pi/3$.

Exercise 5.37

Find the axis and angle for the rotation $X \to X\mathbf{M}$ where M is the matrix

$$\mathbf{M} = \begin{pmatrix} 4/5 & -12/25 & 9/25 \\ 3/5 & 16/25 & -12/25 \\ 0 & 3/5 & 4/5 \end{pmatrix}.$$

Reflections in a Plane

Consider a plane through the origin, $A \cdot P = 0$. Let X be an arbitrary point and let R be the nearest point of the plane to X. Let XX' be the extension of the line XR so that R is the midpoint of XX'. Then the transformation $X \to X'$ is the *reflection* of X in the plane (Fig. 5.6).

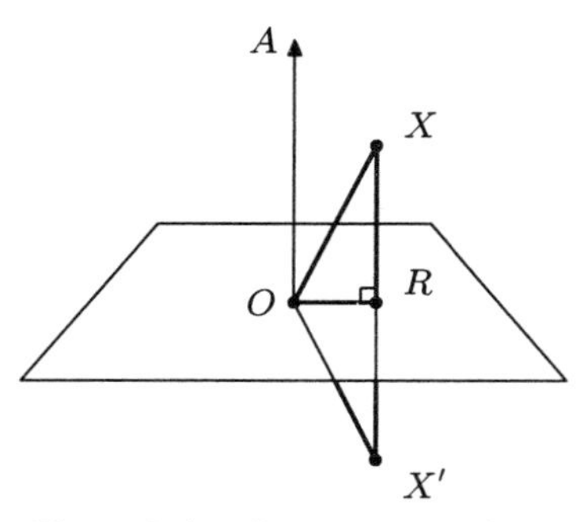

Fig. 5.6 A reflection in a plane

We can find an analytical form for a plane reflection as follows. Note that $X' - X$ is parallel to A and that $X + X'$ is orthogonal to A. The first fact means that $X' = X + cA$ for some number c and the second fact means that $X \cdot A = -X' \cdot A$. We can find the value of c by taking the dot product of the first equation with A and using the second equation. This gives

$$\boldsymbol{X' = X - 2\frac{X \cdot A}{A \cdot A}A.}$$

Exercise 5.38

Find in coordinate form the reflection in the plane $x - y + z = 0$.

Exercise 5.39

Let $X \to X\mathbf{M}$ be a reflection in a plane in space. Show that the plane of the reflection is perpendicular to the direction A where A is an eigenvector corresponding to the eigenvalue -1.

Exercise 5.40

Your image in a mirror has left and right interchanged. Why not up and down?

Consider two reflections T_A and T_B in non-parallel planes $A \cdot P = 0$ and $B \cdot P = 0$. We can assume that both A and B have unit length and this simplifies the calculations. Then the composition is given by

$$T_A T_B(X) = X - 2(X \cdot A - 2X \cdot B\ A \cdot B)A - 2X \cdot B\ B.$$

Then $T_A T_B$ fixes any X orthogonal to A and B (because $X \cdot A = X \cdot B = 0$). Moreover if we act on A by $T_A T_B$ and take the dot product with A then we get $2(A \cdot B)^2 - 1 = \cos 2\theta$ where θ is the angle between A and B. So we conclude

■ The composition of two reflections in non-parallel planes is a rotation about their line of intersection through twice the angle between the planes. □

Exercise 5.41

Show that $T_A T_B = T_B T_A$ if and only if the two planes are orthogonal or coincide.

Exercise 5.42

Show that $T_A T_B = T_{T_A(B)} T_A$.

Any set with a self-distributive binary operation such as the one in the above question is called a rack.

Rotary Reflections

The composition of a reflection in a plane and a rotation about an axis orthogonal to the plane is called a *rotary reflection*. A rotary reflection is determined by the reflection plane (with oriented normal) and an angle of rotation. Notice that we could have reversed the order so that the rotation was first, followed by the reflection.

If the angle is π then the rotary reflection is reflection in the origin O.

Reflections and rotary reflections are both *indirect isometries* because handedness is reversed.

Exercise 5.43

Show that $X \to X\mathbf{M}$ where M is the matrix

$$\mathbf{M} = \begin{pmatrix} \cos\theta & \sin\theta & 0 \\ -\sin\theta & \cos\theta & 0 \\ 0 & 0 & -1 \end{pmatrix}$$

is a rotary reflection in the xy-plane through an angle θ.

Exercise 5.44

Let $X \to X\mathbf{M}$ be a rotary reflection in space. Show that the invariant plane of the rotary reflection is perpendicular to the direction A where A is an eigenvector corresponding to the eigenvalue -1. Show that the angle of the rotary reflection is θ where $-1 + 2\cos\theta$ is the trace of $\mathbf{M}$.

Exercise 5.45

Show that any orthogonal transformation T can be translated to one based at P by the rule $S(X) = T(X - P) + P$.

All the above orthogonal transformations can be illustrated by considering the symmetries of a cube.

Consider a cube $ABCDA'B'C'D'$ with centre O (Fig. 5.7). Let M be the midpoint of the edge AB and let N be the midpoint of the face $ABCD$. Let ϖ be the plane midway between the faces $ABCD$ and $A'B'C'D'$. Let ϖ' be the plane midway between the edges AB and $C'D'$. Let ϖ'' be the plane midway between the vertices A and C'. The plane ϖ meets the cube in a square, the plane ϖ' meets the cube in a rectangle with sides in the ratio $1 : \sqrt{2}$ and the plane ϖ'' meets the cube in a regular hexagon.

There are 48 symmetries of a cube. The identity is one and that leaves 47 to find.

Rotations

There is a rotation about ON through $\pi/2$. If repeated the effect is a rotation about ON through π or a half turn about the line ON. One more repeat gives a rotation about ON through $3\pi/2$ or $-\pi/2$. Since there are 3 axes through the centre of faces there are $3 \times 3 = 9$ of these rotations.

There is a half turn about OM. There are $1 \times 6 = 6$ of these.

There is a rotation about OA through $2\pi/3$. It is convenient to adopt the notation of permutations at this point and note that the permutation of the vertices is $(A'BD)(B'CD')$ where $(A'BD)$ means $A' \to B \to D \to A'$ etc. The vertices A, C' remain fixed. If we repeat this rotation we get a rotation about OA through $-2\pi/3$. There are $2 \times 4 = 8$ of these.

There are $9 + 6 + 8 = 23$ rotational symmetries. Together with the identity there are 24 direct symmetries.

Plane Reflections

There is a reflection in the plane ϖ. There are 3 of these.

There is a reflection in the plane ϖ'. There are 6 of these.

There are $3 + 6 = 9$ plane reflection symmetries.

Rotary Reflections

There is a reflection in the plane ϖ followed by a rotation through $\pi/2$. Repetition gives reflection in the line ON (a half turn). This can be seen by looking at the permutations of the vertices. A further repetition gives a reflection in

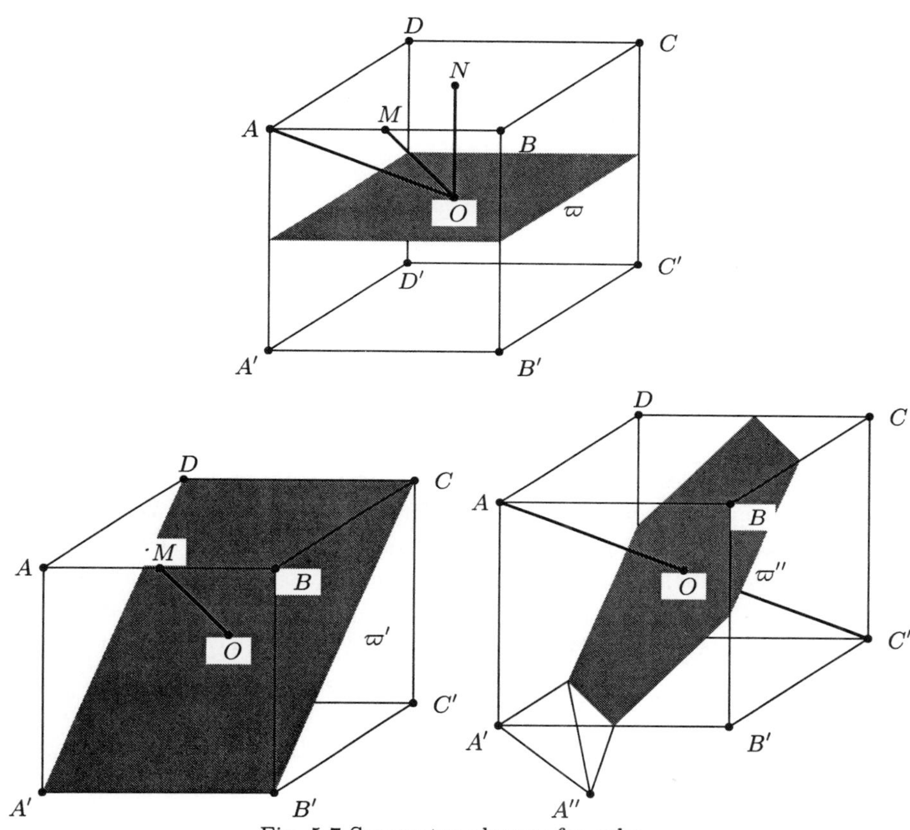

Fig. 5.7 Symmetry planes of a cube

the plane ϖ followed by a rotation through $-\pi/2$. There are $2 \times 3 = 7$ of these rotary reflections.

There is a reflection in the plane ϖ' followed by a rotation through π. This is reflection in O.

There is a reflection in the plane ϖ'' followed by a rotation through $\pi/3$. This rotates the hexagon once and interchages vertices A and C'. The effect on the other vertices is a little tricky to see. The vertex A' goes to the point A'' on reflection and then A'' goes to the vertex B' by the rotation. The corresponding permutation of the vertices is $(AC')(A'B'BCDD')$ in cycle notation.

Let T denote this rotary reflection. Then T^2 is a rotation about OA through $2\pi/3$, T^3 is reflection in O, T^4 is a rotation about OA through $-2\pi/3$, T^5 is a rotatory reflection in ϖ'' with angle $-\pi/3$ and T^6 is the identity. So T and T^5 are the rotatory reflections with symmetry plane ϖ''. Hence there are $2 \times 4 = 8$

of these kinds of rotary reflections.

Hence there are $1 + 6 + 8 = 15$ rotary reflection symmetries. This gives $9 + 15 = 24$ indirect symmetries of the cube.

The 48 symmetries of the cube form a Coxeter group with presentation

$$< x, y, z \mid x^2 = y^2 = z^2 = (xy)^4 = (yz)^2 = (zx)^3 = 1 > .$$

The generators x, y, z are plane reflections and the products xy, yz, zx correspond to the 3 kinds of rotations.

Exercise 5.46

Describe the 24 symmetries of a regular tetrahedron.

The 24 symmetries of a regular tetrahedron form another Coxeter group isomorphic to the symmetry group of four objects, for example the vertices. We can use the geometric picture to prove properties of this group. For example the element (ABC) is conjugate to (ACB) in the full group but not in the subgroup of direct symmetries (even permutations). This is because (ABC) is a rotation through $2\pi/3$ and (ACB) is a rotation through $-2\pi/3$. The signs cannot be reversed with conjugation by direct symmetries.

Orthogonal Transformations and Matrices

Let T be an orthogonal transformation. Since T is linear we can write its action as

$$T(X) = X\mathbf{M}$$

where $\mathbf{M}$ is a 3×3 matrix and X is considered as a 1×3 row vector.

Since orthogonal transformations respect the dot product we must have

$$A \cdot B = AB^T = T(A) \cdot T(B) = A\mathbf{M}(B\mathbf{M})^T = A\mathbf{M}\mathbf{M}^T B^T$$

where $(\ )^T$ denotes transposition. Since this is true for all A, B we have

$$\mathbf{M}\mathbf{M}^T = \mathbf{E} \text{ or equivalently } \mathbf{M}^{-1} = \mathbf{M}^T$$

where $\mathbf{E}$ denotes the identity matrix.

A matrix $\mathbf{M}$ satisfying this condition is called an *orthogonal matrix*. If we write $\mathbf{M}$ in terms of its rows say

$$\mathbf{M} = \begin{pmatrix} A \\ B \\ C \end{pmatrix}$$

then the condition for orthogonality means

$$\begin{pmatrix} A \\ B \\ C \end{pmatrix} (\, A^T \quad B^T \quad C^T \,) = \mathbf{E} = \begin{pmatrix} 1 & 0 & 0 \\ 0 & 1 & 0 \\ 0 & 0 & 1 \end{pmatrix}.$$

Equating the matrix entries on both sides yields

$$|A|^2 = |B|^2 = |C|^2 = 1 \text{ and } A \cdot B = B \cdot C = C \cdot A = 0.$$

In other words $\{A, B, C\}$ is an orthonormal basis.

Since the determinant of $\mathbf{M}$ is $[ABC]$ we deduce

■ The matrix $\mathbf{M}$ with rows A, B, C is orthogonal if and only if $\{A, B, C\}$ is an orthonormal basis. The associated orthogonal transformation is direct if $\{A, B, C\}$ is a right-handed orthonormal basis and indirect otherwise. □

Exercise 5.47

Show that if the matrix $\mathbf{M}$ with rows A, B, C is orthogonal then $C = \pm A \times B$.

Exercise 5.48

Show that the determinant of an orthogonal matrix is ± 1 and the eigenvalues have modulus 1.

Exercise 5.49

Let $\mathbf{S}$ be an anti-symmetric matrix, i.e. $\mathbf{S}^T = -\mathbf{S}$. Show that $(\mathbf{E} - \mathbf{S})(\mathbf{E} + \mathbf{S})^{-1}$ is orthogonal where $\mathbf{E}$ is the identity matrix. Show that the converse is also true provided the orthogonal matrix is not $-\mathbf{E}$.

Points on the unit sphere S^2 are permuted by orthogonal transformations. In particular the unit coordinate points I, J, K are sent to the rows of $\mathbf{M}$. That is

$$I\mathbf{M} = A, \;\; J\mathbf{M} = B, \;\; K\mathbf{M} = C \text{ where } \mathbf{M} = \begin{pmatrix} A \\ B \\ C \end{pmatrix}.$$

So this allows us to find an orthogonal matrix

$$\mathbf{M}^T\mathbf{M}' \text{ where } \mathbf{M} = \begin{pmatrix} A \\ B \\ C \end{pmatrix} \text{ and } \mathbf{M}' = \begin{pmatrix} A' \\ B' \\ C' \end{pmatrix}$$

sending any orthonormal basis $\{A, B, C\}$ to any other $\{A', B', C'\}$. The first matrix $\mathbf{M}^T = \mathbf{M}^{-1}$ in the product sends $\{A, B, C\}$ to $\{I, J, K\}$ and the second matrix $\mathbf{M}'$ sends $\{I, J, K\}$ to $\{A', B', C'\}$.

Since the third row is determined up to sign we can say rather more.

■ Let A, B and A', B' be pairs of unit points subtending the same angle θ at the origin, $\theta \neq 0, \pi$. Then there is a unique direct (indirect) orthogonal transformation taking A to A' and B to B'.

Proof Assume initially that $\theta = \pi/2$ so A, B and A', B' are orthogonal. If

$$\mathbf{M} = \begin{pmatrix} A^T & B^T & (A \times B)^T \end{pmatrix} \begin{pmatrix} A' \\ B' \\ A' \times B' \end{pmatrix}$$

then $T(X) = X\mathbf{M}$ is the required direct orthogonal transformation. The indirect transformation is obtained by negating the bottom row of one matrix factor. If $\theta \neq \pi/2$ replace B by $(A \times B) \times A$ and B' by $(A' \times B') \times A'$. □

Exercise 5.50

Let $\mathbf{M}$ be the matrix

$$\begin{pmatrix} \cos\theta + u^2(1-\cos\theta) & uv(1-\cos\theta) + w\sin\theta & wu(1-\cos\theta) - v\sin\theta \\ uv(1-\cos\theta) - w\sin\theta & \cos\theta + v^2(1-\cos\theta) & vw(1-\cos\theta) + u\sin\theta \\ wu(1-\cos\theta) + v\sin\theta & vw(1-\cos\theta) - u\sin\theta & \cos\theta + w^2(1-\cos\theta) \end{pmatrix}$$

where $A = (u, v, w) \in S^2$ is a unit point. Show that $\mathbf{M}$ is orthogonal and represents a rotation about OA through an angle θ.

5.9 Projections

Orthogonal Projections

If ϖ is a plane in space, say $A \cdot P = d$, then we can define an orthogonal projection p from $\mathbb{R}^3$ to ϖ by sending an arbitrary point X in space to the nearest point of ϖ. A formula for this was given earlier, namely

$$p(X) = X + \frac{d - A \cdot X}{|A|^2} A.$$

Exercise 5.51

Find the orthogonal projection of the line

$$\frac{x-1}{3} = \frac{y-3}{5} = \frac{z-4}{2}$$

in the plane

$$2x - y + z + 3 = 0.$$

This projection is all very well but in order to represent a three-dimensional object on a piece of paper or a computer screen the three-dimensional coordinates must be projected to two-dimensional coordinates. An orthogonal projection is determined by the image plane and a direction in the plane which determines the first coordinate. The second direction is automatically defined by the right-handedness condition.

Let the plane of projection be $A \cdot P = 0$ and let B be a point in the plane chosen so that the line OB is the first coordinate axis or x-axis of the target. Then A, B are orthogonal and we can always assume that they have unit length. A suitable candidate for the second or y-axis in the target plane is $A \times B$. Since $\{A, B, A \times B\}$ is a right-handed basis for $\mathbb{R}^3$ every point $X \in \mathbb{R}^3$ can be written uniquely as $X = xB + yA \times B + zA$. The projection takes this to $p(X) = xB + yA \times B$. The values of x, y can be determined by dot products and so

$$x = X \cdot B, \quad y = X \cdot (A \times B).$$

Hence the corresponding orthogonal projection from $\mathbb{R}^3$ to $\mathbb{R}^2$ is

$$\boldsymbol{p(X) = (X \cdot B, [XAB]).}$$

Example 5.4

Orthogonally project the cube with vertices $(\pm1, \pm1, \pm1)$ onto the plane $x + 2y + 3z = 0$ with x-axis along $(1, 1, -1)$. Then $A = (1, 2, 3)/\sqrt{14}$, $B = (1, 1, -1)/\sqrt{3}$ and $A \times B = (-5, 4, -1)/\sqrt{42}$. So

$$p(x, y, z) = \left((x + y - z)/\sqrt{3}, \quad (-5x + 4y - z)/\sqrt{14} \right).$$

The cube vertices become

$$\left((\pm1 \pm 1 \mp 1)/\sqrt{3}, \quad (\mp5 \pm 4 \mp 1)/\sqrt{14} \right).$$

Stereographic Projections

In order to better mimic what the eye sees we can project spatial bodies into a plane by stereographic projection. This was used in Chapter 4 to represent the extended complex plane as a sphere. The result is more realistic than orthogonal projection since objects nearer to the observer appear larger than further objects of the same size.

Let P, the eye's pupil, be the projection point. If X is a point in $\mathbb{R}^3$ let X' be the point where the line XP meets the plane, ϖ.

To fix coordinates let A, B be orthogonal and of unit length and let ϖ be the plane passing through O and orthogonal to A. So B and $A \times B$ lie in ϖ.

Let $X' = tX + (1-t)P$. Then

$$0 = X'.A = tX \cdot A + (1-t)P \cdot A.$$

So $t = -P \cdot A/(X-P) \cdot A$ and

$$X' = -\frac{P \cdot A}{(X-P) \cdot A} X + \frac{X \cdot A}{(X-P) \cdot A} P.$$

We now want to identify X' with a point of $\mathbb{R}^2$. If $X' = x'B + y'A \times B$ then

$$\boldsymbol{x' = X' \cdot B = \frac{(X \times P) \cdot (A \times B)}{(X - P) \cdot A}}$$

$$\boldsymbol{y' = X' \cdot A \times B = -\frac{(X \times P) \cdot B}{(X - P) \cdot A}}.$$

These are the coordinates of the stereographic projection (Fig. 5.8).

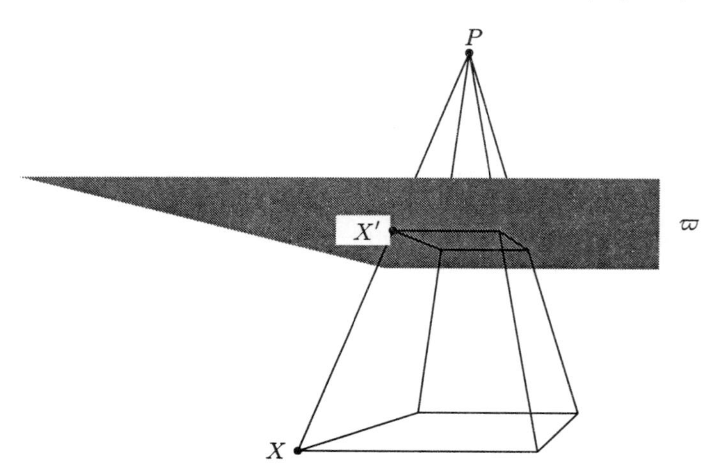

Fig. 5.8 Stereographic projection, $X \to X'$, from P onto the plane ϖ

Example 5.5

Suppose we want to project stereographically the cube with vertices $(\pm 1, \pm 1, \pm 1)$ using the pupil $(-3, -6, -9)$ onto the plane $x + 2y + 3z = 0$ with x'-axis along $(1, 1, -1)$. Then with the notation above $A = (1, 2, 3)/\sqrt{14}$, $B = (1, 1, -1)/\sqrt{3}$ and $A \times B = (-5, 4, -1)/\sqrt{42}$. The coordinates of the stereographic projection are therefore

$$x' = -\frac{\sqrt{42}(x + y - z)}{x + 2y + 3z + 3\sqrt{14}}, \quad y' = -\frac{\sqrt{3}(-5x + 4y - z)}{x + 2y + 3z + 3\sqrt{14}}.$$

The coordinates of the cube vertices are now obtained by substitution.

The appearance of the cubes in the two projections are dramatically different (see Fig. 5.9).

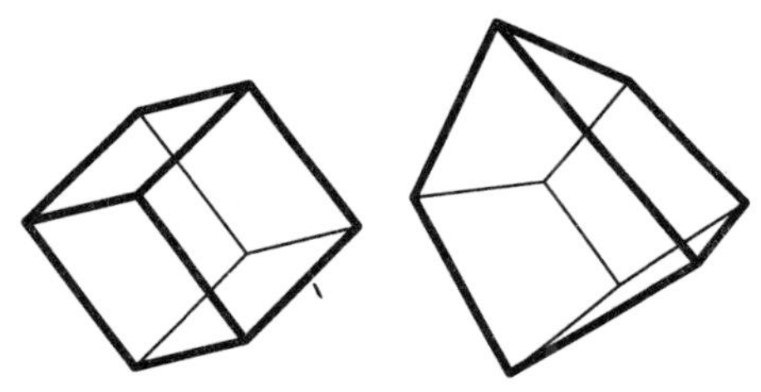

Fig. 5.9 Orthogonal and stereographic projections of a cube

The stereographic projection is not defined on points of the plane $X \cdot A = P \cdot A$. These points may be thought of as going to infinity. Conversely, parallel lines may meet in the projection.

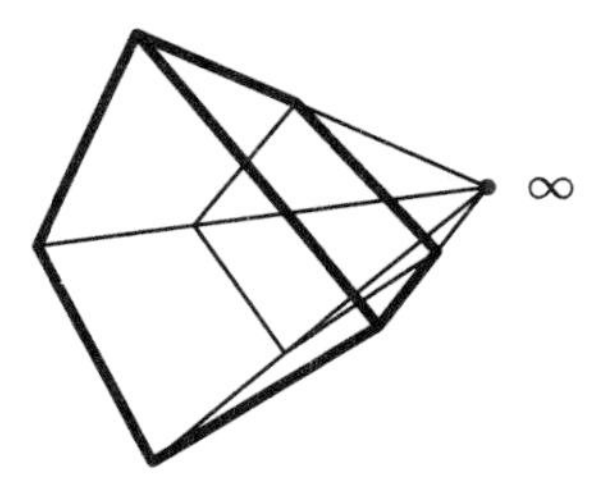

Fig. 5.10 Parallel lines meeting at infinity

For example if we extend the projection of four of the parallel lines of the cube then they converge to an ideal or infinite point (Fig. 5.10). We shall have more to say about this in the next chapter.

5.10 Polyhedra

We now look at polyhedra in $\mathbb{R}^3$. We can think of these as solid bodies with a finite number of polygons making up the faces of the boundary. Each edge of a polygon face will have two polygons on either side and each polygon vertex will be shared by three or more other polygons. Notice that this number is equal to the number of edges which terminate at the vertex. Call this number the *order* of the vertex.

The simplest polyhedron is the tetrahedron which has four vertices and four triangular faces. Each vertex is met by three faces so all vertices have order three. The tetrahedron is called *equilateral* if each face is equilateral.

The cube has six square faces, eight vertices and the order of each vertex is again three. Another name for the cube is equilateral *hexahedron*.

Two regular tetrahedra can be obtained from the face diagonals of a cube. These are illustrated in Fig. 5.11.

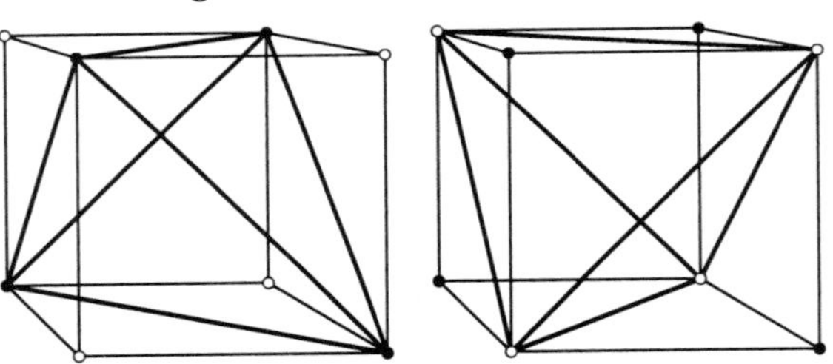

Fig. 5.11 Diagonal tetrahedra of a cube

Exercise 5.52

How many edges has a tetrahedron or a cube?

Exercise 5.53

Let $n: \mathbb{R}^3 \to \mathbb{R}$ be the function defined by

$$n(x, y, z) = |x| + |y| + |z|$$

and let the vertices of a cube be $(\pm 1, \pm 1, \pm 1)$. Show that vertices U, V are

(a) joined by an edge if and only if $n(U - V) = 2$
(b) joined by a face diagonal if and only if $n(U - V) = 4$
(c) joined by a major diagonal if and only if $n(U - V) = 6$.
Deduce that U is a vertex of the diagonal tetrahedron containing the vertex V if and only if $n(U - V)$ is four.

The tetrahedron is an example of a *cone* or *pyramid.* Any face of the tetrahedron can be chosen as *base* and the opposite vertex as *apex.* The cube is an example of a *prism* with base any face.

A prism is called *right* if the segment joining corresponding points on the base and top is orthogonal to the plane of the base. A cone on a circular base or any other centrally symmetric plane object is called right if the segment joining the apex to the centre of the base is orthogonal to the plane of the base. See Fig. 5.12.

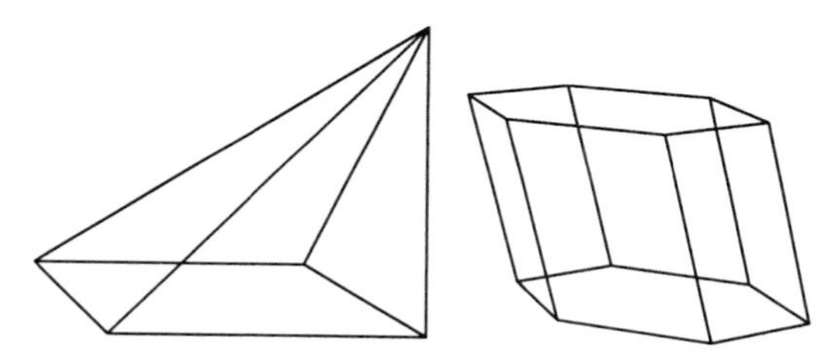

Fig. 5.12 A cone and a prism

Let b be a subset of the plane $\mathbb{R}^2$ and A be a point in space above it. Then points in the cone with base b and apex A lie on a segment joining a point X in b and A. This segment is unique if the point is not the apex and can be written uniquely as $(1-t)X + tA$ where $0 \le t < 1$. So coordinates for the point may be thought of as (X, t). The same thing is almost true for a prism except that $0 \le t \le 1$ and $(X, 0)$ are the coordinates of a point X on the base and $(X, 1)$ is a corresponding point on the top.

Octahedra

Consider two cones with common base a square and apexes on either side of the base. Their union is called an *octahedron.* As the name suggests, the octahedron has eight faces, each of which is a triangle. The octahedron is called equilateral if each face is an equilateral triangle.

Another construction for an equilateral octahedron is to take a vertex for the octahedron in the centre of every face of a cube (Fig. 5.13). This defines the octahedron as the *dual* of a cube.

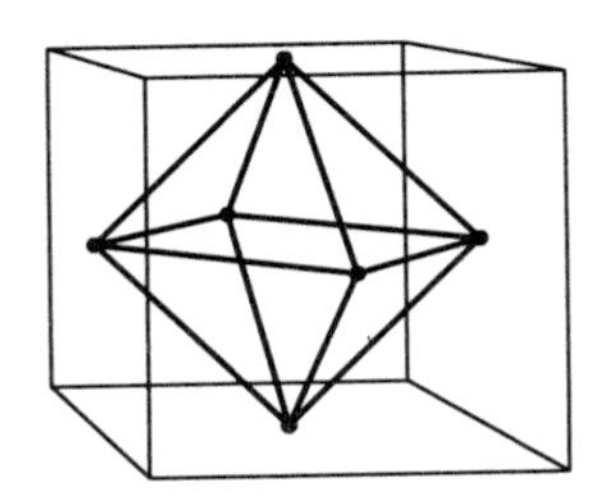

Fig. 5.13 The octahedron as the dual of a cube

Exercise 5.54

How many edges has an octahedron? How many vertices has an octahedron and what are their orders?

Any polyhedron has a dual defined by taking the vertices of the dual as the centres of each face of the original polyhedron. The dual of a tetrahedron is again a tetrahedron and the dual of an octahedron is a cube.

Exercise 5.55

If the coordinates of the vertices of a cube are $(\pm 1, \pm 1, \pm 1)$, show that the coordinates of the vertices of the dual octahedron are $(\pm 1, 0, 0)$, $(0, \pm 1, 0)$, $(0, 0, \pm 1)$.

The double cone construction of the octahedron is called the *suspension* of a square. Here is a similar construction. Take two cubes of the same size. Divide one cube into six pyramids with apex at the centre of the cube and bases the six square faces. Now glue these pyramids via their bases onto the faces of the other cube. The result is called a *rhombic dodecahedron* (Fig. 5.14).

A rhombic dodecahedron gets its name from the fact that it has 12 faces each of which is a rhombus. It might appear that the polyhedron had 24 faces all of which were triangles. However any two triangular faces in the construction which have an edge of the cube in common lie in the same plane and so their union is a rhombus.

If you have difficulty visualising the rhombic dodecahedron why not make one? You need 12 rhombi cut from cardboard with diagonals in the ratio $1 : \sqrt{2}$. Figure 5.21 gives an example to copy.

Exercise 5.56

If the edges of a cube have length 2, what are the lengths of the edges

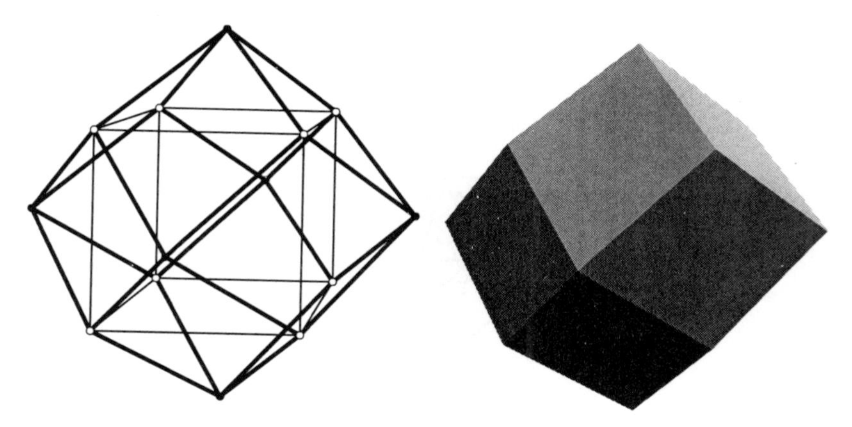

Fig. 5.14 The rhombic dodecahedron

of the associated rhombic dodecahedron?

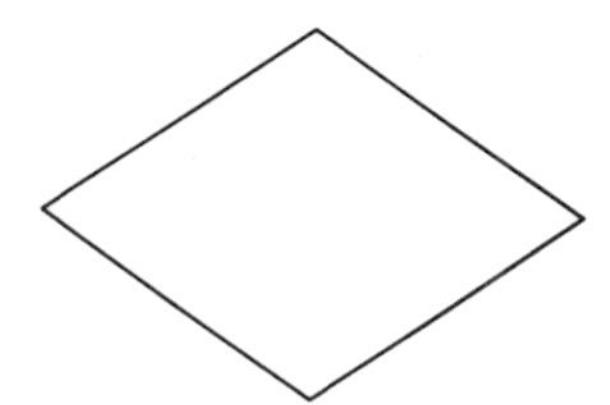

Fig. 5.15 The face of a rhombic dodecahedron

Exercise 5.57

How many edges has a rhombic dodecahedron? How many vertices has a rhombic dodecahedron and what are their orders?

Exercise 5.58

If the coordinates of the vertices of a cube are $(\pm 1, \pm 1, \pm 1)$, show that the coordinates of the extra vertices of the rhombic dodecahedron are $(\pm 2, 0, 0)$, $(0, \pm 2, 0)$ and $(0, 0, \pm 2)$.

Truncation

With the dual of a polyhedron we made a new polyhedron with the centre of each face as vertices. If we do the same with the centre of each edge we get the *truncated* polyhedron. Imagine that at each vertex a small cone is sliced off. Now gradually enlarge the slices until they meet at the midpoints of the edges.

If we truncate a tetrahedron we get an octahedron (Fig. 5.16) and if we truncate a cube we get a *cuboctahedron* (Fig. 5.17).

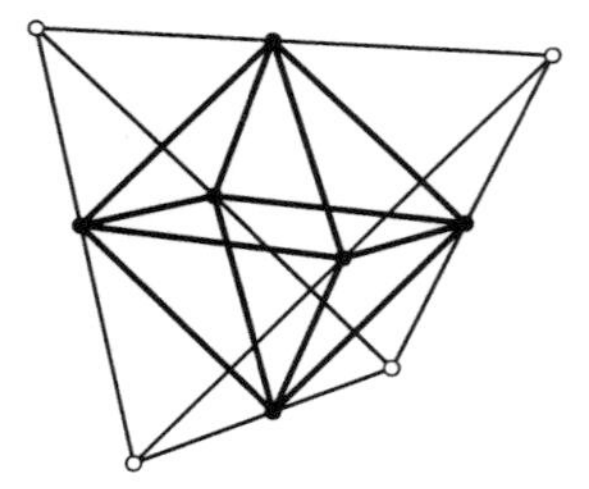

Fig. 5.16 The octahedron as truncated tetrahedron

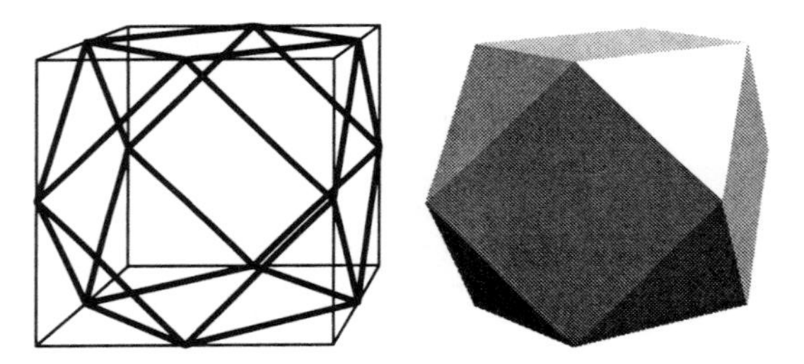

Fig. 5.17 The cuboctahedron as truncated cube

Exercise 5.59

How many faces has a cuboctahedron and what are they? How many vertices has a cuboctahedron and what are their orders?

Exercise 5.60

For all the polyhedra so far encountered count α_0 (number of vertices), α_1 (number of edges) and α_2 (number of faces). Verify in each case that $\alpha_0 - \alpha_1 + \alpha_2 = 2$.

The above result was known to Archimedes. The n-dimensional version was proved by Poincaré in 1893 and is the first theorem of homology theory.

The Five Platonic Polyhedra

We have already met three of them, namely the tetrahedron, the cube and the octahedron. The remaining two are the *icosahedron* which has 20 triangular faces and the *dodecahedron* which has 12 pentagonal faces. Since the icosahedron is the dual of the dodecahedron, constructing one gives a construction of the other.

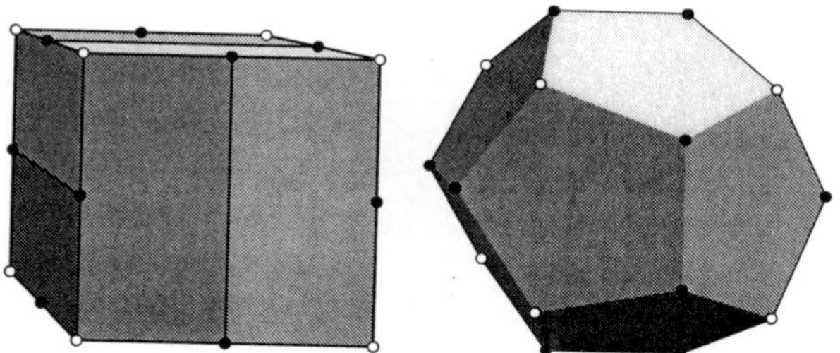

Fig. 5.18 Forming a dodecahedron from a cube

Remarkably the dodecahedron is fabricated by nature from a cube as follows (Fig. 5.18). Certain crystals of iron pyrites or fool's gold form a cube, the faces of which have striations parallel to the edges in sets of three. Over aeons, geological forces push up the central striations like the ridge of a house roof to make a dodecahedral or pyritohedron crystal of pyrites. Note that the cube with the central striation in each face is already a degenerate dodecahedron. The vertices of the cube are indicated by open dots and remain the vertices of a cube. The other vertices which are the endpoints of central striations are indicated by black dots.

Exercise 5.61

Let $ABCDE$ and $ABC'D'E'$ be the vertices of adjacent faces of a regular dodecahedron. Show that $CC'E'E$ form the vertices of a square. Show that there are 30 such squares forming the faces of 5 inscribed cubes.

We will give the details of a dual construction of the icosahedron from an octahedron. In the meantime I strongly urge you to beg, borrow or steal a dodecahedron and an icosahedron to help visualisation. Here is a method of making an icosahedron told to me by Martin Roller.

You will need two sheets of paper (A4 will do nicely), scissors and glue. Masking tape is a help.

1 Fold one sheet down the middle and open up again.

2 Next fold so that corner X lies on the middle line and the new fold runs through corner Y (Fig. 5.19).

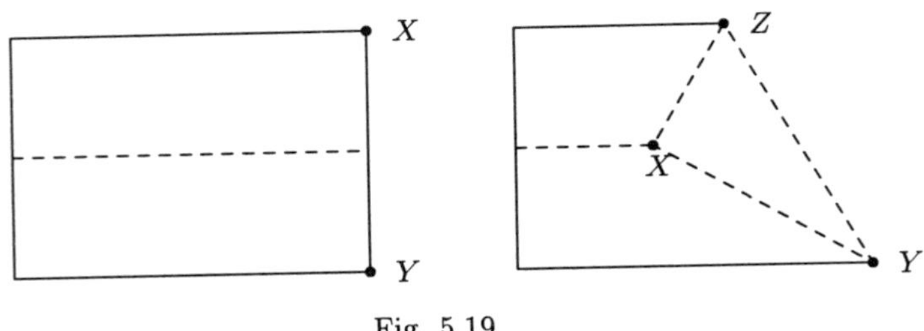

Fig. 5.19

3 Fold along the line from X to Z and open up again.

4 You see an equilateral triangle. Cut it out.

5 Fold the triangle into 16 smaller equilateral triangles labelled a to p as shown in Fig. 5.20.

6 Cut the edge between the triangles d and h, e and f, and n and m.

7 Glue triangle d onto h, e onto f and n onto m. You now have a sort of bowl with three triangular flaps marked a, j and p.

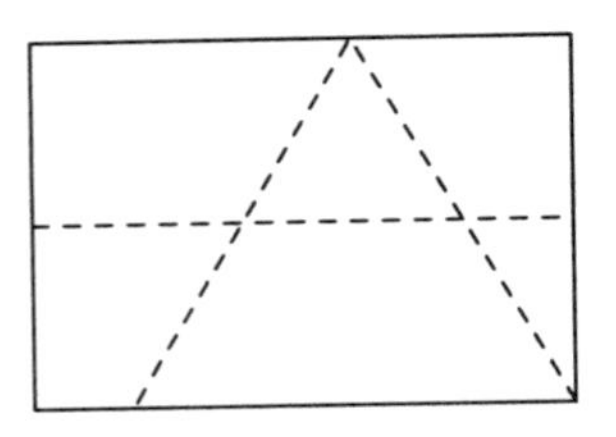

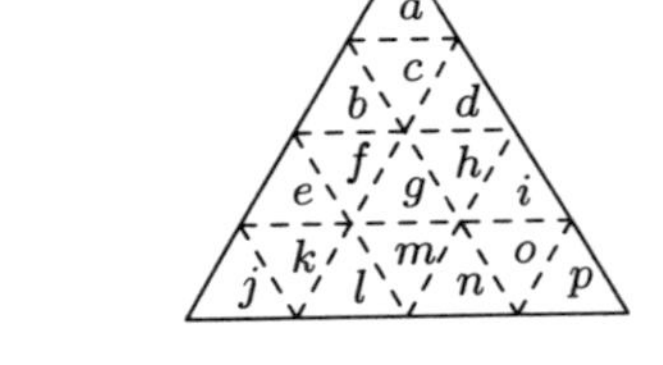

Fig. 5.20

8 Repeat the same with the other sheet of paper but now as a mirror image, i.e. cut the edges between b & f, l & m and i & h, then glue the corresponding triangles.

9 The two halves fit together with a kind of screw motion. You can either fold the flaps of one bowl inside the other or glue them onto the outside.

If you want to make a dodecahedron photocopy the diagram in Fig. 5.21, magnifying if you want, cut out and glue the edges.

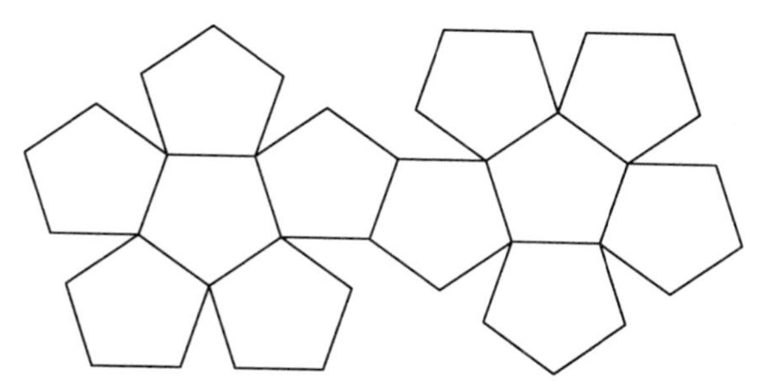

Fig. 5.21

You can skip the gluing by folding the edges, interlacing the flattened dodecahedron with an elastic band and watching the dodecahedron 'pop up'.

Constructing an Icosahedron from an Octahedron

The octahedron has the property that the edges can all be oriented in two ways, so that the boundary of each face is cyclically oriented. Assume that this is done. At each vertex two edges flow in and two edges flow out; see Fig. 5.22.

Pick a number a, where $0 < a < 1$. Then each edge can be divided in the ratio $a : b$, where $a + b = 1$, according to the orientation of the edge.

These 12 new points, one for each edge, define equilateral triangles in each face and isosceles triangles straddling each edge. There are 8 triangles of the first kind exemplified by ABC in the diagram and 12 triangles of the second kind exemplified by ADC in the diagram. These 20 triangles define an irregular icosahedron.

If a is very small then the isosceles triangles are long and thin. If a is nearly 1 then the isosceles triangles are almost right angled like WZY. There must exist an intermediate value of a which makes the isosceles triangles equilateral. It follows that for this value all the triangles are equilateral and the icosahedron has equal equilateral triangular faces.

What is the ratio $a : b$ which gives the equilateral triangles? We need: $AD = CD$. Choose coordinates so that $X = (1,0,0), Y = (0,1,0), Z = (0,0,1)$ and $W = (0,-1,0)$. Then $A = (a,0,b), C = (b,-a,0)$ and $D = (b,a,0)$. The condition then becomes $|(a,0,b) - (b,a,0)| = |(b,-a,0) - (b,a,0)|$ or $(a-b)^2 + a^2 + b^2 = 4a^2$.

Substituting $b = 1 - a$ and solving gives $a = \frac{3-\sqrt{5}}{2}$ and $b = \frac{-1+\sqrt{5}}{2}$. The ratio $b/a = \frac{1+\sqrt{5}}{2}$ is the golden ratio.

It follows that the vertices of a regular icosahedron can be taken as

$$(0, \pm\tau, \pm1), \ (\pm1, 0, \pm\tau), \ (\pm\tau, \pm1, 0)$$

where $\tau^2 = \tau + 1$.

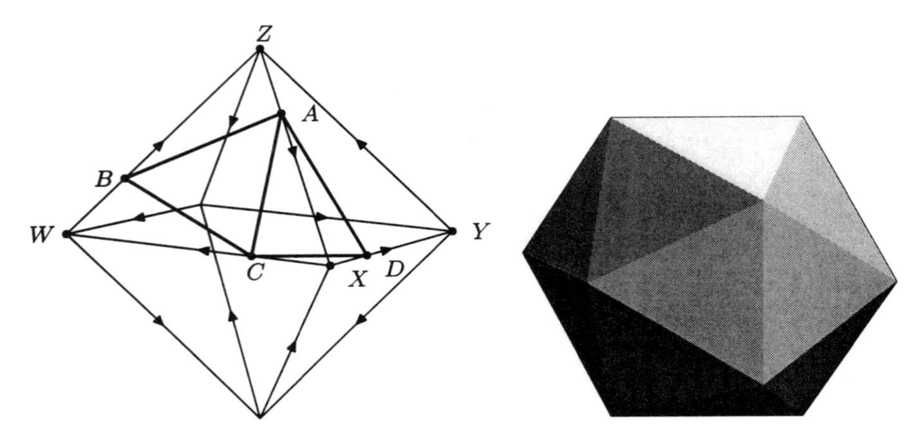

Fig. 5.22 Forming an icosahedron from an octahedron

Exercise 5.62

Show that the points

$$(0, \pm\tau^{-1}, \quad \pm\tau), (\pm\tau, 0, \pm\tau^{-1}), \quad (\pm\tau^{-1}, \pm\tau, 0), \quad (\pm 1, \pm 1, \pm 1)$$

are the vertices of a regular dodecahedron. Hint: the dodecahedron is the dual of the icosahedron.

Exercise 5.63

Show that the angle between adjacent faces of a dodecahedron (icosahedron) is $\pi - \tan^{-1} 2$ ($\pi - \sin^{-1} 2/3$).

The vertices of a regular icosahedron naturally fall into three classes of four, each set of which are the vertices of a rectangle with sides in the golden ratio (Fig. 5.23). The three rectangles are orthogonal and their boundaries have the property that in pairs they are unlinked but as a triple they cannot be separated. Topologically this link of three cycles is called the *Borromean rings*. You can easily make a model from three postcards or train tickets. If you want the result to be equilateral make sure that the sides of the rectangles are in the golden ratio.

Regular Polyhedra

In two dimensions the regular polyhedra are the regular polygons. In three dimensions the regular polyhedra are those that have the same kind of regular

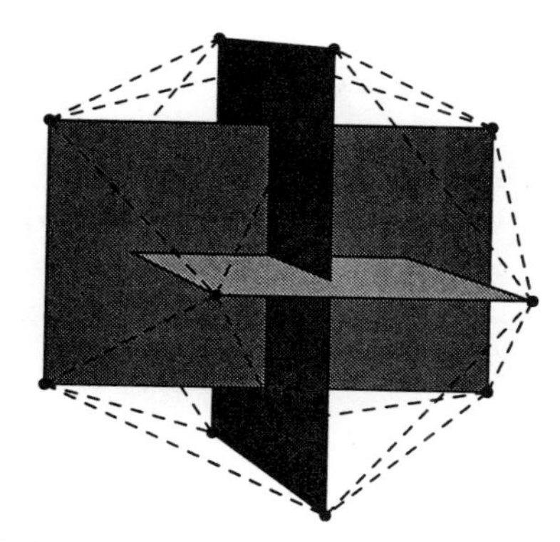

Fig. 5.23 The vertices of a regular icosahedron

polygon as faces and the same order for each vertex. So the five platonic solids are regular in three dimensions. How do we know if there are any others? Let us introduce the following notation due to Schläfli. The symbol $\{p\}$ stands for a regular p-gon and $\{p, q\}$ denotes the solid polyhedron (which may or may not exist) such that each face is a $\{p\}$ and each vertex is surrounded by q $\{p\}$s.

Example 5.6

The tetrahedron is $\{3, 3\}$ and the cube is $\{4, 3\}$.

Exercise 5.64

Find the Schläfli notation for the octahedron, the icosahedron and the dodecahedron.

Exercise 5.65

Show that the dual of a $\{p, q\}$ is a $\{q, p\}$.

We now show that the only regular three-dimensional polyhedra are the platonic solids. The faces of a polyhedron at a vertex all lie on one side of some plane in space. It follows that total angle sum of corners at this vertex is less than 2π. There are q $\{p\}$s at a vertex of a $\{p, q\}$. The angle at the vertex of a $\{p\}$ is

$$2\pi(\frac{1}{2} - \frac{1}{p}).$$

It follows that
$$q2\pi(\frac{1}{2}-\frac{1}{p}) < 2\pi \text{ or } qp - 2q - 2p < 0.$$
Adding 4 to both sides yields: $(q-2)(p-2) < 4$.

The solutions in integers greater than 2 are $\{p, q\} = \{3,3\}, \{3,4\}, \{4,3\}, \{5,3\}$ or $\{3,5\}$. So there are at most five regular three-dimensional polyhedra, corresponding to those we have considered.

Exercise 5.66

Show that the angle at the vertex of a $\{p\}$ is $2\pi(\frac{1}{2}-\frac{1}{p})$.

Exercise 5.67

What do the solutions to $(q-2)(p-2) = 4$ correspond to?

In four dimensions there are 6 regular polyhedra; the analogues of the tetrahedron, cube and octahedron, the analogues of the icosahedron and the dodecahedron and the analogue of the rhombic dodecahedron which happens to be regular in this dimension. In higher dimensions the only regular polyhedra are the analogues of the tetrahedron, cube and octahedron.

The number of regular polyhedra in each dimension leads to the bizarre sequence
$$\infty,\ 5,\ 6,\ 3,\ 3,\ 3,\ \ldots$$
This illustrates a principle, well known to topologists, that there is enough room to screw yourself up in dimensions higher than 2 but it takes at least 5 dimensions to unscrew yourself.

Answers to Selected Questions in Chapter 5

5.1 Any point can be written $(x, y, z) = zI' + (y-z)J' + (x-y)K'$.

5.2 $\hat{P} = (1/3, 2/3, -2/3)$, the angles in degrees up to three places of decimals are $\alpha = 70.529$, $\beta = \gamma = 48.190$.

5.3 $\rho = 3$, $\phi = 131.810$, $\theta = 26.565$.

5.4 $PX = \sqrt{y^2+z^2}$, $\cos\angle GOP = \sqrt{(x^2+y^2)/(x^2+y^2+z^2)}$.

5.5 $\cos^{-1}(1/3)$.

5.6 Let the vertices of the tetrahedron be A, B, C, D. Then $(A-D)\cdot(C-B) = (A-B)\cdot(C-D) + (A-C)\cdot(D-B)$. So if both summands on the right are zero then so is the value on the left.

5.7 $(0, 0, -1)$.

5.8 $O = (A+B) \times (A+B) = A \times A + A \times B + B \times A + B \times B = A \times B + B \times A$.

5.9 One choice is $C = -A \times B = (2, 0, -2)$. Any other choice C' must have $C \cdot C' > 0$.

5.10 Add $(1, 0, 0)$ to all the vertices to make the first vertex O. Then the area is $|(1, 1, 1) \times (1, -1, 1)| = |(2, 0, -2)| = 2\sqrt{2}$.

5.11 The area is half a parallelogram with three of its vertices A, B, C. So the area of the triangle is $|(A - C) \times (B - C)|/2 = \sqrt{26}/2$.

5.13 Add $(-1, 2, 0)$ to each of the points so that the first is $(0, 0, 0)$. Then the volume is

$$\left| \frac{1}{6} \begin{vmatrix} -1 & 3 & 1 \\ 0 & 2 & 1 \\ -3 & 3 & 1 \end{vmatrix} \right| = 1/3.$$

5.14 $A \times (B \times C) = -(B \times C) \times A$. Now use the triple vector product formula.

5.15 Consider $A \times (B \times C) - (A \times B) \times C = (B \cdot C)A - (A \cdot B)C$.
A necessary and sufficient condition for this to be zero is that A is parallel to C.

5.17 $(A \times B) \cdot (C \times D) = A \cdot (B \times (C \times D))$; now expand the right-hand side using the triple vector product identity.

5.19 The volume can be found by the usual methods to be $1/3$. The face opposite $(-2, 1, 1)$ has area $\sqrt{6}/2$. If d is the unknown distance then d times $(\sqrt{6}/2)/6 = 1/3$. So $d = 4/\sqrt{6}$.

5.20

$$\frac{x - 11}{10} = -y + 2 = -\frac{z+1}{3}.$$

5.21 Let O, A, B, C be the vertices of the tetrahedron. Then $(A + B + C)/4$ is the common middle point.

5.22 The line is the diagonal of the face $z = 1$ of a cube with centre O. The direction of the line is the unit vector $A = (1/\sqrt{2}, 1/\sqrt{2}, 0)$ and the nearest point of the line to O is $C = (0, 0, 1)$. So $B = C \times A = (-1/\sqrt{2}, 1/\sqrt{2}, 0)$.

5.25 We know that the line of intersection is in the direction $C \times C'$. If we take the vector product of a point P of the intersection with this and expand we get

$$P \times (C \times C') = (P \cdot C')C - (P \cdot C)C' = p'C - pC'.$$

5.26 Since $(1, 3, 2)$ lies on both lines they must be coplanar and that is their point of intersection. The equation of the plane in which they lie is $17x - 47y - 24z + 172 = 0$.

5.27 If $A \times A' = O$ the lines are parallel and hence not skew. So assume that $A \times A' \neq O$ and $A \cdot B' = -A' \cdot B$. If $A \cdot B' \neq 0$ then the point $B \times B'/(A \cdot B')$ is common to both lines. If $A \cdot B' = A' \cdot B = 0$ then B and B' are both parallel to $A \times A'$. If both B and B' are O then both lines pass through the origin, so assume $B \neq O$. Then both lines lie in the plane $P \cdot B = 0$ and so are not skew.

5.28 The points $(8, 3, 2), (-3, 8, 4)$ are nearest. Their distance apart is $\sqrt{150}$.

5.29 Let $P = P_0 - P_0'$ where P_0 and P_0' are the nearest points on $P \times A = B$ and $P \times A' = B'$ respectively. Then P is perpendicular to both lines and so parallel to $A \times A'$. Now $(A \times A') \cdot P = (P \times A) \cdot A' = A \cdot B' + A' \cdot B$ from which the result follows.

5.30 A plane through the line ℓ_1 has form $\lambda(z-1) + \mu(x+y+1)$ for some real numbers λ, μ. This passes through the origin if $\lambda = \mu$. So the plane has equation $x+y+z = 0$. This meets the other line in the point $P = (0, 1, -1)$. So the required line is OP.

5.31 Since the determinant

$$\begin{vmatrix} 1 & 1 & 1 \\ 1 & -2 & 1 \\ 0 & 1 & -1 \end{vmatrix} = 3 \neq 0$$

the planes have a unique common point. Solving the three equations yields $(1, -2, 1)$ as the triple point.

5.32 The determinant is

$$\begin{vmatrix} 1 & 1 & 1 \\ 5 & -1 & -4 \\ 4 & 2 & 1 \end{vmatrix} = 0$$

so the configuration cannot be of the first kind. Clearly none of the planes are parallel so the configuration must be of type two or three. To find a direction parallel to all the planes take the cross product $(1, 1, 1) \times (5, -1, -4) = (-3, 9, -6)$ which is parallel to $(1, -3, 2)$.

So the plane $\varpi : x - 3y + 2z = 0$ is orthogonal to all three given planes. Finding the common points of this plane with pairs of the given planes yields the three points $\varpi \cap \varpi_1 \cap \varpi_2 = P = (10, -2, -8)/21$, $\varpi \cap \varpi_1 \cap \varpi_3 = Q = (20, -4, -16)/7$, $\varpi \cap \varpi_2 \cap \varpi_3 = R = (10, 6, 4)/7$. Since P, Q, R are all distinct the configuration is of the second kind. That is they form a triangular prism.

The area of the triangle PQR is

$$|(P-R)\times(Q-R)|/2 = |20(1,1,1)/21 \times 10(1,-1,-2)/7|/2$$
$$= 100\sqrt{14}/147.$$

5.33 The determinant is

$$\begin{vmatrix} 8 & 3 & 5 \\ 18 & -3 & 13 \\ 5 & -3 & 4 \end{vmatrix} = 0$$

so the configuration cannot be of the first kind. Clearly none of the planes are parallel so the configuration must be of type two or three. If we solve the equations we see that one is redundant and for general $z = t$ the solution is $(5/13, 4/13, 0)+t(-9/13, 7/13, 1), t \in \mathbb{R}$ which is the parametric form of the common line.

5.35 Clearly any point on the axis is fixed and corresponds to an eigenvector with eigenvalue 1. By a suitable choice of coordinate axes the form of the matrix is as for the previous question where the trace of the matrix is $1 + 2\cos\theta$ and this is an invariant.

5.36 Solving the equation $X\mathbf{M} = X$ yields some multiple X of $(1,1,1)$ which is in the direction of the axis. The angle of rotation satisfies $1 + 2\cos\theta = 0$. So $\theta = \pm 2\pi/3$. By considering the image of the point $(1,0,0)$ which is $(0,1,0)$ the plus sign is taken.

5.37 $(3,1,3)$ and $-\cos^{-1}(31/50)$.

5.38 $(x,y,z) \to (x+2y-2z, 2x+y+2z, -2x+2y+z)/3$.

5.40 To make sense of your image in the mirror you must place yourself mentally as far as possible so as to coincide with the image. Because our usual position is upright this will be achieved by a rotation through a vertical axis through the head.

5.41 $T_AT_B(X) - T_BT_A(X) = 4A \cdot B((X \cdot B)A - (X.A)B)$. This is O if and only if $A \cdot B = 0$ (the planes are orthogonal) or $(X \cdot B)A = (X.A)B$ (the planes coincide).

5.46 The symmetries correspond to all permutations of the four numbers $\{1,2,3,4\}$; the four numbers representing the four vertices. They naturally fall into four types when written as a product of disjoint cycles. For example (12) is the reflection in the plane through the line 34 and the midpoint of 12. The permutation (123) is the rotation through $2\pi/3$ about the line between 4 and the midpoint of 123. The combination (12)(34) is the half twist about the line through the midpoints of 12 and 34. Finally (1234) is the rotary reflection in the plane through the midpoints of 12, 14, 32, 34 through an angle $\pi/2$.

5.48 If $\mathbf{M}$ is orthogonal then $\mathbf{MM}^T$ is the identity matrix. So $\det(\mathbf{MM}^T)$ is 1 which equals $(\det \mathbf{M})^2$. If $U\mathbf{M} = \lambda U$ where λ and U may be complex then $\mathbf{M}^T\overline{U^T} = \overline{\lambda U^T}$. So $U\mathbf{MM}^T\overline{U^T} = \lambda\bar{\lambda}U\overline{U^T}$. It follows that $\lambda\bar{\lambda} = 1$.

5.49 Let $\mathbf{M} = (E-S)(E+S)^{-1}$. Then it is easily checked that $\mathbf{MM}^T = E$ so $\mathbf{M}$ is orthogonal. Conversely $\mathbf{S} = (\mathbf{E}+\mathbf{M})^{-1}(\mathbf{E}-\mathbf{M})$ provided $\mathbf{M} \neq -E$.

5.50 It is easy to check that $\mathbf{M}$ is orthogonal and A is fixed by $\mathbf{M}$. If B is orthogonal to A then a simple calculation shows that

$$B\mathbf{M} = \cos\theta B + \sin\theta A \times B$$

which is the required rotation.

5.51 The line meets the plane in $A = (-5,-7,0)$. The point $B = (1,3,4)$ lies on the line and its projection on the plane is $C = (-3,5,2)$. So the projection of the line in the plane is the line AC.

5.52 6 and 12.

5.54 12, 6 and 4.

5.56 $\sqrt{3}$.

5.57 24, 14, 8 have order 3 and 6 have order 4.

5.59 6 squares corresponding to the faces of a cube and 8 triangles corresponding to the vertices of a cube. There are 12 vertices of order 4.

5.61 See Fig. 5.24.

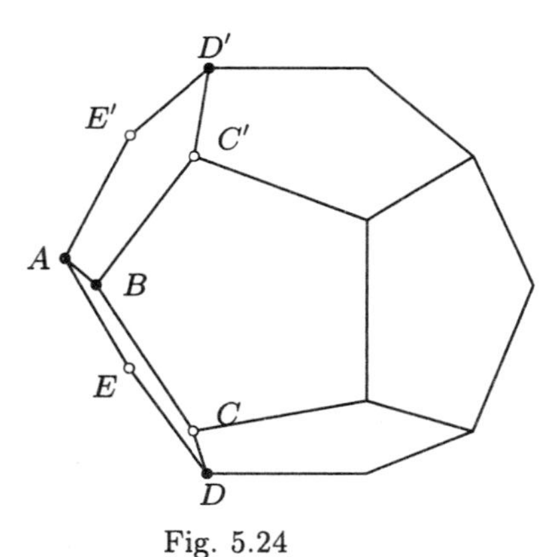

Fig. 5.24

By symmetry $CE = C'E'$ and the lines are parallel. By the same construction $CC' = EE'$ and so $CC'E'E$ is a square. One such square corresponds to each of the 30 edges. Since there are 6 faces to a cube there must be 5 cubes: count them!

5.64 $\{3,4\}, \{3,5\}$ and $\{5,3\}$.

5.66 Create p congruent isosceles triangles by joining the vertices of the $\{p\}$ to O the centre. Let the angles of each triangle be $\alpha,\ \alpha,\ \beta$ say. Then $\beta = 2\pi/p$ and so $2\alpha = \pi - 2\pi/p$.

5.67 In that case

$$q2\pi(\frac{1}{2} - \frac{1}{p}) = 2\pi.$$

So all the faces attached to a vertex lie in a plane. Continuing yields a tiling of the plane by squares, triangles or hexagons according to the three solutions of the equation (Fig. 5.25).

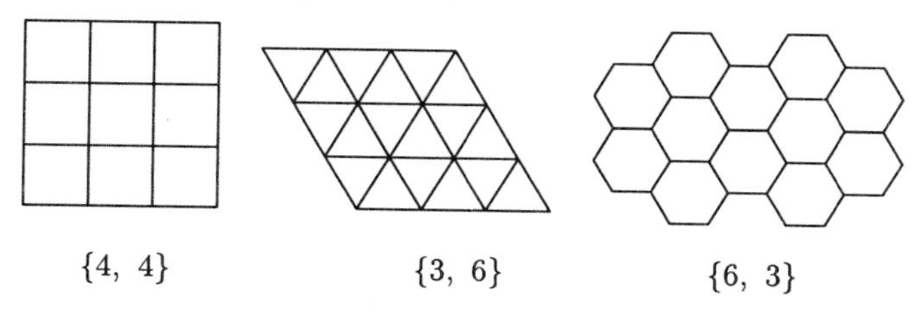

{4, 4} {3, 6} {6, 3}

Fig. 5.25 Tilings of the plane by squares, triangles and hexagons

6
Projective Geometry

The invention of projective geometry led to great improvements in railway safety

Biff

Geometries are defined by their objects of study. Euclidean geometry is naturally preoccupied with distance and angle. Otherwise how could we distinguish between an acute and an obtuse triangle? In the discipline of topology a straight line is as good as a curved one. Projective geometry falls between these two extremes. We have already met projections; projective geometry studies properties unchanged by these.

Under a projection from one plane to another, a straight line will stay a straight line. However the distance between two points or the angle between two lines can vary. Even more disconcertingly, parallel lines may project to lines which meet. To view this, look along a straight railway track or road. The sides will appear to converge at the horizon even though we know they cannot meet. Alternatively look at the sky on a day with broken cloud. Rays of light from the sun are essentially parallel but the clouds, if in the right place, will break the sun's rays into a Λ shape.

In this chapter we consider only the linear properties of the projective plane, that is points, lines and their properties. Conics and higher order curves in the projective plane would take us beyond the scope of this book.

6.1 The Projective Plane

In order to translate the above considerations into a mathematical theory we will describe a model of the real projective plane, written as $\mathbb{R}P^2$.

The *projective plane* $\mathbb{R}P^2$ is the set of lines in space $\mathbb{R}^3$ which pass through the origin.

Since a line in $\mathbb{R}^3$ which passes through the origin is determined by any of its points other than O we see that an element of $\mathbb{R}P^2$ can be considered as the set of points $\{rA\}$ in $\mathbb{R}^3$ where A is a point other than the origin and r varies over all non-zero real numbers. So any non-zero point $A = (x, y, z)$ in $\mathbb{R}^3$ determines a point in the projective plane. We can specify this point by three *homogeneous coordinates* or ratios, written $[x, y, z]$ where $[x', y', z']$ represents the same point provided there is a non-zero coefficient r such that

$$x' = rx, \; y' = ry, \; z' = rz.$$

For example the following,

$$[3, -5, 2], \; [-9, 15, -6], \; [27, -45, 18], \; [-3/5, 1, -2/5]$$

all represent the same point in $\mathbb{R}P^2$.

The projective plane is the set of pairs of antipodal points on the two dimensional sphere. This is because a line in $\mathbb{R}^3$ through the origin meets the two dimensional sphere, S^2, in an antipodal pair A, $-A$ where $|A| = 1$. We can discard the southern point of all antipodal pairs and we can think of the projective plane as the points in the northern hemisphere in S^2 together with pairs of antipodal points on the equator. We may now flatten the northern hemisphere by an orthogonal projection. This means that the projective plane is also the set of points in a plane disc provided pairs of antipodal points on the boundary circle are identified.

The use of homogeneous coordinates also suggests the definition RP^2 for the R projective plane with points specified by three ratios $[x, y, z]$, where x, y, z lie in a division ring R. However some care must be taken if R is non-commutative. In this case the ratios must all be taken on one side.

Exercise 6.1

How many points are represented by $[\pm 1, \pm 1, \pm 1]$?

We shall use the general notation $\mathsf{A} = [a_1, a_2, a_3] = [A]$ for the point in $\mathbb{R}P^2$ represented by the point $A = (a_1, a_2, a_3)$ in $\mathbb{R}^3$. The points

$$\mathsf{X} = [1, 0, 0], \; \mathsf{Y} = [0, 1, 0], \; \mathsf{Z} = [0, 0, 1]$$

are called the vertices of the *triangle of reference*. The point $\mathsf{U} = [1, 1, 1]$ is called the *unit point*.

There is no need to restrict attention to three homogeneous coordinates. Define the projective space of dimension n to be the set of points $[a_1, \dots, a_{n+1}]$ where as before $[a_1, \dots, a_{n+1}] = [ra_1, \dots, ra_{n+1}]$. The coefficients a_i may be real numbers or complex numbers or members of other algebraic systems such as the quaternions.

6.2 Lines in the Projective Plane

What! Will the line stretch
out to the crack of doom?
Macbeth

A *line* in the projective plane is the set of points $[x, y, z]$ satisfying a homogeneous linear equation

$$ax + by + cz = 0.$$

Notice that this equation is well defined in the projective plane because non-zero multiples of coordinates can be cancelled in the above equation.

In space, $\mathbb{R}^3$, the above equation represents a plane, ϖ say, through the origin in space. Any point in the projective plane on the line, $ax + by + cz = 0$, will correspond to a line in space which passes through the origin and lies in the plane ϖ.

We can identify the projective plane as an extension of the euclidean plane as follows. Let $\mathbb{E}$ be the set of points in $\mathbb{R}P^2$ with last coordinate non-zero. Dividing the coefficients by this last element means that any point in $\mathbb{E}$ can be written uniquely as $[x, y, 1]$ and hence can be identified with the point (x, y) in the euclidean plane.

The remaining points in $\mathbb{R}P^2$ have last coordinate zero and so lie on the line $z = 0$. This line is called the *line at infinity* with respect to $\mathbb{E}$. A point on the line at infinity is called a *vanishing point.*

To get a feel for this extension of the euclidean plane, imagine yourself as an observer in a boat on a calm sea far from the sight of land. Think of your eye as the origin O, the sea as the plane $z = -1$ and the sky as the euclidean plane $z = 1$. The sea also corresponds to the euclidean plane because the points $[x, y, 1]$ and $[-x, -y, -1]$ are one and the same. A point in the sky and the corresponding point in the sea lie on a line through your eye.

The horizon corresponds twice over to the line at infinity $z = 0$. Antipodal points on the horizon are identified. In particular north and south are identified and east and west are identified. The line at infinity is a circle on which the horizon circle is wrapped round twice.

Analytically points on the line at infinity may be written

$$[x, y, 0] = [r\cos\theta, r\sin\theta, 0].$$

Since $r \neq 0$ this point corresponds to the point on the unit circle $(\cos\theta, \sin\theta)$ and since $[\cos\theta, \sin\theta, 0] = [-\cos\theta, -\sin\theta, 0]$ the angle θ is an absolute angle modulo π.

Any line may be taken as the line at infinity by a suitable rotation of space. For example the model for the euclidean plane could be $2x - 7y + 345z = 1$. In that case the line at infinity would be $2x - 7y + 345z = 0$.

6.3 Incidence and Duality

Two points determine a unique line containing them. For example the points $[1, 2, 3]$ and $[3, 2, 1]$ lie on the line $x - 2y + z = 0$. Interpreting this in space $\mathbb{R}^3$, the lines through $(1, 2, 3)$ and $(3, 2, 1)$ and the origin lie in the plane $x - 2y + z = 0$.

In addition two lines meet in a unique point. For example the lines $x + 2y + 3z = 0$ and $3x + 2y + z = 0$ meet in the point $[1, -2, 1]$.

The above examples illustrate the notion of *duality*. The line $x - 2y + z = 0$ can be represented by the homogeneous coordinates $[1, -2, 1]$ which represents a point of the projective line called the *dual point*. Conversely the *dual line* of the point $[1, 2, 3]$ is the line $x + 2y + 3z = 0$. The dual of the line containing two points is the point where the dual lines meet. More generally any statement of incidence of points and lines is converted to a statement of incidence of lines and points.

If $A = (a, b, c)$ and $X = (x, y, z)$ then the line $ax + by + cz = 0$ can be written as $A \cdot X = 0$ and the dual point is $[A]$. This is represented in space as the line through the origin at right angles to the plane $ax + by + cz = 0$.

To simplify calculations let us introduce the notation

$$\begin{vmatrix} \boldsymbol{a_1} & \boldsymbol{a_2} & \boldsymbol{a_3} \\ \boldsymbol{b_1} & \boldsymbol{b_2} & \boldsymbol{b_3} \end{vmatrix} = [\boldsymbol{a_2 b_3 - a_3 b_2, a_3 b_1 - a_1 b_3, a_1 b_2 - a_2 b_1}].$$

Then the dual coordinates of the line through $[a_1, a_2, a_3]$ and $[b_1, b_2, b_3]$ are $\begin{vmatrix} a_1 & a_2 & a_3 \\ b_1 & b_2 & b_3 \end{vmatrix}$ which are also the coordinates of the intersection point of the lines $a_1x + a_2y + a_3z = 0$ and $b_1x + b_2y + b_3z = 0$.

Using the notation of the previous chapter the line through the points $[A]$ and $[B]$ has dual coordinates $[A \times B]$. Dually the lines $A \cdot X = 0$ and $B \cdot X = 0$ meet at $[A \times B]$.

Example 6.1

To find the dual coordinates of the line through $[1,1,-1]$ and $[2,0,1]$ consider

$$\begin{vmatrix} 1 & 1 & -1 \\ 2 & 0 & 1 \end{vmatrix} = [1-0,-2-1,0-2] = [1,-3,-2].$$

The equation of the line is $x - 3y - 2z = 0$.

Exercise 6.2

Show that the line through the vertices X and Y of the triangle of reference has equation $z = 0$.

Exercise 6.3

Find the equation of the line through the points $[0,1,1]$, $[2,-1,0]$ and find the line's dual coordinates.

Exercise 6.4

Find the line coordinates of the line through $[1,\theta,\theta^2]$ and $[1,\phi,\phi^2]$, $(\theta \neq \phi)$. What happens when ϕ tends to θ?

If the point $[A]$ lies on the line determined by the points $[B]$ and $[C]$ then $A \cdot (B \times C) = 0$. This can be reinterpreted as

■ The points A, B and C in $\mathbb{R}P^2$ are collinear if and only if

$$\begin{vmatrix} a_1 & a_2 & a_3 \\ b_1 & b_2 & b_3 \\ c_1 & c_2 & c_3 \end{vmatrix} = 0.$$ □

Exercise 6.5

What is the dual of the above result?

Exercise 6.6

Interpret the above result in terms of volume.

Exercise 6.7

Let P be a variable point on the line $z = 0$ and let A, B be fixed points on the line $y = 0$. Let L, M be the intersections of the lines PA, PB with the line $x = 0$ respectively and let P' be the intersection of the lines AM and BL. Show that P' lies on a fixed line through $\mathsf{Y}= [0, 1, 0]$.

6.4 Desargues' Theorem

We now come to the first major theorem of projective geometry. The famous theorem of Girard Desargues (1591–1661) was published in a book by Abraham Bosse (1648). The importance of the theorem was not noted until the 19th century.

■ **Desargues' Theorem** If two triangles are in perspective then the points of intersection of corresponding sides are collinear.

Proof Being in perspective means that the lines defined by corresponding vertices are concurrent. The reader should consult Fig. 6.1 where the triangles are A, B, C and A', B', C' and the perspective point is P. The points of intersection of the corresponding sides are L, M, N and it will be our task to prove that they are collinear using the techniques described above.

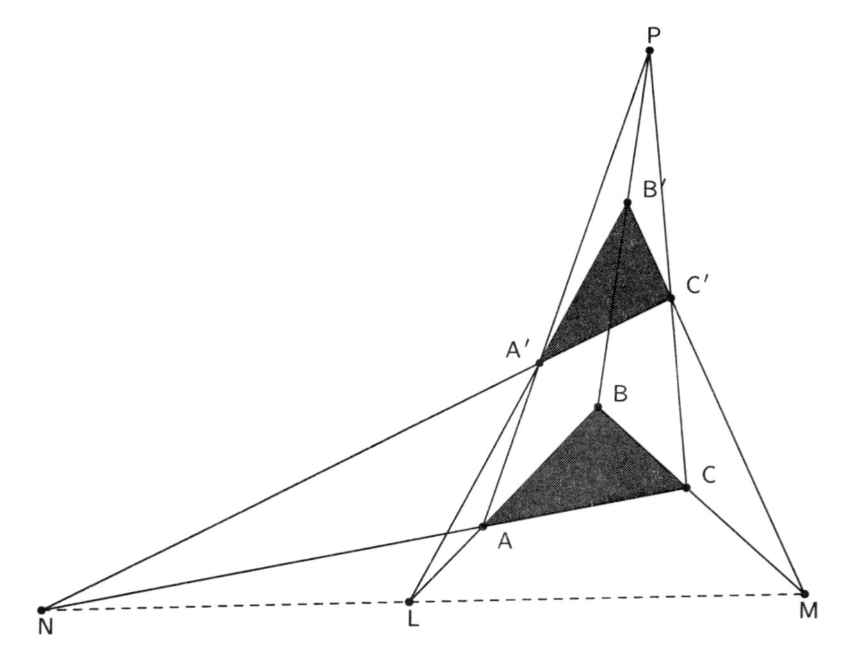

Fig. 6.1

Since $[A], [A'], [P]$ are collinear there are real numbers a, a' (not equal to 0) such that

$$P = aA + a'A'.$$

By choosing the representatives of A, A′ as $aA, a'A'$ we can rewrite this equation as

$$P = A + A'.$$

Similarly

$$P = B + B'$$
$$P = C + C'.$$

So $A - B = B' - A'$. It follows that $[A - B]$ is the intersection of the line $[A][B]$ with the line $[A'][B']$, that is L. Similarly M= $[B - C]$ and N= $[C - A]$. Since $(A - B) + (B - C) + (C - A) = 0$ it follows that L, M, N are collinear. □

The purely algebraic methods of the proof show that Desargues' theorem holds whenever the coefficients lie in a division ring. Plane geometries where Desargues' theorem does not hold are called non-desarguian planes.

By 'lifting' Desargues' theorem into three-dimensional space a much easier proof can be given. Consider Fig. 6.2.

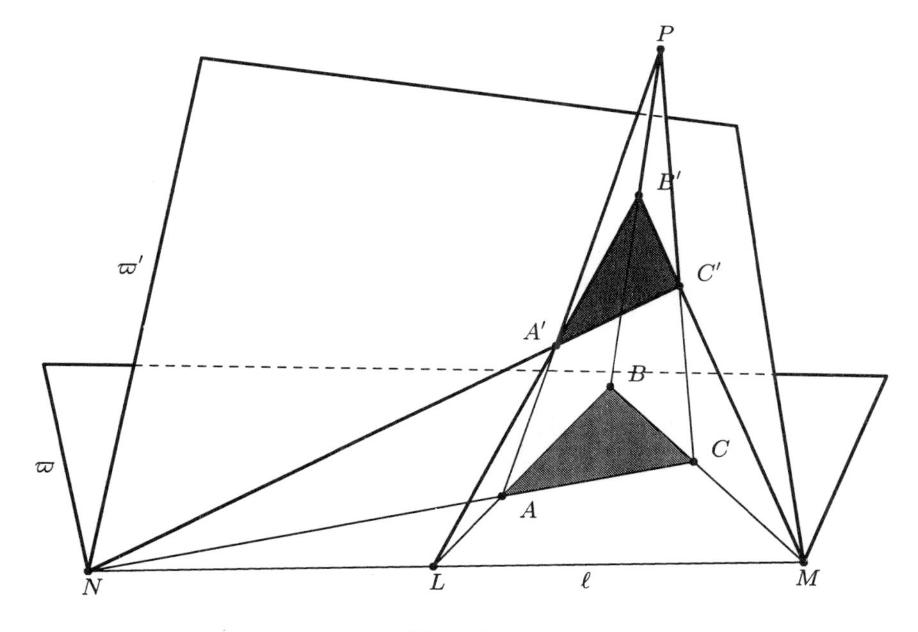

Fig. 6.2

The triangles which are in perspective from P are now assumed to lie in planes ϖ and ϖ' respectively. Let the planes be distinct and meet in the line ℓ. The planes PAB and $PA'B'$ coincide by hypothesis. This common plane meets ϖ in the line AB and meets ϖ' in the line $A'B'$. So these lines, which meet at L must also meet at a point on ℓ. Similarly, so do the points M and N.

If $\varpi = \varpi'$ then the above argument fails completely since the line ℓ common to the two planes is not defined. In order to prove the theorem in this case take a line ℓ through P, the point of perspective, suppose that ℓ is not in ϖ and let P_1, P_2 be points on ℓ other than P.

Then the lines P_1A and P_2A' lie in the plane containing the lines, P_1PP_2 and $PA'A$ and so meet in a point A'' say. Similarly let P_1B and P_2B' meet at B'' and let P_1C and P_2C' meet at C''. Then $A''B''C''$ is a triangle which is in perspective with ABC from P_1 and which is in perspective with $A'B'C'$ from P_2.

Let ϖ' be the plane of $A''B''C''$. Then $\varpi \neq \varpi'$ and we can apply Desargues' theorem in this case. But the line of perspective of the sides is then the line common to ϖ and ϖ' which contains L, M and N. □

The three-dimensional nature of this proof means that non-desarguian planes cannot be embedded in a three-dimensional geometry.

6.5 Cross Ratios Again

It is clear that angles and distance, the familiar invariants of the euclidean plane, are not projective invariants. Projective transformations stretch and change them both. In fact there are precious few numerical projective invariants. However one such is the cross ratio. We will define this initially in terms of the projective plane and then relate this definition to that given in Chapter 4.

Let $[A], [B], [C], [D]$ be four points, no three of which are equal, lying on a projective line. This means that the four points A, B, C, D of $\mathbb{R}^3$ lie in a plane through the origin. If $[A] \neq [B]$ the vectors C, D can be written uniquely as linear combinations of A and B. Suppose

$$C = pA + qB, \;\; D = rA + sB.$$

Then their *cross ratio* is defined to be

$$\text{(AB,CD)} = qr/ps.$$

Before we can be sure that this makes sense we should see if it is unchanged for different representatives of the same points. This is not difficult. If aA, bB, cC

and dD are different representatives of A, B, C, D then

$$cC = (cp/a)(aA) + (cq/b)(bB), \;\; dD = (dr/a)(aA) + (ds/b)(bB)$$

and the new multiples cancel out in the quotient.

Example 6.2

Let $[1, 2, 3]$, $[1, 1, 2]$, $[3, 5, 8]$, $[1, -1, 0]$ be four points in the projective plane. It happens that they all lie on a projective line. Let us find their cross ratio.

The equations

$$(3, 5, 8) = p(1, 2, 3) + q(1, 1, 2), \;\; (1, -1, 0) = r(1, 2, 3) + s(1, 1, 2)$$

have the unique solution $p = 2, \;\; q = 1\, r = -2, \;\; s = 3$ so the cross ratio is

$$\frac{1 \times (-2)}{2 \times 3} = -\frac{1}{3}.$$

Exercise 6.8

Find the cross ratio of the points $[2, 1, 3], [1, 2, 3], [8, 1, 9]$ and $[4, -1, 3]$.

Exercise 6.9

Show that

$$(\mathsf{AB,CC}) = 1, \;\; (\mathsf{AB,AD}) = 0, \;\; (\mathsf{AB,BD}) = \infty.$$

The cross ratio is like a coordinate determining the fourth member from a triple. It is clear that any real number can and ∞ occur. Moreover the answer is unique in the following sense.

■ (Unique fourth point theorem) Let A, B, C, X, Y be collinear points such that

$$(\mathsf{AB,CX}) = (\mathsf{AB,CY}).$$

Then X=Y.

Proof Suppose that $C = sA + tB$, $X = \alpha A + \beta B$ and $Y = \gamma A + \delta B$. Then the left-hand cross ratio is $t\alpha/s\beta$ and the right-hand cross ratio is $t\gamma/s\delta$. So $\alpha/\beta = \gamma/\delta$. It follows that $Y = (\gamma/\alpha)(\alpha A + \beta B) = (\gamma/\alpha)X$ and this represents the same point in the projective plane as X. □

In Chapter 4 the cross ratio was defined in terms of complex numbers. Recall that the cross ratio was real if the points all belonged to a (euclidean) line or circle.

Our next task is to show that the two definitions of cross ratio coincide in any mutually meaningful situation. Take two points A and B in some euclidean space. Then for all real numbers t, the point $C = tA + (1-t)B$ lies on the line through A and B, and has the property that the ratio of distances AC/CB is the fraction $(1-t)/t$. It follows that if $D = sA + (1-s)B$ is another such point then the cross ratio, defined in terms of distances, is

$$(AB, CD) = \frac{AC.BD}{AD.BC} = \frac{(1-t)s}{t(1-s)}.$$

Suppose now that the points A, B, C, D are in $\mathbb{R}^3$. Then the corresponding points A, B, C, D in the projective plane are collinear and have the same cross ratio $(1-t)s/t(1-s)$ according to this chapter's definition.

Exercise 6.10

* Let $[A], [B], [C], [D]$ be four distinct points lying on a projective line. Show that their cross ratio is given by the formula

$$\pm\sqrt{\frac{(A\cdot A)(C\cdot C)-(A\cdot C)^2}{(B\cdot B)(C\cdot C)-(B\cdot C)^2}\cdot\frac{(B\cdot B)(D\cdot D)-(B\cdot D)^2}{(A\cdot A)(D\cdot D)-(A\cdot D)^2}}.$$

As in the complex number definition the cross ratio satisfies a number of identities as the points are permuted. These can be summarised as follows.

■ The cross ratio satisfies,

$$(\mathsf{AB,CD}) = (\mathsf{BA,DC}) = (\mathsf{CD,AB}) = (\mathsf{DC,BA})$$

and

$$\begin{array}{rclrcl} (\mathsf{AB,CD}) & = & c & (\mathsf{AB,DC}) & = & 1/c \\ (\mathsf{AC,BD}) & = & 1-c & (\mathsf{AC,DB}) & = & 1/(1-c)\,. \\ (\mathsf{AD,BC}) & = & 1-1/c & (\mathsf{AD,CB}) & = & c/(c-1) \end{array}$$

Proof For example, to see that $(\mathsf{AB,CD}) = 1-(\mathsf{AC,BD})$ we must interchange the rôle of B and C. Let $C = pA + qB$, $D = rA + sB$. Then $B = (-p/q)A + (1/q)C$ and $D = rA + s((-p/q)A + (1/q)C) = (r - sp/q)A + (s/q)C$ and so

$$(\mathsf{AC,BD}) = \frac{(1/q)(r-sp/q)}{(-p/q)(s/q)} = 1 - qr/ps = 1 - (\mathsf{AB,CD}).$$

The other identities can be safely left as an exercise. □

6.6 Cross Ratios and Duality

A collection of points on a line is called a *range* and the line is called the *axis* of the range. The dual of a range is called a *pencil* of lines. If the range has axis ℓ then the lines of the dual pencil pass through the point dual to the line ℓ. This common point of the lines of the pencil is called the *apex* of the pencil.

Since the dual coordinates of four lines of a pencil are dependent, a cross ratio can be defined for a pencil of four lines in exactly the same way as one was defined for four collinear points.

Exercise 6.11

Show that the four lines

$$x + y + 2z = 0,\ 3x - y + 4z = 0,\ 5x + y + 8z = 0,\ 2x + 3z = 0$$

are concurrent and find their cross ratio.

Any line not through the apex of a pencil of four points (called a *transversal*) will meet the pencil in four collinear points (Fig. 6.3). It turns out that their cross ratio is the same as that of the pencil.

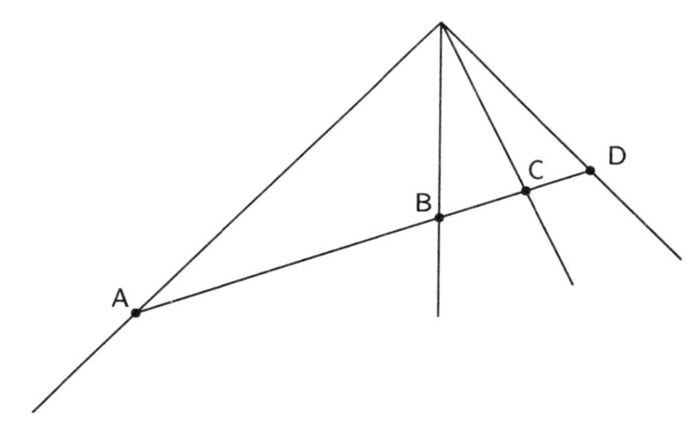

Fig. 6.3 Four collinear points determined by a transversal to a pencil

■ The four collinear points determined by a transversal to a pencil have the same cross ratio as the pencil itself.

Proof Suppose that the lines of the pencil have dual coordinates

$$[A], [B], [C], [D] \text{ where } C = pA + qB,\ D = rA + sB.$$

Let the transversal have dual coordinates $[L]$. Then the points of intersection of the transversal with the pencil have coordinates $[L\times A]$, $[L\times B]$, $[L\times C]$, $[L\times D]$. Since $L \times C = pL \times A + qL \times B$ and $L \times D = rL \times A + sL \times B$ we see that the cross ratio is the same. □

Since the calculation of the cross ratio only depends on the pencil we see that any another transversal will meet the pencil in four points with the same cross ratio. This may be summed up by the following result.

■ Let a transversal meet a pencil in four points A, B, C, D and let another transversal meet the pencil in four points A′, B′, C′, D′ (Fig. 6.4).

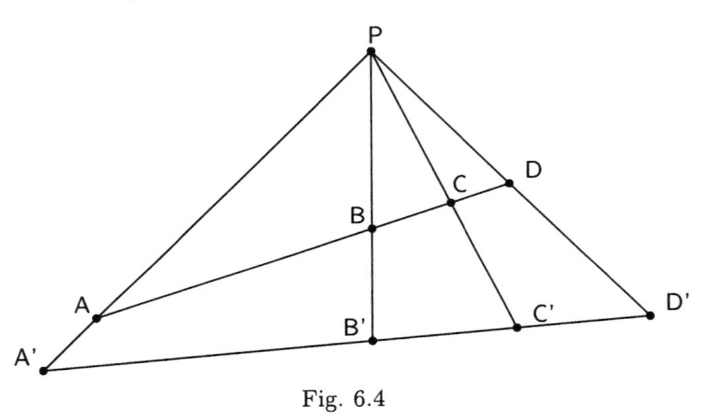

Fig. 6.4

Then (AB,CD)= (A′B′,C′D′). □

Here is a partial converse to the above result.

■ Let ℓ and ℓ' be two lines meeting at a point P. Let A, B, C be three points on ℓ and let A′, B′, C′ be three points on ℓ' such that (PA,BC)=(PA′,B′C′). Then the lines AA′, BB′ and CC′ are concurrent.

Proof Let the lines AA′, BB′ meet at Q and let QC meet ℓ' at C″ (see Fig. 6.5). Our aim will be to show that C″=C′. As P, A, B, C and P, A′, B′, C″ are in perspective from Q it follows that the cross ratios are equal. So (PA,BC)=(PA′,B′C″). But (PA,BC)=(PA′,B′C′) and so (PA′,B′C″)=(PA′,B′C′). By the unique fourth point theorem C″=C′. □

Exercise 6.12

Let A, B, C, X be points on a line ℓ and let A′, B′, C′, X′ be points on a line ℓ' such that the points of intersection of the lines {AB′, BA′}, {AC′, CA′} and {AX′, XA′} are collinear. Show that (AB,CX)=(A′B′,C′X′).

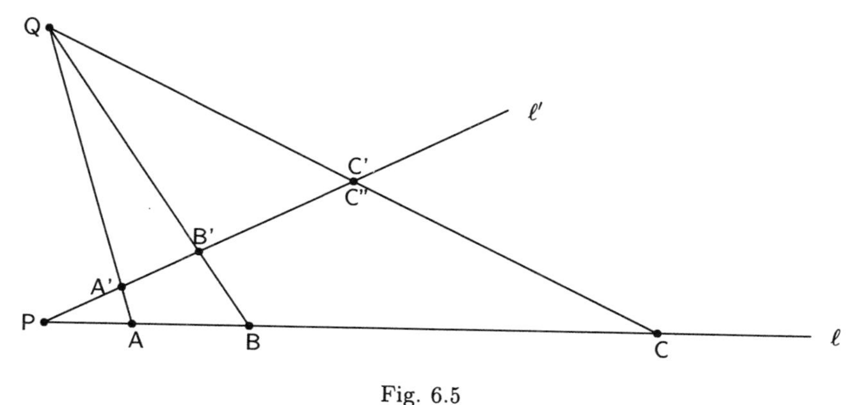

Fig. 6.5

6.7 Projectivities and Perspectivities

Let A_1, A_2, A_3, A be four points on a line a and let B_1, B_2, B_3, B be four points on a line b such that $(A_1A_2, A_3A)=(B_1B_2, B_3B)$. Then A is uniquely determined by B and conversely. The relation between A and B is called a *projectivity* between the points of a and the points of b.

If (A) is a range of points with axis a projectively related to a range of points (B) with axis b then we write

$$(A) \wedge (B).$$

The relation

$$(A_1, A_2, A_3, A_4) \wedge (B_1, B_2, B_3, B_4)$$

is equivalent to the equality

$$(A_1A_2, A_3A_4) = (B_1B_2, B_3B_4).$$

A special example of a projectivity is a *perspectivity* from a point P, since perspective points have the same cross ratio. In that case we write

$$(A) \overset{P}{\wedge} (B).$$

We will now give a geometric construction for any projectivity. Let (A_1, A_2, A_3) be three distinct points on a line a and let (B_1, B_2, B_3) be three distinct points on another line b. Suppose under a projectivity between a and b that

$$(A_1, A_2, A_3, A) \wedge (B_1, B_2, B_3, B).$$

We will assume that A is given and then construct the corresponding point B geometrically.

Let A_1B_1 meet A_2B_2 in the point O and A_3B_3 in the point O'. There are two cases to consider.

(i) If O and O' coincide let B be the point where the line OA meets b.

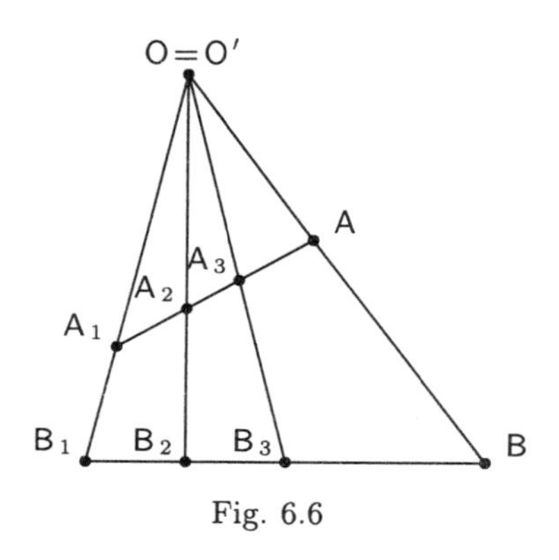

Fig. 6.6

Then

$$(A_1, A_2, A_3, A) \overset{O}{\wedge} (B_1, B_2, B_3, B).$$

That is, the projectivity is a perspectivity from O.

(ii) If O and O' are distinct let OO' meet A_3B_2 in X, let OA meet A_3B_2 in Y and let O'Y meet b in B (see Fig. 6.7).

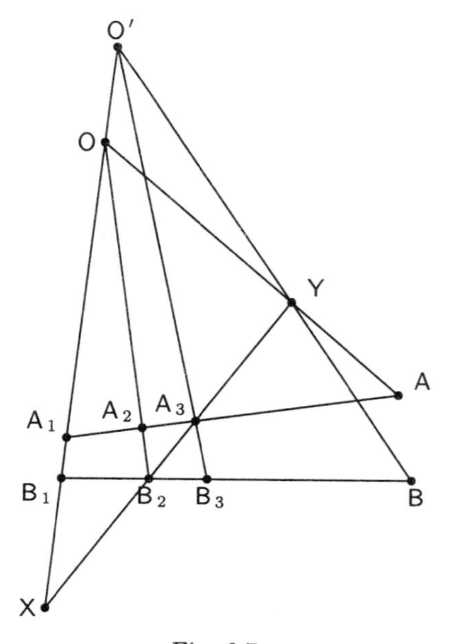

Fig. 6.7

Then

$$(A_1, A_2, A_3, A) \overset{O}{\wedge} (X, B_2, A_3, Y) \overset{O'}{\wedge} (B_1, B_2, B_3, B).$$

This writes the projectivity as the composition of the perspectivities from O and O′. So we have proved the following.

■ Any projectivity between two lines is either a perspectivity or the composition of two perspectivities. □

We can give a simple rule for when a projectivity is a perspectivity as follows.

■ Let a, b be two lines meeting at the point P. Then a projectivity between the points of a and b is a perspectivity if and only if P corresponds to itself under the projectivity.

Proof Clearly under a perspectivity P is self-corresponding. For the converse use the notation of the construction of the two perspectivities given above. Let OP meet B_2A_3 in the point Z. Since P is self-corresponding O′ must lie in the line OZ. But O, O′ also lie on the distinct line A_1B_1. So O, O′ coincide and the projectivity is a perspectivity from O. □

It is useful at this stage to pause and consider the dual of the above results. The dual of a range of points on a line is a pencil of lines through a point. A projectivity between two pencils is a correspondence which preserves the cross ratio. Two pencils, (a) and (b) are in perspective from a line c if any line a from the first pencil meets the corresponding line b from the second pencil on the line c. Any projectivity between two pencils is either a perspectivity or the product of two perspectivities. A projectivity is a perspectivity if and only if the line joining the apexes (common points) of the two pencils is self-corresponding.

The Pappus Line

Let the range of points (A_i) be projectively related to the range of points (B_i) under the correspondence $A_i \leftrightarrow B_i$. Fix i: then the pencil of lines through A_i is projectively related to the pencil of lines through B_i by the sequence

$$(A_iB_j) \overset{A_i}{\wedge} B_j \wedge A_j \overset{B_i}{\wedge} (B_iA_j).$$

The line A_iB_i is self-corresponding and so this projectivity is a perspectivity from a line c_i. If we interchange the rôle of i and j we see that c_i is independent of i. This line is called the *Pappus line* of the projectivity. It has the following property: the intersection of A_iB_j with A_jB_i lies on the Pappus line. In fact once the Pappus line is known then it can be used as an alternative geometric construction of the point B corresponding to A.

Exercise 6.13

If the range of points (A_i) on the line a is projectively related to the range of points (B_i) on the line b under a perspectivity, show that the Pappus line passes through the point common to a and b. What happens otherwise?

Pappus' theorem Let A_1, A_2, A_3 and B_1, B_2, B_3 be two sets of collinear points (lying on distinct lines). Let A_2B_3, A_3B_2 meet in C_1, let A_3B_1, A_1B_3 meet in C_2 and let A_1B_2, A_2B_1 meet in C_3. Then C_1, C_2, C_3 are collinear.

Proof

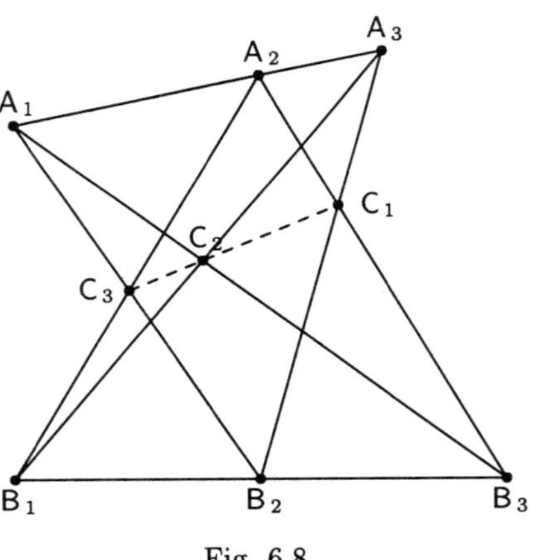

Fig. 6.8

A projectivity is uniquely determined by three pairs of corresponding points. So $A_i \to B_i$, $i = 1, 2, 3$ determines a unique projectivity. The points C_1, C_2, C_3 all lie on the Pappus line of this projectivity (Fig. 6.8). □

Exercise 6.14

State the dual of Pappus' theorem.

6.8 Quadrilaterals

A quadrilateral is defined by four general points A, B, C, D in the projective plane. The four points determine six lines which fall into three pairs of opposite sides $\{AC, BD\}$, $\{AD, BC\}$ and $\{AB, CD\}$ which meet in three diagonal points X, Y, Z (see Fig. 6.9).

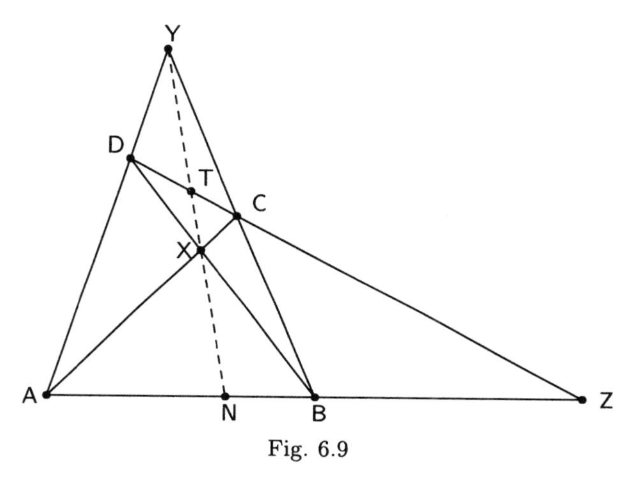

Fig. 6.9

Exercise 6.15

If the coordinates of A, B, C, D are $[\lambda, \mu, \nu]$, $[\lambda, \mu, -\nu]$, $[-\lambda, \mu, \nu]$, $[\lambda, -\mu, \nu]$, show that the three diagonal points are the vertices of the triangle of reference.

There is no particular order to the points of a quadrilateral in the projective plane.

Four collinear points A, B, A', B' are said to be *harmonic* if (AB,A'B')$= -1$.

Exercise 6.16

Show that distinct points A, B, A', B' are harmonic if and only if (AB,A'B')=(BA,A'B'.)

■ In Fig. 6.9 let AB meet XY in N. Then A, B, N, Z are harmonic.

Proof Let DC meet XY in T. Then (AB,NZ)=(DC,TZ) by perspectivity from Y. But (DC,TZ)=(BA,NZ) by perspectivity from X. By the results of Exercise 6.16 A, B, N, Z are harmonic. □

Exercise 6.17

If the coordinates of A, B, C, D are $[\lambda, \mu, \nu]$, $[\lambda, \mu, -\nu]$, $[-\lambda, \mu, \nu]$, $[\lambda, -\mu, \nu]$, find the coordinates of N and show that A, B, N, Z are harmonic by direct calculation.

6.9 Projective Transformations

We have already met projectivities, that is maps from a line to a line which preserve the only one-dimensional projective feature, the cross ratio. Let us now consider transformations of the projective plane which preserve the essential two-dimensional features. Let $T\colon \mathbb{R}^3 \to \mathbb{R}^3$ be an invertible linear transformation. Then T defines a transformation, denoted by $[T]$, of the projective plane, $\mathbb{R}P^2$, by the rule

$$[T]([X]) = [T(X)].$$

Such a transformation is called a *projective transformation.*

If T is defined by a matrix $\mathbf{M}$ then $[T]$ is defined by any non-zero multiple of $\mathbf{M}$. Because the linear transformation is invertible the determinant of $\mathbf{M}$ will be non-zero.

An invertible linear transformation of $\mathbb{R}^3$ takes planes through the origin (subspaces of dimension 2) to planes through the origin. It follows that the corresponding projective transformation will take a line in the projective space to another line. A projective transformation will also preserve incidence. That is, it takes points on a line through two points into points on the line through the corresponding points. As a consequence we say that properties of incidence are projective invariants.

Exercise 6.18

Show that the line $2x+y-3z = 0$ is transformed into the line $5x-5y+z = 0$ under the projective transformation with matrix

$$\mathbf{M} = \begin{pmatrix} 1 & 0 & 1 \\ 1 & 1 & 3 \\ -2 & 0 & 1 \end{pmatrix}.$$

Exercise 6.19

Show that the projective transformation with matrix $\mathbf{M}$ takes the line with dual coordinates A into the line with dual coordinates $A(\mathbf{M}^T)^{-1}$.

We now show that the cross ratio is an invariant of projective transformations.

■ The cross ratio of four collinear points is unchanged by a projective transformation.

Proof Suppose that the four points are $[A], [B], [C], [D]$ where $C = pA + qB,\ D = rA + sB$. Let $[P] \to [P\mathbf{M}]$ be a projective transformation. Then the four collinear points are transformed to

$$[A\mathbf{M}], \quad [B\mathbf{M}], \quad [C\mathbf{M}], \quad [D\mathbf{M}].$$

Since $C\mathbf{M} = pA\mathbf{M} + qB\mathbf{M}$ and $D\mathbf{M} = rA\mathbf{M} + sB\mathbf{M}$ the cross ratio is unchanged. □

Exercise 6.20

* Show that any continuous transformation of $\mathbb{R}P^2$ which takes lines into lines and preserves incidence is determined by a linear transformation of $\mathbb{R}^3$ as above.

In fact the projective transformation is determined by what happens to four general points, that is four points no three of which are collinear.

■ There is a unique projective transformation taking four general points to four other general points.

Proof Since a projective transformation is invertible these points might as well be the vertices of the triangle of reference, $\mathsf{X} = [1,0,0],\ \mathsf{Y} = [0,1,0],\ \mathsf{Z} = [0,0,1]$ and the unit point $\mathsf{U} = [1,1,1]$. Let the other four points be $\mathsf{A} = [A]$, $\mathsf{B} = [B]$, $\mathsf{C} = [C]$, $\mathsf{D} = [D]$.

For arbitrary non-zero real numbers λ, μ, ν, the matrix

$$\mathbf{M} = \begin{pmatrix} \lambda A \\ \mu B \\ \nu C \end{pmatrix}$$

defines a projective transformation $[T]$ taking X to A, Y to B, Z to C and U to $[\lambda A + \mu B + \nu C]$. Since A, B, C form a basis of $\mathbb{R}^3$ we can find unique λ, μ, ν so that $D = \lambda A + \mu B + \nu C$ and then $[T]$ takes U to D. □

Exercise 6.21

Find the matrix (up to a non-zero multiple) of the projective transformation which takes $[1,0,0]$ to $[0,0,1]$, $[0,1,0]$ to $[0,1,1]$, $[0,0,1]$ to $[1,1,1]$ and $[1,1,1]$ to $[3,2,4]$.

6.10 Fixed Points and Eigenvectors

If A is a (real) eigenvector of the matrix $\mathbf{M}$ corresponding to an eigenvalue λ then $A\mathbf{M} = \lambda A$ and so $[A]$ is a fixed point of the corresponding projective transformation and conversely. Since a 3×3 real matrix always has at least one real eigenvalue every projective transformation has at least one fixed point.

Example 6.3

Let us find the fixed points of the projective transformation corresponding to the matrix

$$\begin{pmatrix} 2 & 1 & 0 \\ 0 & 1 & -1 \\ 0 & 2 & 4 \end{pmatrix}.$$

The eigenvalues are given by the equation

$$\begin{vmatrix} \lambda - 2 & -1 & 0 \\ 0 & \lambda - 1 & 1 \\ 0 & -2 & \lambda - 4 \end{vmatrix} = (\lambda - 2)^2(\lambda - 3) = 0.$$

So there are two eigenvalues 2 (twice) and 3. The corresponding eigenvectors are (up to a non-zero multiple) $(0, 2, 1)$ and $(0, 1, 1)$. Hence the fixed points of the projective transformation are $[0, 2, 1]$ and $[0, 1, 1]$. Notice that the line $x = 0$ is invariant in the sense that any point with x-coordinate zero is transformed into another point with x-coordinate zero.

Exercise 6.22

Find the fixed points of the projective transformation corresponding to the matrix

$$\begin{pmatrix} 2 & 0 & 0 \\ 1 & 1 & 2 \\ 0 & -1 & 4 \end{pmatrix}.$$

6.11 Pappus' Theorem

We will now use the above result to prove the famous theorem of Pappus using homogeneous coordinates. Earlier, the proof used the properties of projectivities. Because this proof uses coordinates the dependence of the proof on the commutativity of the real numbers will be made clear.

■ Let A_1, A_2, A_3 and B_1, B_2, B_3 be two sets of collinear points (lying on distinct lines). Let A_2B_3, A_3B_2 meet in C_1; let A_3B_1, A_1B_3 meet in C_2; and let A_1B_2, A_2B_1 meet in C_3. Then C_1, C_2, C_3 are collinear.

Proof By the above discussion we can apply a projective transformation to simplify the algebra. This will not effect the essential features of the situation. In essence we are choosing the triangle of reference and we take $A_1A_2A_3$ as the line $y = 0$ and $B_1B_2B_3$ as the line $z = 0$. Let

$$\begin{array}{lll} A_1 = [p, 0, 1], & A_2 = [q, 0, 1], & A_3 = [r, 0, 1], \\ B_1 = [l, 1, 0], & B_2 = [m, 1, 0], & B_3 = [n, 1, 0]. \end{array}$$

Then the line A_2B_3 has dual coordinates

$$\begin{vmatrix} q & 0 & 1 \\ n & 1 & 0 \end{vmatrix} = [-1, n, q]$$

and the line A_3B_2 has dual coordinates

$$\begin{vmatrix} r & 0 & 1 \\ m & 1 & 0 \end{vmatrix} = [-1, m, r].$$

These lines meet at C_1 which therefore has coordinates

$$\begin{vmatrix} -1 & n & q \\ -1 & m & r \end{vmatrix} = [nr - qm, r - q, n - m].$$

Similarly

$$C_2 = [lp - rn, p - r, l - n], \; C_3 = [mq - pl, q - p, m - l].$$

Since

$$\begin{vmatrix} nr - qm & r - q & n - m \\ lp - rn & p - r & l - n \\ mq - pl & q - p & m - l \end{vmatrix} = 0$$

(the rows sum to zero), it follows that C_1, C_2, C_3 are collinear. □

The alert reader will have noted that the vanishing of the determinant which proves the collinearity of C_1, C_2, C_3 depends on the fact that $nr = rn$ etc. With more general non-commuting algebraic systems such as the quaternions, Pappus' theorem may fail to hold.

6.12 Perspective Drawing: Tricks of the Trade

Before the Renaissance, most paintings were for the use of the church and reflected its priorities. So a saint would be depicted larger than a mere mortal irrespective of their actual positions in space. With the rise of rich and powerful patrons whose interests were more secular a greater realistic representation was necessary. Many artists and mathematicians such as Albrecht Dürer (1471–1528) in Germany and Leonardo da Vinci (1452–1519) in Italy applied themselves to the problem of greater spatial reality in a painting or drawing. The interest that these inquiries generated, fed back to mathematics and to geometry in particular.

The task was to represent three-dimensional space onto a two dimensional space by stereographic projection with the eye as projection point. Dürer had a glass screen with a grid etched on it. The screen was placed between himself and the subject matter so that its appearance on the grid could be transfered to a similar grid on his work surface.

Of course in such a projection parallel lines may appear to meet at a vanishing point. The standard example is railway lines which we know are always a constant distance apart. Nevertheless they appear to meet at a vanishing point on the horizon.

The perspective view of a box shown in Fig. 6.10 is a typical example. The horizontal parallel lines meet at two vanishing points on the horizon.

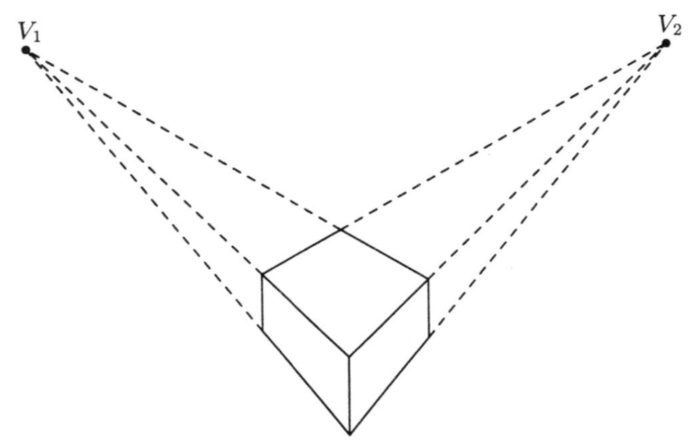

Fig. 6.10 A box in perspective

It must not be thought that paintings with "correct" perspective are necessarily superior to the older type. Perspective was just another technique for the painter to use. With the appearance of photography it was realised that the exact representation of the physical world was not necessary for great art and our views are now in some respects more in tune with medieval ideas.

A Brick in the Wall:

Very often a regular repeating pattern needs to be represented. Obvious examples are telegraph poles, railway sleepers and bricks in a wall. In Figs. 6.11

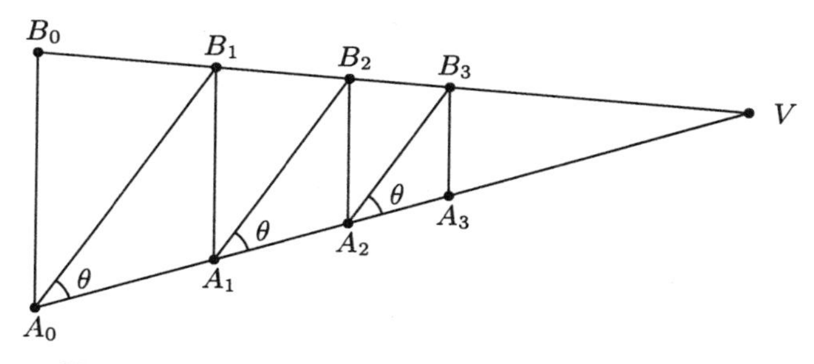

Fig. 6.11 Bricklaying by the constant angle method

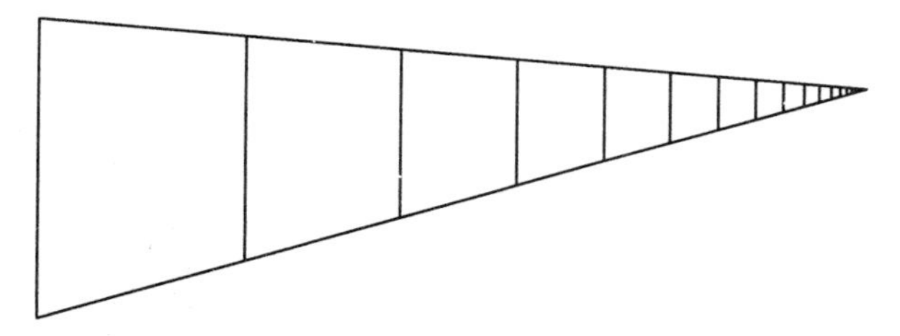

Fig. 6.12 The completed course

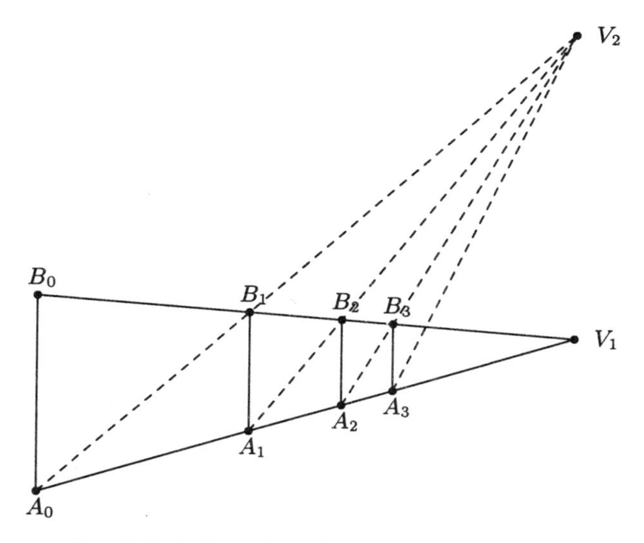

Fig. 6.13 Putting the diagonals in perspective

and 6.12 all the diagonals of the bricks are represented as parallel. This means that they make a constant angle θ with the base line. This means that once the first brick is drawn the positions of the next bricks are determined. Another idea is to make the diagonals meet at a vanishing point V_2 (Fig. 6.13). Now two bricks must be drawn to determine V_2. The rest are determined.

Leon Battista Alberti (1404–1472) described a method for representing a set of squares in a horizontal ground plane, for example a chess board or tiled floor, in the vertical plane of a painting. The initial line AB is divided into equal parts (Fig. 6.14). The sides of the squares are either horizontal or meet at a vanishing point V_2 on the horizon. The diagonals meet at another vanishing point V_1 on the horizon. The method of determining the position of the squares is clear from Fig. 6.14. To find the position of points in general we can use the

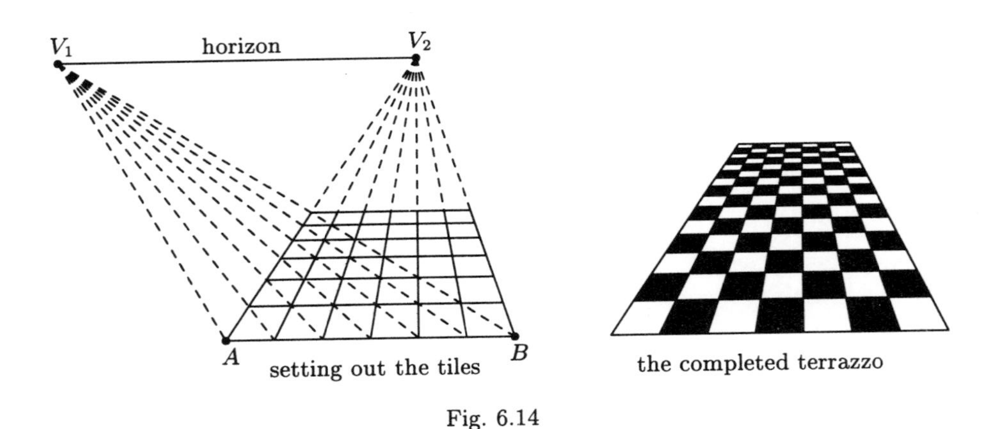

Fig. 6.14

fact that the cross ratio is invariant under a perspectivity. Here is a typical application.

Example 6.4

On a straight road approaching traffic lights there are "slow down" signs, 400 m and 200 m from the traffic lights, and a warning sign 100 m from the traffic lights. A town planner makes a perspective drawing in which the "slow down" signs are 3 cm and 1 cm from the traffic lights. Where should the warning sign be placed on the drawing? Let the slow down signs be S and S' and the warning sign be at W. If T is the traffic lights then the cross ratio is

$$\frac{(S-W)(S'-T)}{(S-T)(S'-W)} = \frac{300 \times 200}{400 \times 100} = \frac{3}{2}.$$

If in the drawing the warning sign is placed x cm from traffic lights then this is equal to

$$\frac{(3-x)(1)}{3(1-x)}.$$

So $9(1-x) = 2(3-x)$ which when solved gives $x = 3/7$ cm.

Exercise 6.23

The Doge's palace in Venice has two pillars and a statue in a line. They are 4 m, 3 m, and 1 metre respectively, from a wall. Leonardo has been asked to make a perspective drawing. He places the pillars 3 cm and 1 cm from the wall in the drawing. Where should the statue be placed in the drawing?

6.13 The Fano Plane

We have previously discussed the possibility of using different algebraic objects for coordinates. Here we illustrate using the integers modulo 2, $\mathbb{Z}_2 = \{0, 1\}$. We can think of 0 as the collection of all even integers and 1 as the collection of all odd integers and add and multiply accordingly. For example an even integer plus an odd integer is odd so $0+1 = 1$ but an even integer times an odd integer is even so $0 \times 1 = 0$.

The *Fano plane*, $\mathbb{Z}_2P^2$, is the projective plane defined by 3 homogeneous $\mathbb{Z}_2$ coordinates. So an element of the Fano plane is specified by a triple (x, y, z) where x, y, z are 0 or 1. Since $(0, 0, 0)$ is excluded $\mathbb{Z}_2P^2$ has seven points. By duality there are seven lines and these are illustrated in Fig. 6.15.

The only possible conceptual difficulty might be the line $x + y + z = 0$ passing through the points $(0, 1, 1), (1, 0, 1), (1, 1, 0)$ and represented by a circle in Fig. 6.15.

Exercise 6.24

What is the maximum number of points in $\mathbb{Z}_2P^2$, no three being collinear? How many (non-degenerate) triangles are there in the Fano plane? How many (non-degenerate) quadrilaterals?

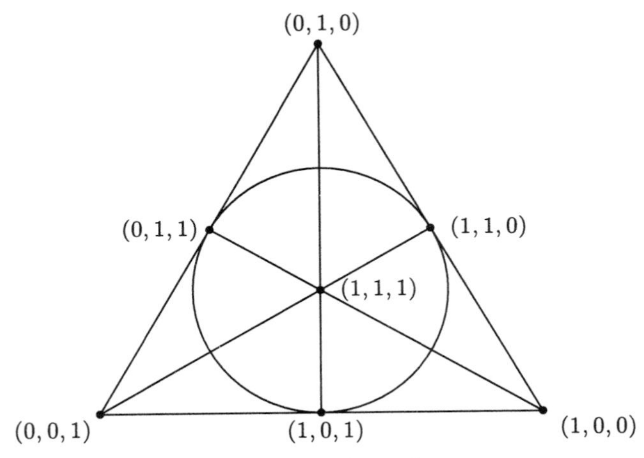

Fig. 6.15 The Fano plane

Answers to Selected Questions in Chapter 6

6.1 4.

6.3

$$\begin{vmatrix} 0 & 1 & 1 \\ 2 & -1 & 0 \end{vmatrix} = [1, 2, -2].$$

So the equation of the line is $x + 2y - 2z = 0$ and the dual coordinates of the line are $[1, 2, -2]$.

6.4

$$\begin{vmatrix} 1 & \theta & \theta^2 \\ 1 & \phi & \phi^2 \end{vmatrix} = [\theta\phi^2 - \theta^2\phi, \theta^2 - \phi^2, \phi - \theta] = [\theta\phi, -\theta - \phi, 1], \qquad \theta \neq \phi.$$

If $\phi = \theta$ then the line has coordinates $[\theta^2, -2\theta, 1]$. (This is the tangent line of a parabola.)

6.7 Let P$= [a, b, 0]$, A$= [a_1, 0, a_2]$, B$= [b_1, 0, b_2]$, then the line through PA has equation

$$\begin{vmatrix} x & y & z \\ a_1 & 0 & a_2 \\ a & b & 0 \end{vmatrix} = -xa_2b + ya_2a + za_1b = 0.$$

It meets the line $x = 0$ at the point L$= [0, a_1b, -a_2a]$. Similarly M is the point given by M$= [0, b_1b, -b_2a]$. So the line AM has equation

$$\begin{vmatrix} x & y & z \\ a_1 & 0 & a_2 \\ 0 & b_1b & -b_2a \end{vmatrix} = -xa_2b_1b + ya_1b_2a + za_1b_1b = 0.$$

Similarly the line BL has equation

$$-xa_1b_2b + ya_2b_1a + za_1b_1b = 0.$$

They meet at the point

$$\begin{vmatrix} -a_2b_1b & a_1b_2a & a_1b_1b \\ -a_1b_2b & a_2b_1a & a_1b_1b \end{vmatrix} = [a_1b_1a, -a_1b_1b, a(a_1b_2 + a_2b_1)]$$

which lies on the line

$$a_1b_1z = (a_1b_2 + a_2b_1)x.$$

This line clearly passes through $[0,1,0]$.

6.8 $(8,1,9) = 5(2,1,3) - 2(1,2,3)$ and $(4,-1,3) = 3(2,1,3) - 2(1,2,3)$. So the cross ratio is $3/5$.

6.10 Use the angle formula given in Chapter 4 and the fact that, for example, $B \cdot C = |B||C|\cos\phi$ where ϕ is the angle $\angle BOC$.

6.11 The dual coordinates are

$$[1,1,2],\ [3,-1,4],\ [5,1,8],\ [2,0,3].$$

Since $(5,1,8) = 2(1,1,2) + (3,-1,4)$ and $(2,0,3) = \frac{1}{2}(1,1,2) + \frac{1}{2}(3,-1,4)$ the lines are concurrent with cross ratio $\frac{1\times 1/2}{2\times 1/2} = 1/2$.

6.12 Let the points of intersection of the lines {AB', BA'}, {AC', CA'} and {AX', XA'} be L, M, N and let A" be the point of intersection of the line containing L, M, N with the line AA'. Perspective from A' gives (AB,CX)=(A"L,MN) and perspective from A gives (A"L,MN)=(A'B'C'X').

6.13 Let the point common to a and b be P. Then by construction the Pappus line passes through P since it is self-corresponding. Otherwise suppose as a point of a the point P corresponds to L and as a point of b it corresponds to M. Then the Pappus line is LM.

6.14 Let a_1, a_2, a_3 and b_1, b_2, b_3 be two sets of concurrent lines. Let the points of intersection a_2b_3, a_3b_2 define the line c_1. Define c_2 and c_3 similarly. Then c_1, c_2, c_3 all pass through a common point.

6.15 The dual coordinates of the line AC are $[0,-\nu,\mu]$ and the dual coordinates of the line BD are $[0,\nu,\mu]$. These lines meet in the point $X = [1,0,0]$. A similar calculation proves that for the other diagonal points $Y = [0,1,0]$ and $Z = [0,0,1]$.

6.16 If (AB,A'B')$= c$ then (BA,A'B')$= 1/c$. So $c^2 = 1$. Since $c \neq 1$ if the points are distinct we must have (AB,A'B')$= -1$.

6.17 The point N is where the line $z = 0$ meets the line AB which has dual coordinates $[-\mu, \lambda, 0]$. It follows that N= $[\lambda, \mu, 0]$. So N= $[A + B]$ and Z= $[A - B]$. Harmony is now clear.

6.18 The point with coordinates $[x, y, z]$ is transformed into $[a, b, c] = [x + y - 2z, y, x + 3y + z]$ by the matrix. If we write $[x, y, z]$ in terms of $[a, b, c]$ we get $[x, y, z] = [a - 7b + 2c, 3b, c - a - 2b]/3 = [a - 7b + 2c, 3b, c - a - 2b]$. So $2x + y - 3z = 0$ becomes $5a - 5b + c = 0$ after simplification.

6.19 Suppose $Y = X\mathbf{M}$. Then $X = Y\mathbf{M}^{-1}$. The line $AX^T = 0$ with dual coordinates A becomes $A(\mathbf{M}^T)^{-1}Y = 0$ with dual coordinates $A(\mathbf{M}^T)^{-1}$. To see how this compares with the previous question note that

$$\mathbf{M}^{-1} = \begin{pmatrix} 1 & 0 & -1 \\ -7 & 3 & -2 \\ 2 & 0 & 1 \end{pmatrix} /3.$$

So $(2, 1, -3) \to (2, 1, -3)(\mathbf{M}^T)^{-1} = (5, -5, 1)/3$.

6.21 Since

$$(3, 2, 4) = 2(0, 0, 1) - (0, 1, 1) + 3(1, 1, 1)$$

the matrix

$$\begin{pmatrix} 0 & 0 & 2 \\ 0 & -1 & -1 \\ 3 & 3 & 3 \end{pmatrix}$$

defines the transformation which does the job.

6.22 Note that the matrix is the transpose of the previous matrix and so has the same eigenvalues 2 and 3. The corresponding eigenvectors are (up to a non-zero multiple) $(1, 0, 0)$ and $(1, 1, -2)$.

6.23 Suppose in the drawing he places the statue x cm from the wall. The important thing to remember is that the cross ratio is preserved so

$$\frac{4-1}{4} \cdot \frac{3}{3-1} = \frac{3}{4} \cdot \frac{3}{2} = \frac{9}{8} = \frac{3-x}{3} \cdot \frac{1}{1-x},$$

giving $x = 3/19$.

6.24 Four.

6.25 The number of triangles is $\binom{7}{3} = 35$. Of these, 7 lie in a line. So the number of non-degenerate triangles is 28. The are $\binom{7}{4} = 35$ quadrilaterals. The degenerate ones consist of three points in a line and another point. There are $7 \times (7-3) = 28$ of these. So there are 7 non-degenerate quadrilaterals. Alternatively: a non-degenerate quadrilateral is the complement of a line and there are seven of these.

7
Conics and Quadric Surfaces

If we think of lines, planes and general affine subspaces as sets of points satisfying a linear equation then circles and spheres are examples of sets of points which satisfy a quadratic equation. The solutions to a quadratic equation in the plane are called *conic sections* or conics for short. These were known to the ancient Greeks and were given this name because they can be thought of as the intersection of a plane with a circular cone. This is the definition we shall start with and we shall end with their definition in terms of a focus and directrix. The latter, introduces us to all sorts of pretty properties of a conic. The focus, as the name suggests, involves rays of light reflected by the conic curve with important practical consequences.

We look at the paths of planets along elliptical orbits. It is an amusing fact that the conic curves studied by the ancient Greeks, solely for their interesting and æsthetic contribution to intellectual life, should govern the motion of planets and comets: objects whose motion was such a mystery to them. Studying subjects solely for their utilitarian aspects is not only boring but carries the danger that important discoveries may be missed. There should always be room for blue sky research in any decent society.

In three-dimensional space the solution set of a quadratic equation defines a surface called a *quadric surface.* These surfaces have the property that any plane cuts them in a conic. Since a circular cone is a (degenerate) example of a quadric surface this leads us back to the beginning of the chapter.

7.1 Conic Sections

Consider a (double) circular cone defined as follows. In $\mathbb{R}^3$ let a line g meet another line a (the z-axis say) in O. Suppose that the two lines make an angle α at O. Rotating the line g about the line a will trace out the cone. The line g is called the *generator* of the cone and the line a is called the *axis* of the cone. We call α the *semi-angle* of the cone and the point O the *apex* or the *vertex*.

Exercise 7.1

Show that the cone is the set of points satisfying the equation

$$x^2 + y^2 = z^2 \tan^2 \alpha.$$

To make definite, consider the axis of the cone to be vertical. Then a horizontal plane will cut the cone in a circle (Fig. 7.1). If this plane is tilted slightly then the cut will be an *ellipse*. Notice that the circular cuts will appear elliptical because their appearance is in a plane (of the paper) cutting a cone with apex the eye.

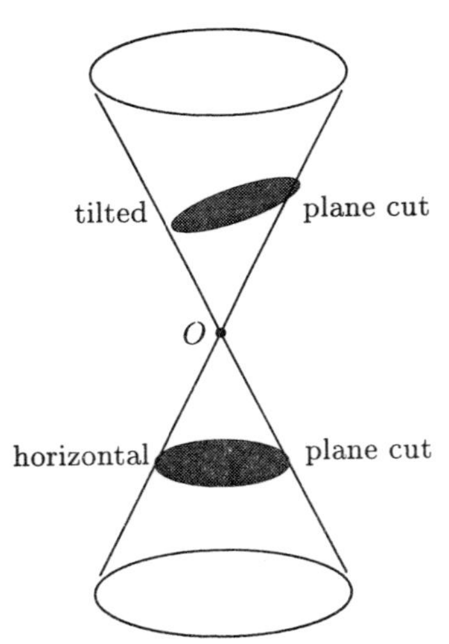

Fig. 7.1 Elliptic cuts of a cone

If the plane is further tilted so that it is parallel to a generator of the cone then the plane cuts the cone in a curve called a ***parabola*** (Fig. 7.2). Unlike the ellipse, the parabola is unbounded and has two ends which go off to infinity.

A further tilt of the cutting plane results in a ***hyperbola*** (Fig. 7.3). Since the plane now meets the double cone in both top and bottom bits the hyperbola is in two pieces. Like the parabola the hyperbola is unbounded.

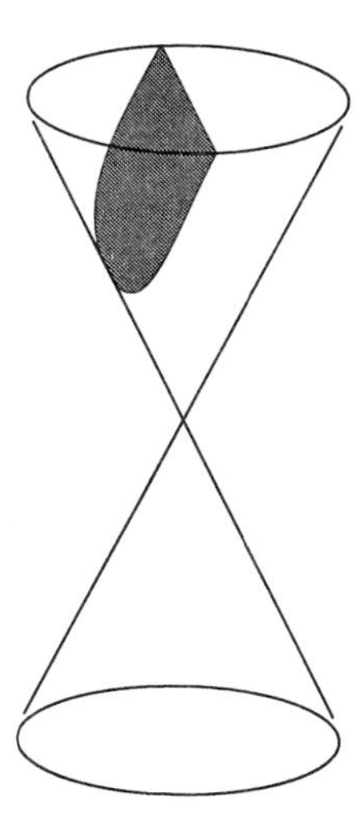

Fig. 7.2 Parabolic cut of a cone

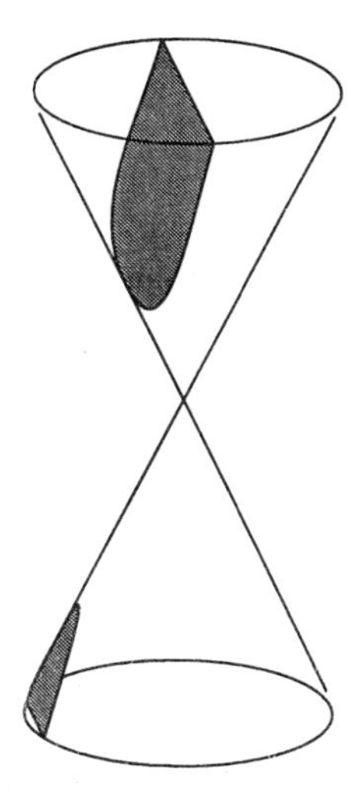

Fig. 7.3 Hyperbolic cut of a cone

If the plane passes through O, the apex of the cone, then the intersection is called a *degenerate conic.* These consist of a point (the apex) or a pair of lines through O (two generators) or a line (one generator where the plane is tangent).

Conic sections arise all the time in the real world. If we look at a circular plate on a table it will only appear circular if we are directly above the plate.

Otherwise it will appear elliptical. The tip of the shadow of a vertical stake in the ground will follow the circular path of the sun through the sky. Normally the tip of the shadow will trace out a hyperbolic curve being the horizontal section of the resulting cone, although in the land of the midnight sun the path will be elliptical.

7.2 The Conic as Quadratic Curve

The cone in space is the solution set of a quadratic equation. The plane, being a linear object, will meet it in a curve also given by a quadratic equation. The most general quadratic equation in the plane has the form

$$\boldsymbol{ax^2 + 2hxy + by^2 + 2fx + 2gy + c = 0.}$$

We shall see that for suitable choice of the coefficients a, b, c, f, g, h, the solution set is one of the conic sections above. The only difference is that the empty set can occur if the equation above has no solutions. Before proving this we shall look at how various conics can appear as particular quadratic equations.

Exercise 7.2

Show that the general quadratic equation above can be written

$$\begin{pmatrix} x & y & 1 \end{pmatrix} \begin{pmatrix} a & h & g \\ h & b & f \\ g & f & c \end{pmatrix} \begin{pmatrix} x \\ y \\ 1 \end{pmatrix} = 0.$$

Exercise 7.3

Show that the general quadratic curve through the five points (x_i, y_i), $i = 1, \dots 5$ has equation

$$\begin{vmatrix} x^2 & xy & y^2 & x & y & 1 \\ x_1^2 & x_1y_1 & y_1^2 & x_1 & y_1 & 1 \\ x_2^2 & x_2y_2 & y_2^2 & x_2 & y_2 & 1 \\ x_3^2 & x_3y_3 & y_3^2 & x_3 & y_3 & 1 \\ x_4^2 & x_4y_4 & y_4^2 & x_4 & y_4 & 1 \\ x_5^2 & x_5y_5 & y_5^2 & x_5 & y_5 & 1 \end{vmatrix} = 0.$$

A convenient condition for degeneracy is

$$\begin{vmatrix} a & h & g \\ h & b & f \\ g & f & c \end{vmatrix} = 0.$$

The Circle

The circle centre (p, q) and radius r has equation

$$(x-p)^2 + (y-q)^2 = r^2.$$

Writing out in full this becomes

$$\boldsymbol{x^2 + y^2 - 2px - 2qy + p^2 + q^2 - r^2 = 0.}$$

So the conditions for the general quadratic to represent a circle are

$$a = b(= 1), \qquad h = 0, \qquad f^2 + g^2 - c > 0.$$

The last inequality ensures that the radius, $r = \sqrt{f^2 + g^2 - c}$, is defined and positive.

Exercise 7.4

Find the equation of the circle centre $(2, 1)$ and radius 3. What is the centre and radius of the circle $x^2 + y^2 + 2x - 6y + 7 = 0$?

Two circles may meet in two points, be disjoint or touch at one point where they have a common tangent. If

$$\begin{aligned} f(x,y) &= x^2 + y^2 - 2p_1x - 2q_1y + p_1^2 + q_1^2 - r_1^2 = 0 \\ g(x,y) &= x^2 + y^2 - 2p_2x - 2q_2y + p_2^2 + q_2^2 - r_2^2 = 0 \end{aligned}$$

are two circles, the line $f(x, y) = g(x, y)$ is called the ***radical axis*** of the two circles. The equation of the radical axis can be written

$$\boldsymbol{-2p_1x - 2q_1y + p_1^2 + q_1^2 - r_1^2 = -2p_2x - 2q_2y + p_2^2 + q_2^2 - r_2^2.}$$

■ The radical axis of two circles is orthogonal to their line of centres. If the circles meet in two points, the radical axis is the line through these two points. If the circles touch at P, the radical axis is the common tangent at P.

Proof Use the above notation. The slope of the radical axis is

$$\frac{p_2 - p_1}{q_1 - q_2}.$$

On the other hand, the slope of the line of centres is

$$\frac{q_1 - q_2}{p_1 - p_2}.$$

Since the product of the slopes is -1 the lines are orthogonal.

Any point satisfying $f(x,y) = 0$ and $g(x,y) = 0$ must lie on the line $f(x,y) = g(x,y)$. The common tangent is a limiting case. □

Exercise 7.5

Show that the circles $x^2 + y^2 - 4 = 0$ and $x^2 + y^2 + 2x - 2y - 7 = 0$ meet in two points and find the equation of the line through these two points.

Two circles are said to be *orthogonal* if they meet at two points where their tangents are orthogonal. The following theorem gives a condition for this to happen.

■ Two circles

$$x^2 + y^2 + 2f_1x + 2g_1y + c_1 = 0$$
$$x^2 + y^2 + 2f_2x + 2g_2y + c_2 = 0$$

are orthogonal if and only if

$$\mathbf{2(f_1f_2 + g_1g_2) = c_1 + c_2.}$$

Proof The centres of the two circles are at

$$C_1 = (-f_1, -g_1) \text{ and } C_2 = (-f_2, -g_2).$$

Suppose that the two circles meet orthogonally at P. Then C_1C_2P is a right-angled triangle with hypotenuse C_1C_2. By Pythagoras' theorem if d is the distance between C_1 and C_2 then

$$d^2 = r_1^2 + r_2^2$$

where r_1, r_2 are the radii. So we have

$$\begin{aligned} d^2 &= (f_2 - f_1)^2 + (g_1^2 - g_1)^2 \\ &= r_1^2 + r_2^2 \\ &= f_1^2 + g_1^2 - c_1 + f_2^2 + g_2^2 - c_2. \end{aligned}$$

Simplification yields the equality.

The converse is also true since if the equality holds there is a right-angled triangle with hypotenuse C_1C_2 and two sides equal to the radii r_1 and r_2. The vertex opposite C_1C_2 must be a point where the circles meet orthogonally. □

Exercise 7.6

Are the circles

$$x^2 + y^2 - 2x + 3y - 7 = 0$$
$$x^2 + y^2 + x + 4y + 12 = 0$$

orthogonal?

Exercise 7.7

Show that for all values of f, g, c the circles

$$x^2 + y^2 - 2fx + c = 0 \text{ and } x^2 + y^2 - 2gy - c = 0$$

are orthogonal.

The Ellipse

The equation of the standard ellipse as a quadratic curve is

$$\frac{x^2}{a^2} + \frac{y^2}{b^2} = 1$$

and it is illustrated in Fig. 7.4.

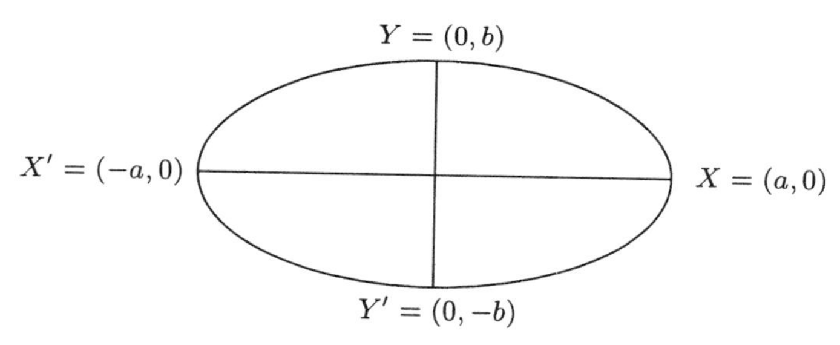

Fig. 7.4 The ellipse $x^2/a^2 + y^2/b^2 = 1$

If $a > b$ then a is called the (length of) the major semi-axis and b is the minor semi-axis. The corresponding lines XX' and YY' in the diagram are axes of symmetry. Of course if $a = b$ then the equation defines a circle which is a special kind of ellipse and there are no axes, major or minor.

A point on the ellipse above is conveniently given as $P = (a\cos\theta, b\sin\theta)$ where θ is an angle parameter. If we put $t = \tan\theta/2$ we obtain the rational parameterisation

$$x = a\frac{1-t^2}{1+t^2}, \qquad y = b\frac{2t}{1+t^2}.$$

For any real value of the parameter t there corresponds a unique point on the ellipse. To get all points we need $t = \infty$ corresponding to $(-a, 0)$.

The *auxiliary circle* to the ellipse has the major axis as diameter. Its area, πa^2, can be used to find the area of the ellipse. If we think of this area as the sum of very thin vertical strips parallel to the minor axis then the area of such a strip is b/a times the area of a longer strip used to find the area of the auxiliary circle. It follows that the area of the ellipse is b/a times the area of the auxiliary circle. Hence

$$\textbf{area of the ellipse} = \pi ab.$$

Exercise 7.8

Show that the chord of the ellipse with end-points $P = (a\cos\theta, b\sin\theta)$ and $Q = (a\cos\phi, b\sin\phi)$ has slope

$$\left(\frac{b}{a}\right)\left(\frac{\sin\theta - \sin\phi}{\cos\theta - \cos\phi}\right).$$

By letting ϕ tend to θ, show that the tangent line to the ellipse at the point (x_0, y_0) has equation

$$\frac{xx_0}{a^2} + \frac{yy_0}{b^2} = 1.$$

Exercise 7.9

Let PQ be a chord of an ellipse with centre O. If M is the midpoint of PQ show that

$$\text{slope } (PQ) \times \text{slope } (OM)$$

is constant. If PQ is orthogonal to OM show that the ellipse is a circle.

The Parabola

The equation of the standard parabola as a quadratic curve is

$$y^2 = 4ax$$

where a is some constant.

The parabola is symmetric about the line $y = 0$, which is called the *axis* of the parabola (Fig. 7.5). Note that any two parabolas are similar. If their equations are $y^2 = 4ax$ and $y^2 = 4a'x$ then the change of scale,

$$x \to ax/a', \ y \to ay/a'$$

takes the first parabola to the second. Points on the parabola may be represented rationally by the parameter t as

$$P_t = (at^2, 2at).$$

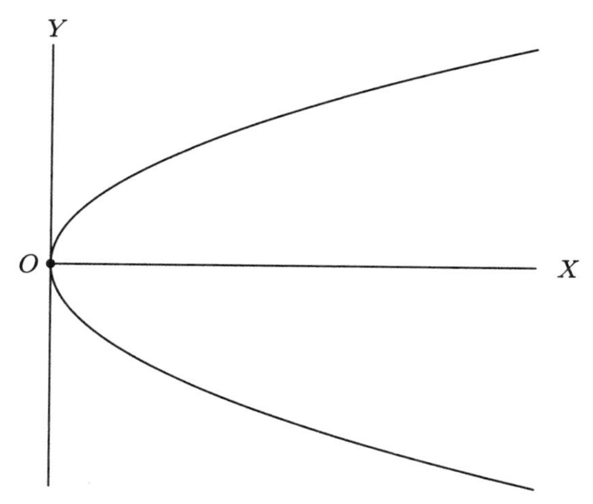

Fig. 7.5 The parabola $y^2 = 4ax$

Exercise 7.10

Show that the chord $P_tP_{t'}$ of the parabola has equation $2x - y(t+t') + 2att' = 0$. By letting t' tend to t, show that the tangent at P_t has slope $1/t$ and equation $x - yt + at^2 = 0$. If $P_t = (x_0, y_0)$, show that the tangent may be written $yy_0 = 2ax + 2ax_0$.

Exercise 7.11

Find the coordinates of the midpoint of the chord $P_tP_{t'}$. Show that the midpoints of parallel chords of a parabola lie on a fixed line parallel to the axis.

The Hyperbola

The equation of the standard hyperbola as a quadratic curve is

$$\boldsymbol{\frac{x^2}{a^2} - \frac{y^2}{b^2} = 1.}$$

The hyperbola consists of two pieces or *branches*. These are separated by the two lines

$$\frac{x^2}{a^2} - \frac{y^2}{b^2} = 0$$

or

$$\frac{x}{a} = \pm\frac{y}{b}$$

which are called the *asymptotes* of the hyperbola (Fig. 7.6). These lines never meet the hyperbola but appear to touch at infinity.

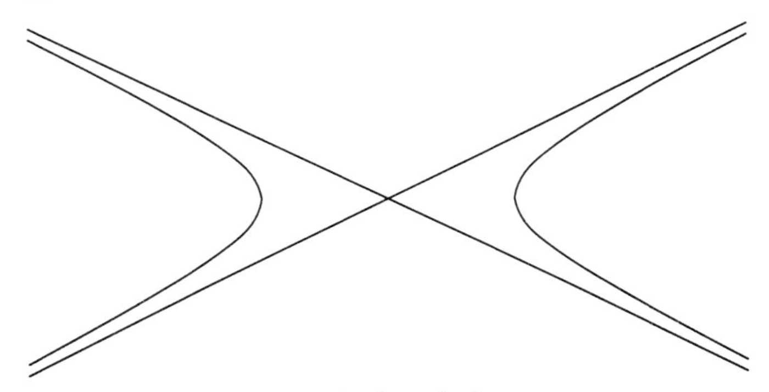

Fig. 7.6 The hyperbola $x^2/a^2 - y^2/b^2 = 1$ with asymptotes

If $a = b$ these lines are perpendicular and the hyperbola is called *rectangular*. So a rectangular hyperbola is to a general hyperbola as a circle is to an ellipse. A convenient equation to represent a rectangular hyperbola is $xy = c^2$. In this case the asymptotes are the coordinate axes.

The hyperbolic functions provide a partial parametric representation of the points on a hyperbola by

$$\boldsymbol{x = a\cosh\theta, \quad y = b\sinh\theta.}$$

However since the value of cosh is always positive this only represents one branch of the hyperbola. A full parameterisation by trigonometric functions is given by

$$\boldsymbol{x = a\sec\theta, \quad y = b\tan\theta.}$$

To get a rational parameterisation of $x^2/a^2 - y^2/b^2 = 1$ consider the factorisation

$$\frac{x^2}{a^2} - \frac{y^2}{b^2} = \left(\frac{x}{a} + \frac{y}{b}\right)\left(\frac{x}{a} - \frac{y}{b}\right).$$

Put $x/a + y/b = t$. Then on the hyperbola, $x/a - y/b = 1/t$. So

$$\boldsymbol{P_t = \left(\frac{a}{2}\left(t + \frac{1}{t}\right), \frac{b}{2}\left(t - \frac{1}{t}\right)\right)}$$

is a point on the hyperbola $x^2/a^2 - y^2/b^2 = 1$. This is valid for all non-zero t. The right-hand branch is $t > 0$ and the left-hand branch is $t < 0$.

A convenient rational parameterisation for the rectangular hyperbola $xy = c^2$ is

$$\boldsymbol{x = ct, \quad y = c/t.}$$

Exercise 7.12

Show that the tangent to the hyperbola at the point (x_0, y_0) has equation

$$\frac{xx_0}{a^2} - \frac{yy_0}{b^2} = 1.$$

Exercise 7.13

Show that the triangle formed by a tangent to the hyperbola together with the asymptotes has constant area.

Exercise 7.14

* Let $P_1P_2P_3$ be a triangle formed by three points on a rectangular hyperbola. Show that the orthocentre also lies on the hyperbola.

The general quadratic curve

We now show that the general quadratic equation having the form

$$ax^2 + 2hxy + by^2 + 2fx + 2gy + c = 0$$

represents, after a change of coordinates, one of the standard conics considered earlier or the empty set or a degenerate conic.

The change of coordinates considered will be of two types

(1) **Translations**

$$x \to x + u, \quad y \to y + v$$

where u, v are constants or

(2) **Rotations**

$$x \to x\cos\theta - y\sin\theta, \quad y \to x\sin\theta + y\cos\theta$$

where θ is the angle of rotation.

These coordinate changes are euclidean in form and so do not change the type or shape of the conics.

If we are not fussed about the size of the conic then we can make a scale change

$$x \to rx, \quad y \to ry$$

where the constant r is chosen to eliminate tiresome scale constants.

The easiest coordinate change is the translation and this is the only one needed in the case where the general equation has the coefficient of the cross term xy zero. The translations are those needed to "complete the square". This is best illustrated by an example.

Example 7.1

Consider the curve

$$x^2 + 2y^2 - 4x + 4y + 4 = 0.$$

This can be written as

$$(x-2)^2 + 2(y+1)^2 = 2.$$

The translation $x \to x+2,\ y \to y-1$ transforms the equation to

$$\frac{x^2}{2} + y^2 = 1,$$

an ellipse. So the curve is an ellipse with centre $(2, -1)$ and semi-axes of length $\sqrt{2}$ and 1, the axes being parallel to the original coordinate axes.

Exercise 7.15

Show that if the constant term in Example 7.1 had been 6 instead of 4 then the curve would have consisted of the single point $(2, -1)$. What would have happened if the constant term had been greater than 6?

Exercise 7.16

Show that the curve

$$2x^2 - 2y^2 + 4x + 12y = 19$$

is a rectangular hyperbola with asymptotes

$$x - y + 4 = 0, \qquad x + y - 2 = 0.$$

Show that the curve is a pair of lines if the constant term 19 is replaced by 16.

Exercise 7.17

Show that the curve

$$y^2 + 2x + 4y + 7 = 0$$

is a parabola and find its axis.

Exercise 7.18

Identify the conics with equations,
(a) $2x^2 + y^2 - 12x - 4y + 22 = 0$
(b) $2x^2 + y^2 - 12x - 4y + 23 = 0$
(c) $4x^2 - y^2 + 16x + 4y + 3 = 0$
(d) $-y^2 + x + 4y = 1$.

Exercise 7.19

Show that the graph of the quadratic function

$$y = ax^2 + bx + c$$

is a parabola and find the axis and the point where the axis meets the parabola.

If the xy-term is present then a rotation is needed to eliminate it and reduce to the first case. In other words, the rotation will make the axes parallel to the coordinate axes.

The rotation changes the coefficient of the xy-term to

$$-2a\cos\theta\sin\theta + 2b\cos\theta\sin\theta + 2h(\cos^2\theta - \sin^2\theta),$$

which can be rewritten as

$$\boldsymbol{(b-a)\sin 2\theta + 2h\cos 2\theta.}$$

The angle θ is now chosen to make this zero.

Example 7.2

Consider

$$x^2 - 6xy - 7y^2 + 10x + 2y + 9 = 0.$$

The rotation through θ changes the cross term to $-8\sin 2\theta - 6\cos 2\theta$ So if $\tan 2\theta = -3/4$ this is zero. We can look up θ using a calculator or argue as follows. Since $\tan 2\theta = 2\tan\theta/(1-\tan^2\theta)$ we have

$$3\tan^2\theta - 8\tan\theta - 3 = 0.$$

For an acute positive θ we have $\tan\theta = 3$ so $\cos\theta = 1/\sqrt{10}$ and $\sin\theta = 3/\sqrt{10}$.

A suitable combination of rotation and scale change is therefore.

$$x \to x - 3y, \qquad y \to 3x + y.$$

This changes the equation to

$$-80x^2 + 20y^2 + 16x - 28y + 9 = 0.$$

This can be seen to be a hyperbola by the translation

$$x \to x + 1/10, \qquad y \to y + 7/10.$$

Exercise 7.20

What is the conic with equation

$$5x^2 + 6xy + 5y^2 - 4x + 4y - 4 = 0?$$

Exercise 7.21

Determine the nature of the conics
(a) $3x^2 - 10xy + 3y^2 + x - 32 = 0$
(b) $41x^2 - 84xy + 76y^2 = 168$
(c) $16x^2 + 24xy + 9y^2 - 30x + 40y = 2$
(d) $x^2 - 4xy + 4y^2 = 9$.

An alternative method for finding the conic is to write the general conic

$$ax^2 + 2hxy + by^2 + 2fx + 2gy + c = 0$$

by putting $X = (x, y)$ as

$$X\mathbf{M}X^T + 2fx + 2gy + c = 0$$

where $\mathbf{M}$ is the matrix

$$\mathbf{M} = \begin{pmatrix} a & h \\ h & b \end{pmatrix}.$$

Find the eigenvalues and eigenvectors of $\mathbf{M}$. Let A and B be unit eigenvectors forming a right-handed orthogonal basis of $\mathbb{R}^2$ and let $\mathbf{P}$ be the 2×2 matrix with rows A and B. Then the rotation

$$X \to X\mathbf{P}$$

eliminates the cross term and we continue as before. The reason that this works will be given later.

Exercise 7.22

Use the above method for

$$x^2 - 6xy - 7y^2 + 10x + 2y + 9 = 0.$$

Exercise 7.23

For the general conic show that if $ab - h^2 > 0$ the conic is either an ellipse, a point or the empty set. If $ab - h^2 < 0$ show that the conic is either a hyperbola or a pair of lines and if $ab - h^2 = 0$ show that it is either a parabola or two parallel lines (may be equal).

Exercise 7.24

For the general central conic

$$ax^2 + 2hxy + by^2 = 1$$

show that if $ab - h^2 > 0$ the conic is either an ellipse, a point or the empty set. If $ab - h^2 < 0$ show that the conic is a hyperbola and if $ab - h^2 = 0$ show that it is a pair of parallel lines.

7.3 Focal Properties of Conics

We have seen the conic as a cross-section of a cone and a quadratic curve in the plane. Here is another definition. The conic is the set of points P whose distance FP from a fixed point F is ε times its distance from a fixed line d. The point F is called a *focus*, the line d is called a *directrix* and the number ε is a positive constant called the *eccentricity*. We will see that the curve is an ellipse if $\varepsilon < 1$, a parabola if $\varepsilon = 1$ and a hyperbola if $\varepsilon > 1$. The condition even makes sense if $\varepsilon = 0$ provided we push the directrix to infinity. Then FP is constant and the curve is a circle. Or, putting it another way, a circle is an ellipse with eccentricity zero.

In Fig. 7.7 P is an arbitrary point of some conic, K is the nearest point of the directrix d to P and J is the nearest point of d to F. The line LL' through F and perpendicular to FJ rejoices in the name *latus rectum*. The point H is the nearest point of d to L and X is the nearest point of the conic to d.

Let r be the distance FP, θ be the angle $\angle JFP$ and l be the distance FL, called the *semi-latus rectum*. Then

$$r = \varepsilon PK = \varepsilon(LH - r\cos\theta) = l - \varepsilon r \cos\theta$$

so that

$$\boldsymbol{\frac{l}{r} = 1 + \varepsilon \cos\theta}$$

which is the polar equation of a conic with eccentricity ε, semi-latus rectum l, focus at the origin and x-axis in the direction FJ. Squaring the formula for r

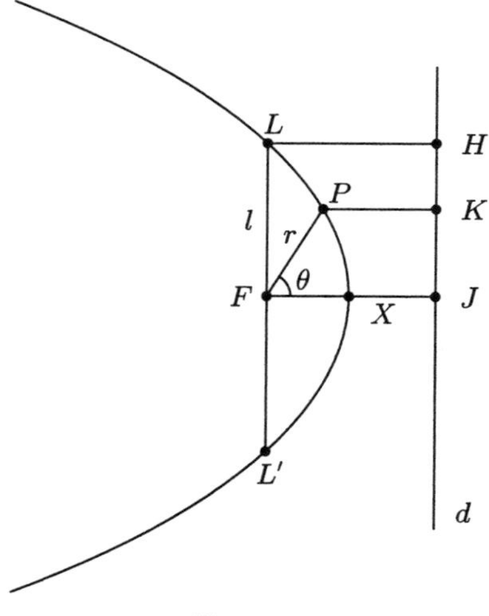

Fig. 7.7

we obtain

$$r^2 = x^2 + y^2 = (l - \varepsilon r \cos\theta)^2 = (l - \varepsilon x)^2$$

or

$$(1 - \varepsilon^2)x^2 + 2\varepsilon l x + y^2 = l^2.$$

We now consider the three cases.

Eccentricity, $0 \le \varepsilon < 1$, the Ellipse

Since $\varepsilon \neq 1$ we can complete the square in the above equation to obtain

$$(1 - \varepsilon^2)\left(x + \frac{\varepsilon l}{1 - \varepsilon^2}\right)^2 + y^2 = \frac{l^2}{1 - \varepsilon^2}.$$

This represents an ellipse with centre and semi-axes

$$\left(-\frac{\varepsilon l}{1 - \varepsilon^2}, 0\right), \quad \frac{l}{1 - \varepsilon^2} \text{ and } \frac{l}{\sqrt{1 - \varepsilon^2}}.$$

Rewriting in terms of the standard ellipse $x^2/a^2 + y^2/b^2 = 1$ we have

eccentricity: $\boldsymbol{\varepsilon = \sqrt{1 - b^2/a^2}}$

foci: $\boldsymbol{(\pm\sqrt{a^2 - b^2}, 0)}$

directrices: $\boldsymbol{x = \pm a^2/\sqrt{a^2 - b^2}}$

semi-latus rectum: $\boldsymbol{l = b^2/a}$.

By symmetry the ellipse (and hyperbola) has two foci and two directrices.

The equation

$$r = l - \varepsilon r \cos\theta = l - \varepsilon x$$

where x is the projection of the point P on the line FX can be written

$$r' = l + \varepsilon x'$$

where r' is the distance from the other focus F' and x' is the projection. Notice that the sign changes because the point is on the opposite side of the focus.

Adding the two equations gives

$$r + r' = 2l + \varepsilon FF'$$

which is a constant (equal to the length of the major axis). It follows that a method of drawing the ellipse is to attach the ends of a piece of string to two fixed points (the foci). Keeping the string taut with the end of your pencil draw the ellipse (Fig. 7.8).

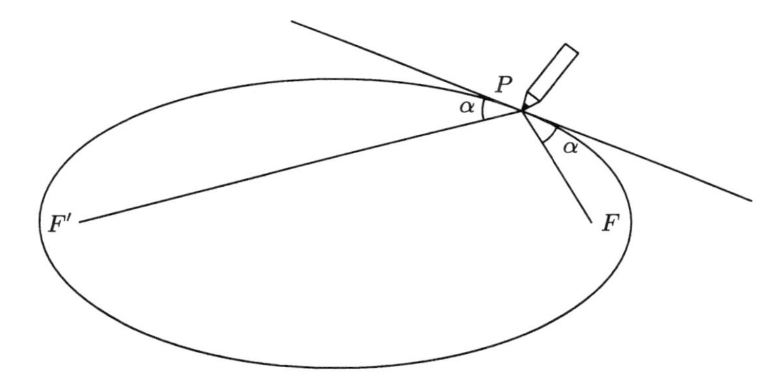

Fig. 7.8 Drawing an ellipse

It follows that the tangent at any point on the ellipse makes equal angles with the two foci. This can be verified directly or by the following argument. The force which prevents the pencil tip at P moving away from the ellipse consists of two equal tensions in the string directed towards the foci. Therefore the resultant force, normal to the curve, bisects the angle $\angle FPF'$.

This means that a billiard ball projected from a focus of an elliptical billiard table will pass though a focus after bouncing off the cushion.

A rather more important application occurs if the inside of the ellipse is silvered to reflect light. Then a ray of light from one focus will be reflected to a ray passing through the other. So a light source at F' will produce an image at F.

Exercise 7.25

If an ellipse with minor axis 3 cm is drawn with a string of length 5 cm attached to the foci, what is the eccentricity? What is the distance between the two ends of the string?

Exercise 7.26

Show that the distances from the foci to the point (x, y) on $x^2/a^2 + y^2/b^2 = 1$ are $a \pm \varepsilon x$.

Exercise 7.27

Let P be a point on an ellipse with major axis XX' and foci FF'. Let R, R' be points on the tangent at P so that XR, $X'R'$ are parallel to FP, $F'P$ respectively. Show that $XR + X'R' = XX'$.

Eccentricity, $\varepsilon = 1$, the Parabola

If the eccentricity is 1 the equation for the conic reduces to

$$y^2 = l^2 - 2lx$$

which is a parabola with axis the x-axis and focus (just one) at $(l, 0)$. The focus is distant $l/2$ from the point X where the curve crosses its axis.

Comparing with the standard parabola $y^2 = 4ax$ we see that the semi-latus rectum is $2a$, the focus is at the point $F = (a, 0)$ and the directrix is the line $x + a = 0$.

If we think of the parabola as the limit of an ellipse with one focus sent to infinity then the tangent at a point P makes an equal angle with the line from P to the focus F and a line from P parallel to the axis (Fig. 7.9). So a parabolic mirror will take light from infinity, say a star, and its image will be at the focus of the parabola.

Exercise 7.28

Show that the tangents at the ends of the latus rectum of a parabola are orthogonal and make an angle of $\pi/4$ with the axis.

Exercise 7.29

Find the coordinates of the point where the tangent at $(at^2, 2at)$ meets the directrix.

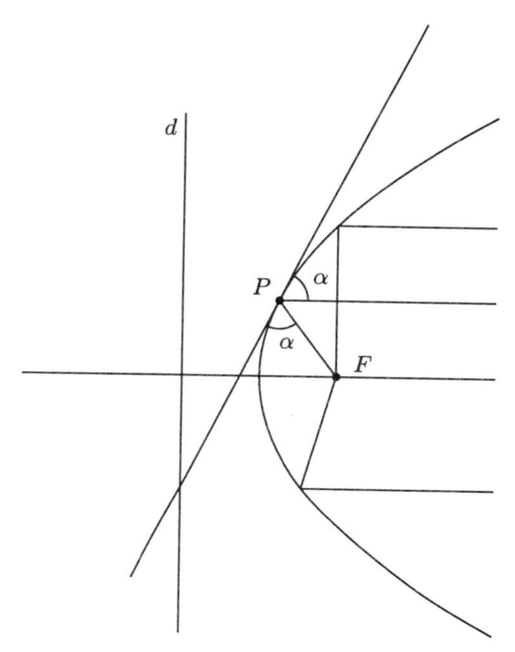

Fig. 7.9 Focal properties of the parabola

Exercise 7.30

Let D be the nearest point on the directrix to a general point P on the parabola. Show that DF is orthogonal to the tangent at P.

Exercise 7.31

Let the tangent to P on a parabola meet the axis at T and let the normal to P meet the axis at N. Show that the midpoint of TN is the focus F.

Eccentricity, $\varepsilon > 1$, the Hyperbola

As in the case of the ellipse, $\varepsilon \neq 1$, so we can complete the square in the equation of the general conic to get

$$(\varepsilon^2 - 1)\left(x + \frac{\varepsilon l}{1 - \varepsilon^2}\right)^2 - y^2 = \frac{l^2}{\varepsilon^2 - 1}.$$

This is now a hyperbola. Rewriting in terms of the standard hyperbola $x^2/a^2 - y^2/b^2 = 1$ we have

$$\textbf{eccentricity: } \varepsilon = \sqrt{1 + b^2/a^2}$$
$$\textbf{foci: } (\pm\sqrt{a^2 + b^2}, 0)$$
$$\textbf{directrices: } x = \pm a^2/\sqrt{a^2 + b^2}$$
$$\textbf{semi-latus rectum: } l = b^2/a.$$

This time the tangent at a general point P on a hyperbola bisects the angle $\angle FPF'$ (Fig. 7.10). It follows that if the right-hand branch of the hyperbola is silvered on the outside, a light source at F' will produce a *virtual* image at F.

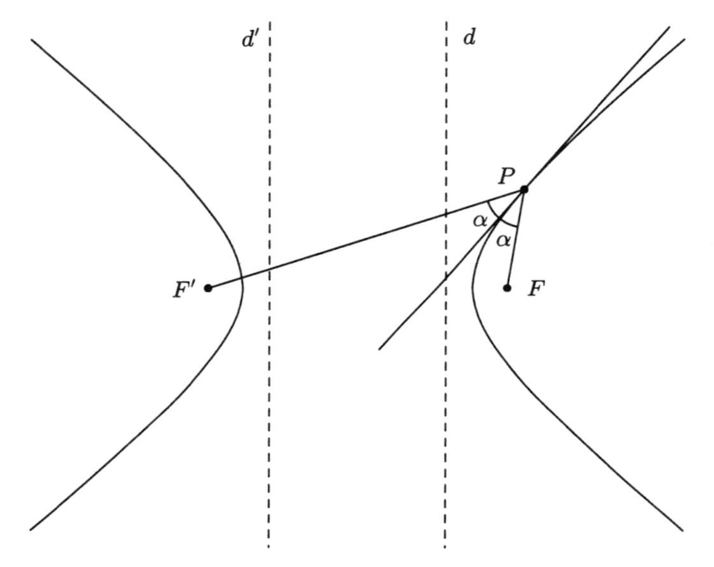

Fig. 7.10 Focal properties of the hyperbola; directrices shown dashed

Exercise 7.32

Show that the difference $PF - PF'$ of the two distances of an arbitrary point on a hyperbola from the two foci is constant.

Exercise 7.33

Let D be the point where the directrix d meets an asymptote of a hyperbola. Show that DF is orthogonal to the asymptote.

Exercise 7.34

* Let AB be a chord of a circle centre C. Suppose a hyperbola with

eccentricity 2, focus A and directrix BC meets AB at D. Show that CD trisects $\angle ACB$.

The focal properties of two different conics can be utilised to make a compound reflecting telescope. Consider two points F, F' and a parabolic mirror with axis FF' and focus F (Fig. 7.11). Light from a star is reflected by the parabola and is directed towards the focus F. It is intercepted by a hyperbolic mirror with axis FF' and foci F, F'. This reflects the light again and directs it through a hole in the parabolic mirror to an image at F'. The point F' is called the *Cassegrain focus*.

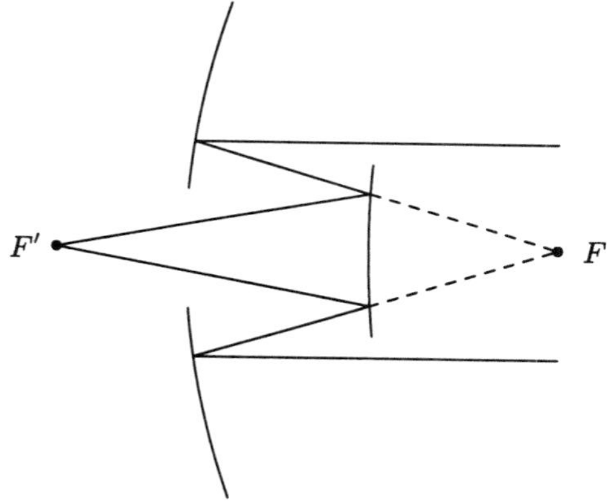

Fig. 7.11 Parabolic/hyperbolic compound mirror (not to scale)

7.4 The Motion of the Planets

> *HELENA: The wars have so kept you under that you must needs be born under Mars.*
> *PAROLLES: When he was predominant.*
> *HELENA: When he was retrograde, I think, rather.*
>
> All's well that ends well

The notion that the Sun and not the Earth is the centre of the cosmos had the most devastating effect upon the world of thought and eventually upon the world of actions when it was first proposed by Copernicus in 1543. The motion of the stars across the night sky does indeed suggest a geocentric universe. Their regular ordered pattern of repetition once a year suggests that the celestial sphere is rotating about the Earth.

However we now know that this is an illusion due to the huge distances of the stars and that nearer objects such as the planets, the Sun, Moon and comets do not exhibit this neat periodic behaviour. Indeed the very name, planet, comes from the Greek word for wanderer.

Ptolemy in about AD 150 attempted to reconcile the eccentric motion of planets with an Earth-centred universe by the use of complicated epicycloids and hypocycloids. This worked well enough until the more accurate measurements of Tycho Brahe at the end of the 16th century pushed this system to its limits. After careful consideration of Brahe's observations Kepler deduced his famous three laws of planetary motion. These are

I Each planet describes an ellipse with the Sun in one focus.

II The radius vector drawn from the Sun to a planet sweeps out equal areas in equal times.

III The squares of the periodic times of the planets are proportional to the cubes of the major semi-axes of their orbits.

With this information the motion of the planets can be determined with great accuracy as we will now see.

Newton was able to use Kepler's laws, which were entirely empirically deduced, to formulate theoretically his inverse square law of gravitational attraction. Because this law is universal it applies to other solar system objects such as comets which to a first approximation move along orbits which are conic sections. If the orbit is an ellipse then the comet will return and good guesses can be made for when this will happen. On the other hand, if the path is parabolic or hyperbolic the comet will return to the depths of space, never to reappear.

Luckily Brahe's observations were not sufficiently accurate to detect slight deviations from Kepler's law in the motion of Jupiter or they might never have been deduced. This is because the great mass of Jupiter causes the focus of its orbit to be shifted from the centre of the Sun. Newton's theory is able to account for this.

The days of the week are named after the seven solar system objects known before the discovery of Uranus. Saturday, Sunday and Monday are obvious. Tuesday is named after the Nordic god of war, Tiw (Roman equivalent Mars, e.g. French *mardi*). Wednesday, Thursday and Friday are named after Woden, Thor and Freya (Mercury *mercredi*, Jupiter *jeudi* and Venus *vendredi*).

The Planetary Setup

Assume the planet P moves along an ellipse, $x^2/a^2 + y^2/b^2 = 1$ with major semi-axis a and focus F occupied by the Sun. The coordinates of F are $(\varepsilon a, 0)$ where ε is the eccentricity. Then the minor semi-axis is related to the major by the equation $b = a\sqrt{1-\varepsilon^2}$. Let $X = (a, 0)$, called the *perihelion*, be the nearest point of the orbit to the Sun and let $X' = (-a, 0)$, called the *aphelion* be the furthest point of the orbit from the Sun. It follows from Kepler's second law that the planet is moving fastest at perihelion and slowest at aphelion.

This is illustrated by Fig. 7.12 where the shaded areas have the same area and so P takes the same time to move along the edge of both.

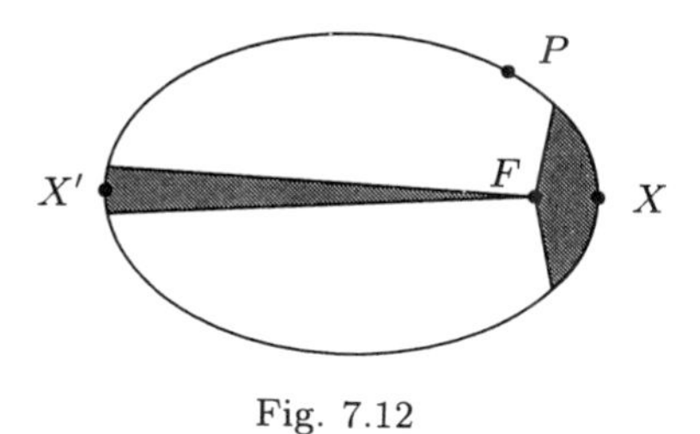

Fig. 7.12

Let α denote the angle $\angle XFP$. Then α is called the *true anomaly.* Let P be given parametrically by $(a\cos\beta, b\sin\beta)$. The angle β is called the *eccentric anomaly.*

The eccentric anomaly is constructed as follows. Draw the auxiliary circle, with the major axis as diameter (Fig. 7.13). Let Q be the point where the line through P at right angles to the major axis meets the circle. Then $\beta = \angle XOQ$.

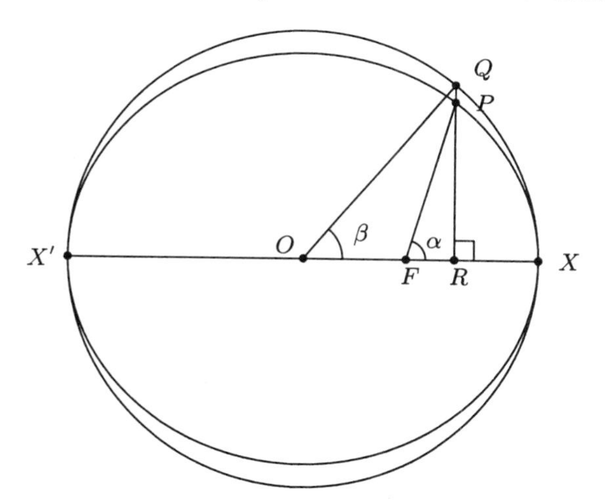

Fig. 7.13 Position of the planet P and the anomalies α, β

To find the relationship between the two anomalies let O be the centre of the ellipse and let R be the foot of the perpendicular from P to OX. The distance FP, denoted by r, satisfies the polar equation

$$r = \frac{a(1-\varepsilon^2)}{1+\varepsilon\cos\alpha}.$$

Eliminating r between this equation and the relation

$$FR = r\cos\alpha = a\cos\beta - a\varepsilon$$

gives α in terms of β. That is

$$\cos\alpha = \frac{\cos\beta - \varepsilon}{1 - \varepsilon\cos\beta}.$$

Exercise 7.35

If $\beta = 45°$ and $\varepsilon = 1/2$, find α.

Exercise 7.36

Show that the true anomaly is equal to the eccentric anomaly only if the ellipse is a circle or the planet is at perihelion or aphelion.

Exercise 7.37

Show that

$$\tan^2\frac{\alpha}{2} = \left(\frac{1+\varepsilon}{1-\varepsilon}\right)\left(\frac{1-\cos\beta}{1+\cos\beta}\right).$$

In order to exploit Kepler's second law we need a formula for the area swept out by the radius FP from the Sun to the planet. We will twice use the fact that $RP/RQ = b/a$ with a corresponding ratio for areas. The sector swept out after perihelion is

$$FPX = FPR + RPX.$$

The formula for a triangle gives

$$FPR = \frac{1}{2}FR \cdot RP = \frac{a}{2}(\cos\beta - \varepsilon)b\sin\beta.$$

The formula for the sector is determined by

$$RPX = \frac{b}{a}RQX$$

and

$$\begin{aligned} RQX &= OQX - OQR \\ &= \frac{a^2}{2}(\beta - \cos\beta\sin\beta). \end{aligned}$$

Putting these two together gives

$$FPX = \frac{ab}{2}(\beta - \varepsilon\sin\beta).$$

Let t be the time for the planet to reach P after perihelion and let p be the time for a complete period. Then comparing areas and using Kepler's second law we have

$$FPX/\pi ab = t/p$$

or

$$\boldsymbol{\beta - \varepsilon \sin\beta = nt}$$

where $n = 2\pi/p$. This is called *Kepler's equation.*

A Word about Units

In the above equation β is measured in radians. In practical situations angles are usually measured in degrees. Remember that one radian is $180/\pi$ degrees. A convenient approximation to this if using integer arithmetic is $4068/71$. The period p is usually measured in Earth years (approx. 365.25 days). The rate n is usually measured in degrees per day. The major semi-axis a is usually measured in astronomical units. One astronomical unit is equal to the major semi-axis for the Earth, being about 149,597,900 km.

Solving Kepler's Equation

Although Kepler's equation is *transcendental* and so has no convenient analytical solution, nevertheless for particular values it can be solved by approximation techniques. Here is one such method. Assume n, t and ε are given. If $\varepsilon < 1$ there is just one solution, β, to Kepler's equation. Let $\beta_0 = nt$ be a first approximation. Define the sequence $\beta_1, \beta_2, \beta_3, \ldots$ by the iterative rule

$$\boldsymbol{\beta_n = nt + \varepsilon \sin\beta_{n-1}}.$$

The sequence (β_n) will tend to a limit β and this will be a solution to Kepler's equation.

Example 7.3

Let us calculate the eccentric anomaly of Mars 200 days after perihelion passage, given that $\varepsilon = 0.0933$ and $n = 0.52403$ degrees per day. Then $\beta_0 = 104.806$ degrees. The sequence is defined by $\beta_n = 104.806 + \varepsilon(180/\pi)\sin\beta_{n-1}$, giving

$$\begin{aligned}\beta_1 &= 109.96921\\ \beta_2 &= 109.82513\\ \beta_3 &= 109.82971\\ \beta_4 &= 109.82954.\end{aligned}$$

The last calculation is correct to 4 decimal places.

Exercise 7.38

Calculate the eccentric anomaly of Venus 100 days after perihelion passage, given that $\varepsilon = 0.006785$ and $n = 1.60213$.

Exercise 7.39

* A planet moving in an elliptical orbit with eccentricity ε moves from perihelion to a point with eccentric anomaly β in the same time as it moves from a point with eccentric anomaly β' to aphelion. Find the relationship between β and β'. If $\beta = \pi/4$ and $\varepsilon = 0.055$ find β' to 4 decimal places.

The mathematics which proves that this method works is quite simple. Consider the graphs of the functions x and $nt + \varepsilon \sin x$. Where they meet will provide a solution to Kepler's equation. If $\varepsilon < 1$ it is clear that there is one and only one solution. Consider the sequence (β_n) defined above. Then

$$\begin{aligned}\beta_n - \beta_{n-1} &= \varepsilon \sin \beta_{n-1} - \varepsilon \sin \beta_{n-2} \\ &= 2\varepsilon \sin \frac{\beta_{n-1} - \beta_{n-2}}{2} \cos \frac{\beta_{n-1} + \beta_{n-2}}{2}.\end{aligned}$$

We now use the inequalities

$$|\sin \theta| \leq |\theta| \text{ and } |\cos \theta| \leq 1$$

to obtain the inequality

$$|\beta_n - \beta_{n-1}| < |\beta_{n-1} - \beta_{n-2}|$$

if $\varepsilon < 1$. So the difference in value between successive terms is always decreasing. Since the sequence is bounded it must converge to a limit. However beware of rounding errors for large values of ε.

Exercise 7.40

Sketch graphs of $y = x$ and $y = nt+\varepsilon \sin x$ for $\varepsilon > 1$ and convince yourself that in this case Kepler's equation can have more than one solution.

Finally consider Kepler's third law. This implies that p^2/a^3 is the same for all planets. If we measure p in years and if we measure a in astronomical units then the law becomes the equation

$$\boldsymbol{p^2 = a^3.}$$

Exercise 7.41

If Jupiter takes 11.86224 years to complete its orbit of the Sun find the length of its major semi-axis.

Retrograde Motion

One of the most mysterious properties of the planets to pre-Copernican astronomers was their change of direction or *retrograde motion* against the background of the stars.

Let P, Q be two planets with the orbit of P smaller than that of Q. Let a (b) be the major semi-axis of P (Q) and let the period be p (q). By Kepler's third law, $p = a^{3/2}$ and $q = b^{3/2}$. Let P move to P' in the same small time as Q moves to Q'. Then

$$\begin{aligned} PP'/QQ' &= \frac{OP \times \text{ angular speed of } P}{OQ \times \text{ angular speed of } Q} \\ &= \frac{OP}{p} \Big/ \frac{OQ}{q} \\ &= \sqrt{\frac{b}{a}}. \end{aligned}$$

To simplify the algebra in the above we have assumed that the eccentricity is small and the orbit is nearly circular. The upshot is that PP' is longer than QQ' because b is greater than a.

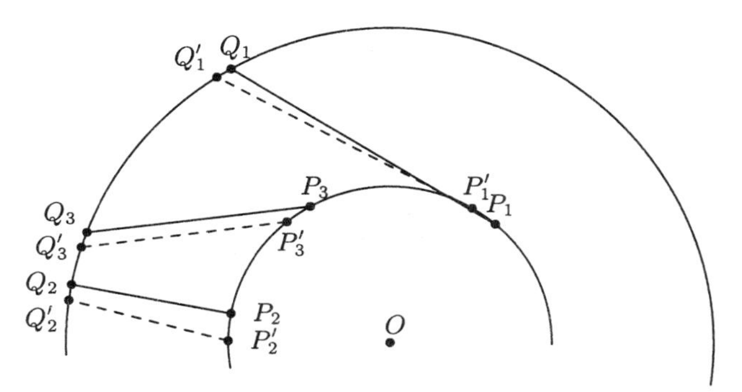

Fig. 7.14 Explanation of retrograde motion

Consider in Fig. 7.14 the various positions of P, Q which again for simplicity we represent on circular orbits about the Sun, O, without regard to scale. This will not affect the qualitative description of the phenomena. At P_1 the motion of P is tangential with respect to Q and consequentially the transverse motion of Q is large and in the expected direction. However at inferior conjunction P_2 the transverse motion of P is greater than the transverse motion of Q which therefore appears to an observer at P to be moving backwards or retrograde. It follows that at some intermediate position P_3 the planet Q at Q_3 will appear stationary. The same analysis holds for an observer of P at Q.

7.5 Quadric Surfaces

In this section we look at the surfaces defined in space by equations of the second degree. We start by considering the various types and some of their properties and finish by describing how to the recognise the kind of quadric surface given by a general quadratic equation.

A good way to visualise these surfaces, and this also applies to higher dimensional objects, is to consider their intersections with families of parallel planes (or hyperplanes in higher dimensions). If we think of the planes as ordered by height then the intersections will be contour lines. Since planes are linear, the contours will be conics, possibly degenerate. We will call the x, y-plane the *equatorial* plane and think of this and any parallel section to be horizontal. Planes through the z-axis will be called *polar* planes and parallel planes will be called *vertical.*

Ellipsoids

These are the two-dimensional analogue of the ellipse (Fig. 7.15). The standard example has equation

$$\frac{x^2}{a^2} + \frac{y^2}{b^2} + \frac{z^2}{c^2} = 1.$$

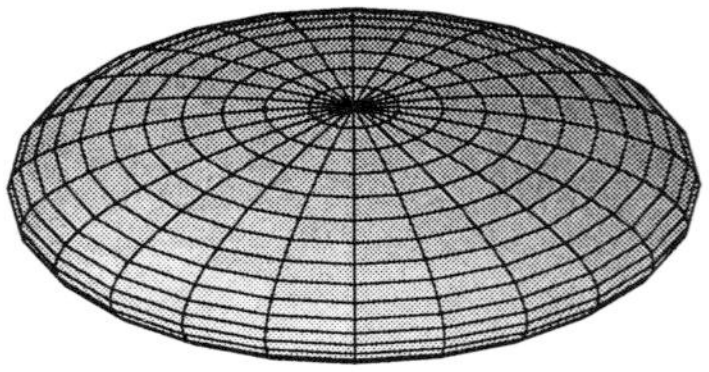

Fig. 7.15 The ellipsoid

All sections are ellipses. This is the only quadric surface which is bounded. So any plane sufficiently far from the origin will not meet the ellipsoid.

If $a = b > c$ the surface is called a *geoid* or *oblate spheroid.* This is the surface formed by rotating an ellipse about its minor axis. Many planets and other rotating bodies under the influence of their own gravitational forces form geoids.

Exercise 7.42

The distance from the centre of the Earth to any point on the equator is 6,378,166 m and the polar radius is 6,356,784 m. Find the eccentricity of a polar section.

A trigonometric parametric formula for a general point on the ellipsoid is

$$(a\cos\theta\sin\phi, b\sin\theta\sin\phi, c\cos\phi),$$

where $0 \le \theta \le 2\pi$ and $-\pi/2 \le \phi \le \pi/2$.

Exercise 7.43

Show that on a geoid, lines of longitude (θ = constant) and lines of latitude (ϕ = constant) are orthogonal. At what points does this occur if the ellipsoid is not a geoid?

A rational parameterisation involving s, t is given by

$$\left(a\frac{2s}{1+s^2}\left(\frac{1-t^2}{1+t^2}\right), b\left(\frac{1-s^2}{1+s^2}\right)\left(\frac{1-t^2}{1+t^2}\right), c\frac{2t}{1+t^2}\right).$$

Hyperboloids

These are the two-dimensional analogue of the hyperbola. They come in two varieties.

Hyperboloids of one sheet: This is the shape that launched a thousand ethnic bamboo waste paper baskets and hundreds of power station cooling towers (Fig. 7.16).

The standard equation is

$$\frac{x^2}{a^2} + \frac{y^2}{b^2} - \frac{z^2}{c^2} = 1.$$

Horizontal sections are ellipses, all having the same eccentricity. The equatorial ellipse is smallest, being like a waist band. Vertical sections are hyperbolae.

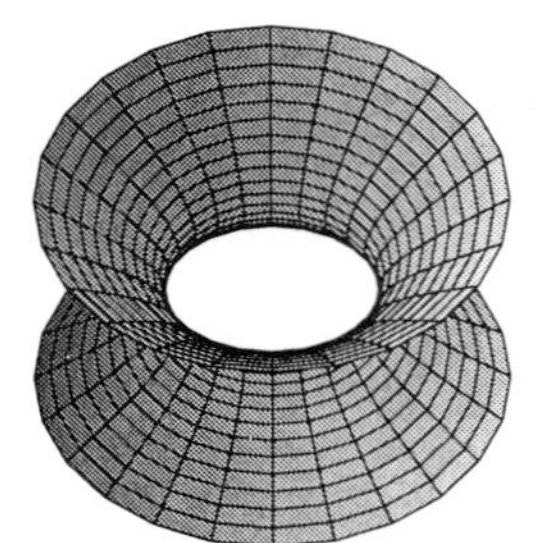

Fig. 7.16 The hyperboloid of one sheet

Exercise 7.44

Show that the plane $ux + vy + wz = 0$ meets the standard hyperbola of one sheet in a pair of parallel lines if

$$w^2c^2 = a^2u^2 + b^2v^2.$$

This surface has the curious property of being made up of a collection of straight lines. This will be clear if you have ever tried to make one or seen one in a craft shop. Consider two ellipses, one above the other. Let $X_+ = (u\cos(\theta + \alpha), v\sin(\theta+\alpha), 1)$ be a point on the top one and $X_- = (u\cos(\theta - \alpha), v\sin(\theta - \alpha), -1)$ be a point on the bottom one. Then a general point on the line joining X_+ and X_- is of the form $X = tX_+ + (1-t)X_-$. This simplifies to

$$X = (u(\cos\theta\cos\alpha + (1-2t)\sin\theta\sin\alpha), \\ v(\sin\theta\cos\alpha + (2t-1)\cos\theta\sin\alpha), 2t-1).$$

Letting θ and t vary, the lines carve out a surface of which X is a general point. If $X = (x, y, z)$ then the coordinates satisfy the equation of the standard hyperboloid if

$$\boldsymbol{a = u\cos\alpha, \;\; b = v\cos\alpha, \;\; c = \cot\alpha.}$$

Exercise 7.45

Show that a point on the hyperboloid of one sheet has a parametric representation in terms of $\theta (0 \le \theta \le 2\pi)$ and $t, t \in \mathbb{R}$ where

$$x = a(\cos\theta + \frac{(1-2t)}{c}\sin\theta)$$
$$y = b(\sin\theta - \frac{(1-2t)}{c}\cos\theta)$$
$$z = 2t - 1.$$

Another parametric representation by trigonometric and hyperbolic functions is

$$x = a\cos\theta\cosh u,\ \ y = b\sin\theta\cosh u,\ \ z = c\sinh u.$$

Exercise 7.46

Show that another family of lines defining the same surface can be obtained by changing the sign of α.

A surface made up of a bundle of straight lines is called a *ruled surface*. Another example is a cylinder. In fact if we allow complex numbers then any quadric surface is ruled. For example an "ellipsoid" is the union of "lines" joining $X_+ = (u\cos(\theta + \alpha), v\sin(\theta+\alpha), i)$ to $X_- = (u\cos(\theta-\alpha), v\sin(\theta-\alpha), -i)$ where $i = \sqrt{-1}$. In this case the "ellipsoid" has four real dimensions and the "lines" are copies of $\mathbb{C}$ and so have two real dimensions. The standard work on this subject was written by the mathematician with the serendipitous name, W.L. Edge.

The hyperboloid $x^2 + y^2 - z^2 = 1$ is locally isometric to the hyperbolic plane.

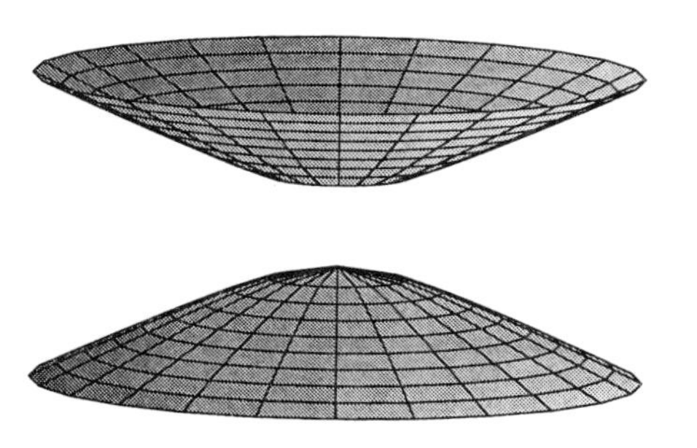

Fig. 7.17 The hyperboloid of two sheets

Hyperboloids of two sheets: This is the only disconnected non-degenerate quadric surface. Its standard equation is

$$\frac{x^2}{a^2} + \frac{y^2}{b^2} - \frac{z^2}{c^2} = -1$$

and it is illustrated in Fig. 7.17. Horizontal cross-sections are ellipses unless $-c \le z \le c$. If $-c < z < c$ then the cross-sections are empty. Vertical cross-sections are hyperbolæ.

Parameterisation in terms of trigonometric and hyperbolic functions is given by

$$x = a\cos\theta\sinh u,\ \ y = b\sin\theta\sinh u,\ \ z = c\cosh u.$$

The asymptotic surface for both hyperbolæ is the cone (Fig. 7.18)

$$\frac{x^2}{a^2} + \frac{y^2}{b^2} - \frac{z^2}{c^2} = 0.$$

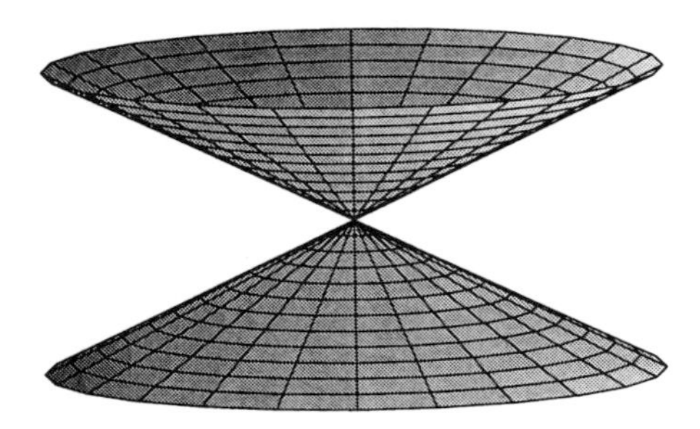

Fig. 7.18 The cone as asymptotic surface of the hyperboloid

Paraboloids

As the name suggests these are the analogue of the parabola. Like the hyperbolæ they come in two varieties.

Elliptic paraboloids: The standard equation is

$$\frac{x^2}{a^2} + \frac{y^2}{b^2} = z.$$

Its shape is like a bowl (Fig. 7.19). Horizontal sections are ellipses if $z > 0$ and vertical sections are parabolas.

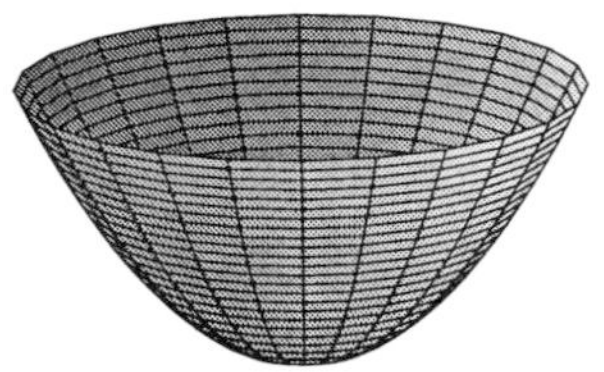

Fig. 7.19 The elliptic paraboloid

Parameterisation is given by

$$x = at\cos\theta,\ \ y = bt\sin\theta,\ \ z = t^2.$$

Hyperbolic paraboloids: Perhaps the most interesting quadric surface. The standard equation is

$$\frac{x^2}{a^2} - \frac{y^2}{b^2} = z.$$

Its shape is like a saddle or a mountain pass (Fig. 7.20). Horizontal sections are hyperbolæ unless $z = 0$ when the cross-section is a pair of lines which form asymptotes to the other contours. Vertical sections are parabolas except the polar sections given by $x/a = \pm y/b$ which determine the asymptotes considered above.

This is also the regulus encountered in Chapter 5.

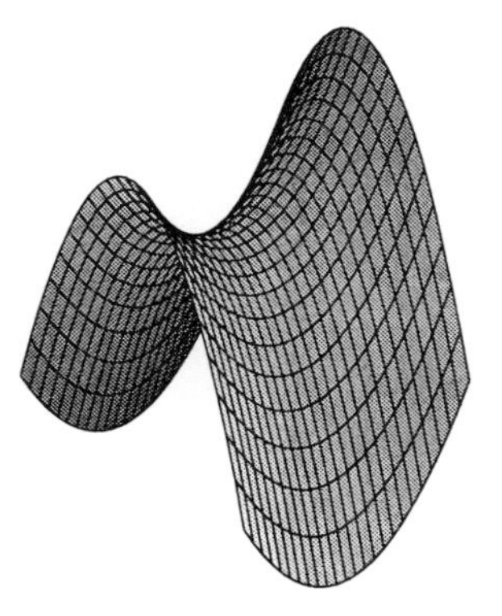

Fig. 7.20 The hyperbolic paraboloid

Parameterisation is given by

$$\boldsymbol{x = at\cos\theta,\ \ y = bt\sin\theta,\ \ z = t^2\cos 2\theta.}$$

Degenerate quadric surfaces

The degenerate quadric surfaces are of the following types: two planes (which may intersect in a line, be parallel or coincide), a point, a cylinder or a cone. The cylinder and the cone have conics as cross-sections.

Exercise 7.47

The quadric surface $2x^2 - y^2 + xy + 2yz - 4zx + 8x - y - 4z + 6 = 0$ is the union of two planes. Find their equations.

7.6 The General Quadric Surface

In this section we shall show that the general quadratic equation in three unknowns x, y, z is one of the types considered above. The most general such equation has the form.

$$ax^2 + by^2 + cz^2 + 2exy + 2fyz + 2gzx + 2px + 2qy + 2rz + d = 0.$$

As in the case of a general conic the easiest situation is where the cross term coefficients e, f, g vanish. Then by completing squares and applying a translation one of the standard types is reached.

Example 7.4

The quadric

$$4x^2 + 3y^2 - z^2 + 2x + y + 2z - 1 = 0$$

can be written

$$4(x + 1/4)^2 + 3(y + 1/6)^2 - (z - 1)^2 - 1/3 = 0.$$

This is a hyperboloid of one sheet with centre at $(-1/4, -1/6, -1)$.

Exercise 7.48

Describe the following quadric surfaces:
(a) $3x^2 + y^2 + 2z^2 + 3x + 3y + 4z = 0$
(b) $2x^2 + 4z^2 - 4x - y - 24z + 36 = 0$.

In general we have to eliminate the cross term coefficients. The quadratic equation can be written

$$X\mathbf{M}X^T + 2X \cdot N + d = 0,$$

where

$$\mathbf{M} = \begin{pmatrix} a & e & g \\ e & b & f \\ g & f & c \end{pmatrix} \qquad N = (p, q, r), \qquad X = (x, y, z).$$

Example 7.5

Consider the quadric surface whose equation is

$$4x^2 + 9y^2 + 5z^2 - 4xy + 8yz + 12xz + 9z - 3 = 0.$$

This can be written

$$(x,y,z)\begin{pmatrix} 4 & -2 & 6 \\ -2 & 9 & 4 \\ 6 & 4 & 5 \end{pmatrix}(x,y,z)^T + (0,0,9)\cdot(x,y,z) - 3 = 0.$$

The matrix $\mathbf{M}$ is symmetric and we can use the following well-known theorem.

■ Let $\mathbf{M}$ be an $n \times n$ real symmetric matrix. Then there is a right-handed orthonormal basis of $\mathbb{R}^n$ whose elements are eigenvectors of $\mathbf{M}$. Moreover if $\mathbf{U}$ is the matrix whose rows are these eigenvectors then $\mathbf{UMU}^T$ is a diagonal matrix whose diagonal elements are the corresponding eigenvalues.

Proof Right handed in this context means that $\det \mathbf{U} = 1$. Because the basis is orthonormal the matrix $\mathbf{U}$ is orthogonal and $\mathbf{U}^{-1} = \mathbf{U}^T$.

The proof of the above result can be found in any good book on linear algebra.

□

Example 7.6

Consider the symmetric matrix in Example 7.5. Then the eigenvalues are the roots of

$$\begin{vmatrix} \lambda - 4 & 2 & -6 \\ 2 & \lambda - 9 & -4 \\ -6 & -4 & \lambda - 5 \end{vmatrix} = \lambda^3 - 18\lambda^2 + 45\lambda + 324.$$

These are 12, 9 and -3. The eigenvector (e_1, e_2, e_3) corresponding to λ is the solution of

$$(e_1, e_2, e_3)\begin{pmatrix} \lambda - 4 & 2 & -6 \\ 2 & \lambda - 9 & -4 \\ -6 & -4 & \lambda - 5 \end{pmatrix} = (0,0,0).$$

If $\lambda = 12$ a solution giving an eigenvector of unit length is $(1,2,2)/3$. Similarly $(2,-2,1)/3$ and $(-2,-1,2)/3$ are unit eigenvectors corresponding to $\lambda = 9$ and -3 respectively. So if

$$\mathbf{U} = \frac{1}{3}\begin{pmatrix} 1 & 2 & 2 \\ 2 & -2 & 1 \\ 2 & 1 & -2 \end{pmatrix},$$

then

$$\mathbf{UMU}^T = \begin{pmatrix} 12 & 0 & 0 \\ 0 & 9 & 0 \\ 0 & 0 & -3 \end{pmatrix}.$$

Notice that we have changed the sign of the last eigenvector (the bottom row) so that $\det \mathbf{U} = 1$.

Exercise 7.49

* If $\mathbf{M}$ is a real symmetric matrix with eigenvalue λ and corresponding eigenvector E, show that λ is real ($\bar{\lambda} = \lambda$) by expanding $E\overline{\mathbf{M}}\bar{E}^T$ in two ways.

Exercise 7.50

* If $\mathbf{M}$ is a real symmetric matrix with distinct eigenvalues λ and μ with corresponding eigenvectors E and F, show that $E \cdot F = 0$ by expanding $E\mathbf{M}F^T$ in two ways.

Exercise 7.51

* Use the above two results to prove the diagonalisation theorem above.

This method allows us to eliminate the coefficient of the cross terms, since the rotation, $X \to X\mathbf{U}$, changes $X\mathbf{M}X^T$ to $X\mathbf{U}\mathbf{M}\mathbf{U}^T X$ which has no cross terms as $\mathbf{U}\mathbf{M}\mathbf{U}^T$ is diagonal.

Applying this to our quadric $4x^2 + 9y^2 + 5z^2 - 4xy + 8yz + 12xz + 9z - 3 = 0$ we get

$$12x^2 + 9y^2 - 3z^2 + 6x + 3y + 6z - 3 = 0.$$

Cancelling the common factor 3 leads to our example above which is a hyperboloid of one sheet.

If one of the eigenvalues is repeated then we must find an orthonormal basis of the corresponding eigenspace.

Example 7.7

Consider the equation

$$4x^2 + 4y^2 + 4z^2 + 4xy + 4yz + 4zx - 3 = 0.$$

Then the equation can be written $X\mathbf{M}X^T - 3 = 0$ where

$$\mathbf{M} = \begin{pmatrix} 4 & 2 & 2 \\ 2 & 4 & 2 \\ 2 & 2 & 4 \end{pmatrix}.$$

The eigenvalues are 2 (twice) and 8. So the rotation matrix is

$$\mathbf{U} = \begin{pmatrix} -1/\sqrt{2} & 1/\sqrt{2} & 0 \\ -1/\sqrt{6} & -1/\sqrt{6} & 2/\sqrt{6} \\ 1/\sqrt{3} & 1/\sqrt{3} & 1/\sqrt{3} \end{pmatrix}.$$

The first two rows are an orthonormal basis of the two-dimensional eigenspace corresponding to 2.

The rotation $X \to X\mathbf{U}$ transforms the equation to

$$2x^2 + 2y^2 + 8z^2 = 3$$

which is an ellipsoid.

Exercise 7.52

Show that the equation $2xy - 6x + 10y + z - 31 = 0$ defines a hyperbolic paraboloid.

Exercise 7.53

For the following quadrics find a rotation $X \to X\mathbf{U}$ that eliminates the cross terms. Write down the new equation and name the quadric.
(a) $2x^2 + 3y^2 + 23z^2 + 72xz + 150 = 0$
(b) $2x^2 + 2y^2 + 2z^2 + 2xy + 2yz + 2zx = 3$
(c) $144x^2 + 100y^2 + 81z^2 - 216xz - 540x - 720z = 0$
(d) $2xy + z = 0$.

Exercise 7.54

Find the equation of the set of points, the difference of whose distances from $(0, 0, 3)$ and $(0, 0, -3)$ is 4.

Exercise 7.55

The plane curve $x = 4 - y^2$ is rotated about the x-axis. What is the resulting surface?

Answers to Selected Questions in Chapter 7

7.1 Let $P = (x, y, z)$ be a general point on the cone and let $Q = (0, 0, z)$ be a point on the axis of symmetry. Then OQP is a right angled triangle with hypotenuse OP. The angle $\angle POQ$ is α, so $\tan\alpha = QP/OQ = \sqrt{x^2 + y^2}/z$.

7.3 If we substitute the six values (x, y) and (x_i, y_i), $i = 1, \ldots 5$ into the general equation we get six homogeneous linear equations in six unknowns,

a, b, c, f, g, h. The condition for a non zero solution is

$$\begin{vmatrix} x^2 & 2xy & y^2 & 2x & 2y & 1 \\ x_1^2 & 2x_1y_1 & y_1^2 & 2x_1 & 2y_1 & 1 \\ x_2^2 & 2x_2y_2 & y_2^2 & 2x_2 & 2y_2 & 1 \\ x_3^2 & 2x_3y_3 & y_3^2 & 2x_3 & 2y_3 & 1 \\ x_4^2 & 2x_4y_4 & y_4^2 & 2x_4 & 2y_4 & 1 \\ x_5^2 & 2x_5y_5 & y_5^2 & 2x_5 & 2y_5 & 1 \end{vmatrix} = 0.$$

Cancelling factors of 2 gives the answer.

7.4 $x^2 + y^2 - 4x - 2y - 4 = 0$, $(-1, 3)$, $\sqrt{3}$.

7.5 The radical axis is $2x - 2y - 3 = 0$ and if the circle met in two points this would be the equation of the line through them. To show that this is indeed the case it is probably easiest to use geometric methods. The first circle has centre $(0, 0)$ and radius 2. The second circle has centre $(-1, 1)$ and radius 3. If a circle centre $(-1, 1)$ is to be disjoint from the first circle (or tangent) its radius must either be less than or equal to $2 - \sqrt{2}$ or greater than or equal to $2 + \sqrt{2}$. Since this is not the case, the result follows.

7.6 No.

7.9 M is the point $M = (a(\cos\theta + \cos\phi)/2, (\sin\theta + \sin\phi)/2)$. So the products of the slopes are $b^2(\sin^2\theta - \sin^2\phi)/a^2(\cos^2\theta - \cos^2\phi) = -b^2/a^2$. This is -1 if and only if $a = b$, i.e. the ellipse is a circle.

7.11 The mid-point is $(a(t_1^2 + t_2^2)/2, a(t_1 + t_2))$. The chord has slope $2/(t_1 + t_2)$. If this is a constant k say then the mid-point always lies on the line $y = 2a/k$ which is parallel to the axis $y = 0$.

7.13 Let the point on the hyperbola be (x_0, y_0) where $x_0^2/a^2 - y_0^2/b^2 = 1$. Then the tangent at this point meets the asymptotes $y = \pm bx/a$ in the points $(a^2b/(bx_0 - ay_0), ab^2/(bx_0 - ay_0))$ and $(a^2b/(bx_0 + ay_0), -ab^2/(bx_0 + ay_0))$. Using the formula $(x_1y_2 - x_2y_1)/2$ for the area of the triangle with vertices O, (x_1, y_1) and (x_2, y_2) gives the answer ab which is independent of (x_0, y_0).

7.14 Let $P_i = (ct_i, c/t_i), i = 1, 2, 3$ be three points on the rectangular hyperbola, $y = c^2/x$. Then the equation of the altitude through P_3 (perpendicular to P_1P_2), is

$$\frac{y - c/t_3}{x - ct_3} = -\left(\frac{c/t_1 - c/t_2}{ct_1 - ct_2}\right)^{-1} = t_1t_2.$$

This line meets the hyperbola $y = c^2/x$ in the point P_3 and the point $H = (-c/t_1t_2t_3, -ct_1t_2t_3)$. Since H is symmetric in t_1, t_2 and t_3 it must be the orthocentre of the triangle $P_1P_2P_3$.

7.15 If the constant term had been greater than 6 then there would have been no real solutions to the equation.

7.19 By completing the square a translation can be found which changes the curve to $y = dx^2$. This is clearly a parabola. The axis is parallel to the y-axis and passes through the minimum or maximum of the function. By differentiation this is found to be the line $x = -b/2a$. The point where it meets the parabola is $(-b/2a, (4ac - b^2)/4a)$.

7.20 To find θ use $\tan 2\theta = 6/(5-5) = \infty$. So $2\theta = \pi$ and $\theta = \pi/2$. The substitution $x \to (x-y)/\sqrt{2}$ and $y \to (x+y)/\sqrt{2}$ and a translation $x \to x+1$ and $y \to y-1$ changes the equation to

$$4x^2 + y^2 = 4,$$

an ellipse with semi-major axes 2 and semi-minor axes 1. The centre of the ellipse was at the point $(1, -1)$ and the axes were rotated through $-45°$ with respect to the original coordinates.

7.21 (a) Hyperbola, (b) ellipse, (c) parabola, (d) pair of parallel lines.

7.22 This can be written as

$$(x, y)\begin{pmatrix} 1 & -3 \\ -3 & -7 \end{pmatrix}\begin{pmatrix} x \\ y \end{pmatrix} + 10x + 2y + 9 = 0$$

The eigenvalues of the matrix $\begin{pmatrix} 1 & -3 \\ -3 & -7 \end{pmatrix}$ are $2, -8$ and the unit eigenvectors are $(3, -1)/\sqrt{10}$ and $(1, 3)/\sqrt{10}$. So the transformation

$$(x, y) \to (x, y)\begin{pmatrix} 3 & -1 \\ 1 & 3 \end{pmatrix}$$

which is a rotation of the plane with a scale multiplication changes the equation to

$$-80x^2 + 20y^2 + 16x - 28y + 9 = 0$$

which is a hyperbola as we have seen.

7.25 $4/5$, $4\,\text{cm}$

7.26 The squared distance to one focus is $(x - \sqrt{a^2 - b^2})^2 + y^2 = a^2 - 2a\varepsilon x + \varepsilon^2 x^2 = (a - \varepsilon x)^2$. Similarly for the other focus.

7.28 Let the ends of the latus rectum be L, L'. Consider a line LM parallel to the axis. Since LL' passes through the focus F, by focal properties LM is orthogonal to LL' and both make equal angles with the tangent at L. The result now follows easily.

7.29 $(-a, a(t-1/t))$.

7.30 Let $P = (at^2, 2at)$ then $D = (-a, 2at)$. The focus $F = (a, 0)$ so the slope of DF is $(0-2at)/(a-(-a)) = -t$. Since the slope of the tangent is $1/t$ the result follows.

7.31 Let $P = (at^2, 2at)$ then $T = (-at^2, 0)$. The normal has equation $(y - 2at)/(x - at^2) = -t$ and meets the axis at $N = (at^2 + 2a, 0)$. The midpoint is $(a, 0) = F$.

7.33 The point D has coordinates $(a^2/\sqrt{a^2+b^2}, ab/\sqrt{a^2+b^2})$. The line DF has slope $(ab/\sqrt{a^2+b^2})/(a^2/\sqrt{a^2+b^2} - \sqrt{a^2+b^2}) = -a/b$. This is perpendicular to the slope of the asymptote $y = bx/a$.

7.35 $\cos^{-1}(3\sqrt{2}-2)/7 = 71.31°$.

7.37 Use the trigonometric identity $\tan^2\frac{\alpha}{2} = (1-\cos\alpha)/(1+\cos\alpha)$ and the earlier formula for $\cos\alpha$.

7.38 160.34377 to five decimal places

7.39 With the usual notation $FPX = ab(\beta - \varepsilon\sin\beta)/2$ and $FP'X = ab(\beta' - \varepsilon\sin\beta')/2$. So

$$\begin{aligned} FP'X' &= \pi ab/2 - ab(\beta' - \varepsilon\sin\beta')/2 \\ &= ab(\pi - \beta' + \varepsilon\sin\beta')/2. \end{aligned}$$

Equating the two equal areas gives

$$\beta - \varepsilon\sin\beta = \pi - \beta' + \varepsilon\sin\beta'.$$

If $\beta = \pi/4$ and $\varepsilon = 0.055$ then β' satisfies the equation

$$\beta' = 2.3951 + 0.055\sin\beta'.$$

This has the solution 2.3974 rad or 137.3609 degrees.

7.41 5.2012908 au

7.42 0.08181.

7.43 Along the equator $\phi = 0$ and along lines of longitude $\theta = 0$ and $\theta = \pi$.

7.47 The planes have equations $x + y - 2z + 3 = 0$ and $2x - y + 2 = 0$.

7.48 (a) Ellipsoid centre $(-1/2, -3/2, -1)$, (b) elliptic paraboloid.

7.53 (a) $25x^2 - 3y^2 - 50z^2 = 150$, hyperboloid of two sheets
(b) $x^2 + y^2 + 4z^2 = 3$, ellipsoid
(c) $9x^2 + 4y^2 - 36z = 0$, elliptic paraboloid
(d) $x^2 - y^2 + z = 0$, hyperbolic paraboloid.

7.54 $5x^2 - 4y^2 - 4z^2 = 20$, a hyperboloid of two sheets.

7.55 $x = 4 - y^2 - z^2$, an elliptic paraboloid.

8
Spherical Geometry

... I could be bounded in a nut shell and
count myself a king of infinite space...
Hamlet

In this chapter we will mostly consider the two-dimensional sphere. For definiteness we will take the unit sphere of unit radius in three-dimensional space with centre the origin, O. This is the set of points S^2 in $\mathbb{R}^3$ satisfying

$$x^2 + y^2 + z^2 = 1.$$

For clarity of exposition we consider the xy-plane, called the *equatorial plane*, as horizontal and the z-axis as vertical. The equatorial plane meets the sphere in a circle called the *equator*. Any plane passing through the origin cuts the sphere in a circle called a *great circle*. So the centre of a great circle and the centre of the sphere coincide. The equator is an example of a great circle. The line through the centre of the sphere perpendicular to the plane of a great circle meets the sphere in two points called the *poles* of the great circle. The poles of the equator are the *north pole* $N = (0, 0, 1)$ and the *south pole* $S = (0, 0, -1)$.

The *antipode* or *antipodal point* of a point X on the sphere is the point $-X$ and is the reflection of X in the centre. The antipode of X is often denoted X^*. The poles of a great circle form an antipodal pair.

Any other plane intersecting the sphere in a circle but not passing through the centre cuts the sphere in a *small circle*.

A plane which touches a sphere in one point is called a *tangent plane* at that point. Any vector parallel to that plane is a tangent vector at the point of contact. Note that tangent vectors are orthogonal to the vector from the centre to the point of contact.

Exercise 8.1

Let

$$X = (1/\sqrt{2}, 1/\sqrt{2}, 0), \qquad Y = (0, 1/\sqrt{2}, -1/\sqrt{2})$$
$$Z = (1/\sqrt{14}, 3/\sqrt{14}, -2/\sqrt{14}), \quad T = (-1/\sqrt{2}, 1/\sqrt{2}, 0)$$

be four points on the unit sphere. Show that X, Y, Z all lie on a great circle and that T is a unit tangent vector at X.

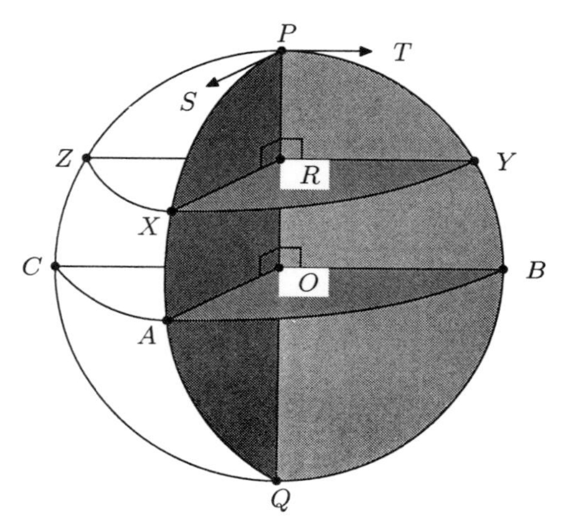

Fig. 8.1 Circles, great and small

The reader should now look at Fig. 8.1. ABC is a great circle and XYZ is a small circle parallel to ABC with centre R. Note that R is the nearest point of the plane XYZ to the origin. The poles of ABC are P, Q. The great circles $PYBQCZ$ and $PXAQ$ are orthogonal to ABC. The tangent T at P points along the great circle $PYBQCZ$ and the tangent S at P points along the great circle $PXAQ$.

The $(n-1)$-dimensional unit sphere, S^{n-1} is the set of points in n-dimensional space distance 1 from the origin. So S^0 is a pair of points, S^1 is a circle, S^2 is a sphere as above, etc.

The $(n-1)$-dimensional sphere has great S^p's and corresponding poles which are S^q's where $p + q = n - 2$.

Points X on a general sphere with radius r and centre P satisfy the equation

$$|X - P| = r.$$

If we square both sides we get

$$(X - P) \cdot (X - P) = r^2.$$

Expanding and rearranging we have

$$X \cdot X - 2X \cdot P = r^2 - P \cdot P$$

for the equation of the sphere.

Writing out the resulting equation in terms of coordinates we get

$$(x_1 - p_1)^2 + (x_2 - p_2)^2 + \cdots + (x_n - p_n)^2 = r^2.$$

Consider a hyperplane with equation in X given by $A \cdot X = b$. The point $Q = \frac{b - A \cdot P}{|A|^2} A - P$ lies on the plane and is the nearest point to P. If $r = |b - A \cdot P|/|A|$ then the plane just meets the sphere $|X - P| = r$ in the one point Q and is the tangent plane at Q.

In that case the vector $P - Q$ is at right angles to the tangent plane. So if X is an arbitary point on the tangent plane we have

$$(Q - P) \cdot (Q - X) = 0.$$

Eliminating $Q \cdot Q$ using the equation of the sphere, we see that the equation of the tangent plane at the point Q is

$$(Q - P) \cdot X = r^2 + P \cdot Q - P \cdot P.$$

Exercise 8.2

Show that $z = 2$ is a tangent plane to the sphere $(x+1)^2+(y-2)^2+z^2 = 4$ and find its point of contact.

8.1 Geodesics

An arc of a great circle is called a *geodesic*. If X, Y are the endpoints of a geodesic on the unit sphere then the length of the geodesic is

$$\psi = \cos^{-1} \mathbf{X} \cdot \mathbf{Y}$$

provided the geodesic does not contain an antipodal pair. This is the path of shortest distance on the sphere between X and Y.

Note that ψ must be measured in radians. If the circle has radius r then this angle must be multiplied by r to get the distance.

With reference to Fig. 8.1,

$$\psi = \angle AOB = \angle XRY = \angle SPT.$$

The length of the small circle arc XY is

$$XR\psi = \cos \angle AOX \psi = AB \cos AX.$$

We can find an arbitary point on the great circle containing X, Y by the following. The vector $Y - X\cos\psi$ is orthogonal to X and lies in the plane of X, Y. The length of this vector is $\sin\psi$. Therefore a unit vector in the plane defined by X, Y which is at right angles to X is given by the formula

$$\boldsymbol{Y'} = \frac{(\boldsymbol{Y} - \cos\psi \boldsymbol{X})}{\sin\psi} = \boldsymbol{Y}\,\mathrm{cosec}\,\psi - \boldsymbol{X}\cot\psi.$$

This is also the tangent vector based at X in the direction of the geodesic from X to Y.

It follows that the general point on the great circle containing X, Y and making an angle θ with OX is given by $X\cos\theta + Y'\sin\theta$ or

$$\begin{aligned}\boldsymbol{X}\cos\boldsymbol{\theta} + (\boldsymbol{Y}\,\mathrm{cosec}\,\psi - \boldsymbol{X}\cot\psi)\sin\boldsymbol{\theta} &= \\ &\boldsymbol{X}(\cos\boldsymbol{\theta} - \cot\psi\sin\boldsymbol{\theta}) + \boldsymbol{Y}\,\mathrm{cosec}\,\psi\sin\boldsymbol{\theta}.\end{aligned}$$

The tangent vector based at this general point is given by $-X\sin\theta + Y'\cos\theta$ or

$$\begin{aligned}-\boldsymbol{X}\sin\boldsymbol{\theta} + (\boldsymbol{Y}\,\mathrm{cosec}\,\psi - \boldsymbol{X}\cot\psi)\cos\boldsymbol{\theta} &= \\ &\boldsymbol{X}(-\sin\boldsymbol{\theta} - \cot\psi\cos\boldsymbol{\theta}) + \boldsymbol{Y}\,\mathrm{cosec}\,\psi\cos\boldsymbol{\theta}.\end{aligned}$$

Exercise 8.3

Find the general point on the great circle containing the two points $X = (1/2, 1/2, 1/\sqrt{2})$ and $Y = (-1/2, -1/2, 1/\sqrt{2})$ on the sphere. Find the point on this great circle with greatest second coordinate. Find the initial direction of the great circle arc from this point to Y.

Exercise 8.4

Find the general point on the great circle containing the two points $X = (1/2, 1/2, 1/2, 1/2)$ and $Y = (-1/2, -1/2, -1/2, 1/2)$ on the three-dimensional sphere S^3. Find the point on this great circle with greatest third coordinate. Find the initial direction of the great circle arc XY at X and at Y.

8.2 Geodesic Triangles

Let A, B, C be three points on the sphere. These points together with the shortest geodesic arcs joining them are said to form a *geodesic* or *spherical triangle* (Fig. 8.2). Denote the distance (on the sphere) between A and B by

$c = AB$. Similarly denote $a = BC$ and $b = CA$. Let α be the angle at A between the geodesic to B and the geodesic to C. Define β, γ similarly.

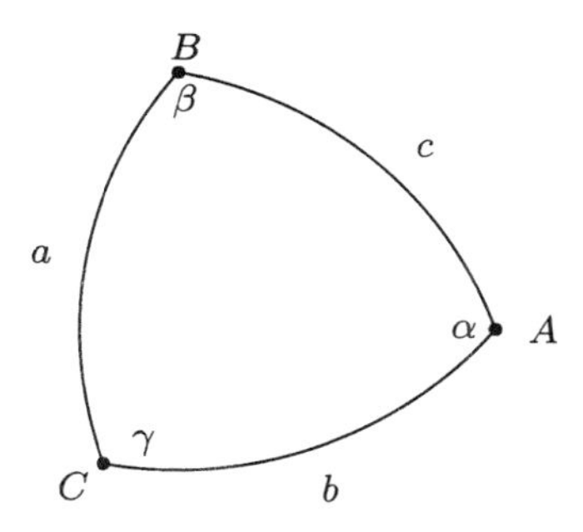

Fig. 8.2 A triangle on a sphere

The relationships and properties enjoyed by $a, b, c, \alpha, \beta, \gamma$ are quite different from those of a euclidean triangle. For example $\alpha + \beta + \gamma > \pi$ and the values of α, β, γ determine the size of the triangle, as we will see.

Since the distance $a = BC$ is the angle between the unit vectors B and C we have

$$\boldsymbol{\cos a = B \cdot C \text{ and } \sin a = |B \times C|}$$

with similar formulæ for the other sides.

The angle between the planes through the origin containing A, B and A, C is the angle between the vectors $A \times B$ and $A \times C$. So

$$\boldsymbol{\cos \alpha = \frac{(A \times B) \cdot (A \times C)}{\sin c \sin b}}$$

The triple scalar product can be expanded by the usual rules, giving

$$\begin{aligned}(A \times B) \cdot (A \times C) &= ((A \times C) \times A) \cdot B \\ &= ((A \cdot A)C - (C \cdot A)A) \cdot B \\ &= C \cdot B - (C \cdot A)(A \cdot B) \\ &= \cos a - \cos b \cos c.\end{aligned}$$

Hence we obtain the *cosine rule* for spherical triangles

$$\boldsymbol{\cos a = \cos \alpha \sin b \sin c + \cos b \cos c.}$$

Alternatively

$$\sin \alpha = \frac{|(A \times B) \times (A \times C)|}{\sin b \sin c} = \frac{(A \times B) \cdot C}{\sin b \sin c}.$$

So

$$\frac{\sin \alpha}{\sin a} = \frac{(A \times B) \cdot C}{\sin a \sin b \sin c}.$$

Because this expression is unaltered by permuting A, B, C we get the *sine rule* for spherical triangles

$$\frac{\sin \alpha}{\sin a} = \frac{\sin \beta}{\sin b} = \frac{\sin \gamma}{\sin c}.$$

Exercise 8.5

Find a spherical triangle ABC such that $\alpha = \beta = \gamma = \pi/2$. What is its area?

Exercise 8.6

In the spherical triangle ABC, $a = 57^\circ\ 22'\ 11''$, $b = 72^\circ\ 12'\ 19''$ and $\gamma = 94^\circ\ 1'\ 49''$. Calculate the values of c, α and β.

Exercise 8.7

* Using the formula $B \operatorname{cosec} c - A \cot c$ for a unit tangent at A in the direction of B and $C \operatorname{cosec} b - A \cot b$ for a unit tangent at A in the direction of C, show that

$$\cos \alpha = \frac{a}{\sin c \sin b} - \frac{b}{\tan c \sin b} - \frac{c}{\sin c \tan b} + \frac{1}{\tan c \tan b}.$$

Areas

The surface area of a sphere of radius r is easily found by the methods of the calculus to be $4\pi r^2$. It follows that the area of the unit sphere is 4π.

We have mentioned that the sum of the angles of a spherical triangle is greater than π. It turns out that this "excess" is equal to the area of the triangle. We first find the area of a simpler object.

A *lune* is the region enclosed by two great circles meeting at an angle α, called the angle of the lune (Fig. 8.3). The two vertices of the lune are antipodal pairs. Let $L(\alpha)$ be the area of a lune of angle α. Because two lunes can be placed next to one another to make a third we have

$$L(\alpha + \beta) = L(\alpha) + L(\beta)$$

so L is a linear function. A lune with angle 2π is the whole sphere and so $L(2\pi) = 4\pi$. It follows that the area of a lune with angle α is

$$\boldsymbol{L(\alpha) = 2\alpha.}$$

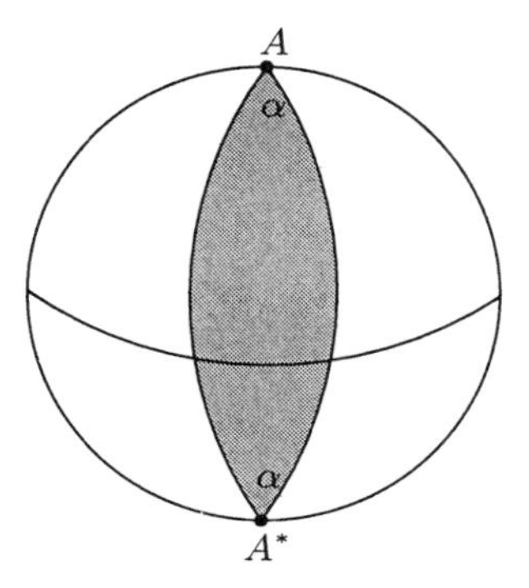

Fig. 8.3 A lune of angle α

Consider a spherical triangle ABC. Denote its area by Δ. Let A^*, B^*, C^* be the antipodal points to A, B, C. The triangle A^*CB is antipodal to the triangle AC^*B^* and so they have the same area x say. The configurations of regions and their areas formed by the great circles through A, B, C are indicated diagramatically in Fig. 8.4.

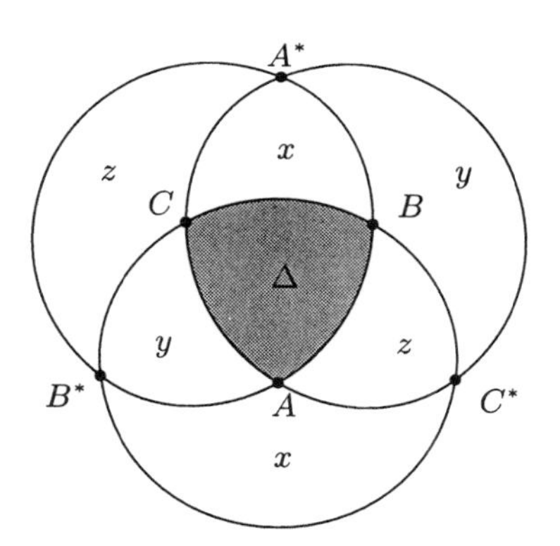

Fig. 8.4 Areas formed by the triangle ABC

The triangles ABC and A^*CB together form a lune of angle α. So

$$\Delta + x = 2\alpha.$$

Similarly

$$\Delta + y = 2\beta, \qquad \Delta + z = 2\gamma.$$

Adding these together we get the equation of the three lunes

$$3\Delta + x + y + z = 2(\alpha + \beta + \gamma).$$

The great circle BCB^*C^* bounds a hemisphere containing Δ, x, y and z so

$$\Delta + x + y + z = 2\pi.$$

Taking this from the equation of the three lunes yields the following.

■ A spherical triangle ABC has area $\alpha + \beta + \gamma - \pi$. In particular the area only depends on the triangle's angles. □

A lune is the shape of the illuminated Moon visible from the Earth.

A similar formula holds for the area of a triangle in the hyperbolic plane, namely

$$\Delta = \pi - \alpha - \beta - \gamma.$$

In this case $\alpha + \beta + \gamma < \pi$.

Because of these formulæ there are no similarities in elliptic or hyperbolic geometry.

8.3 Latitude and Longitude

With instruments that made the ancient astronomers illustrious, the altitude of a star above the horizon is established, its distance from the zenith is deduced, and knowing its declination, and since zenith distance plus or minus declination determines latitude, you know immediately on which parallel you are, that is, how much you are north or south of a known point. That is clear I think.

Umberto Eco: The Island of the Day Before

The surface of the Earth can be thought of as a sphere in $\mathbb{R}^3$ of radius 3963 miles or 6377.5 kilometres. (In actuality the radius at the Equator is slightly more than the radius at the poles but we shall ignore this difference.)

We shall represent the surface of the Earth as the unit sphere S^2 and will multiply by a scaling factor whenever we want to find a numerical answer.

For a traveller the two most important questions are

Where am I? and *How do I get somewhere else?*

In answer to the first question the position on the surface of the Earth is specified by two angles ϕ, θ being *latitude, longitude* respectively. Figure 8.5 gives a picture of the Earth showing circles of constant latitude (called *parallels*) and great semicircles of constant longitude (called *meridians*).

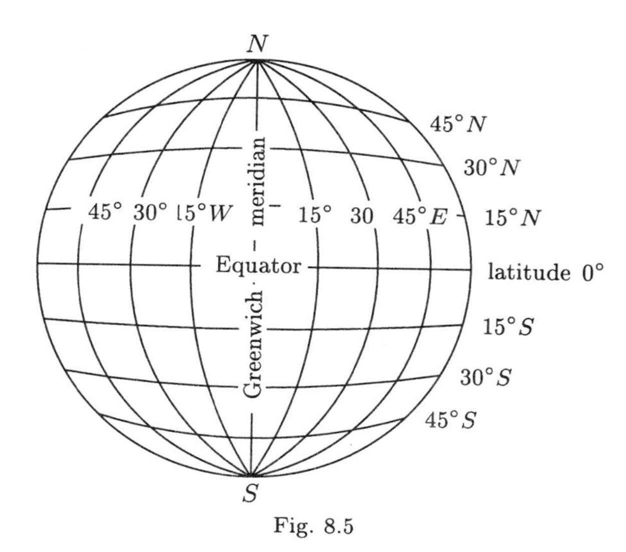

Fig. 8.5

The Earth rotates about an axis through the two poles N and S. The poles are antipodal points and they are the endpoints of a line through the centre of the Earth. The *Equator* is the circle of points equidistant from N and S. It is the parallel of latitude 0° and is the only parallel which is a great circle. Other important parallels are the tropics of Capricorn and Cancer and the Arctic and Antarctic circles. Their significance will be explained later. Important meridians are the Greenwich, which has longitude 0° and the date line, which has longitude 180°. (Actually the date line has a few kinks in it to avoid inhabited islands.)

Figure 8.6 tells us how latitude ϕ and longitude θ of a general point P are defined. It is convenient to consider the equatorial plane containing the Equator $\mathcal{E}$ as being horizontal and the perpendicular axis through the poles as vertical.

The meridian through P meets the Equator at E say. Then the angle $\angle EOP$ is called the *latitude* of P and is denoted by ϕ. The value of ϕ is taken between $\pm\pi/2$ in radians and $\pm 90°$ in degrees. Points north of the Equator have positive latitude whilst points south of the Equator have negative values.

The Arctic circle has constant latitude $66°32'$ and the tropic of Cancer has constant latitude $90 - 66°32' = 23°27'$. The corresponding values for the Antarctic circle and the tropic of Capricorn are negative.

The latitude at any point on the Earth north of the Equator is easily found by calculating the angle between a plumb line pointing to the centre of the Earth and a line pointing at the Pole star which (roughly) lies on the Earth's axis of rotation.

In Fig. 8.6, G represents Greenwich in London with latitude 51.4°. Let the Greenwich meridian meet the Equator at X. Then the angle $\angle XOE$ is called

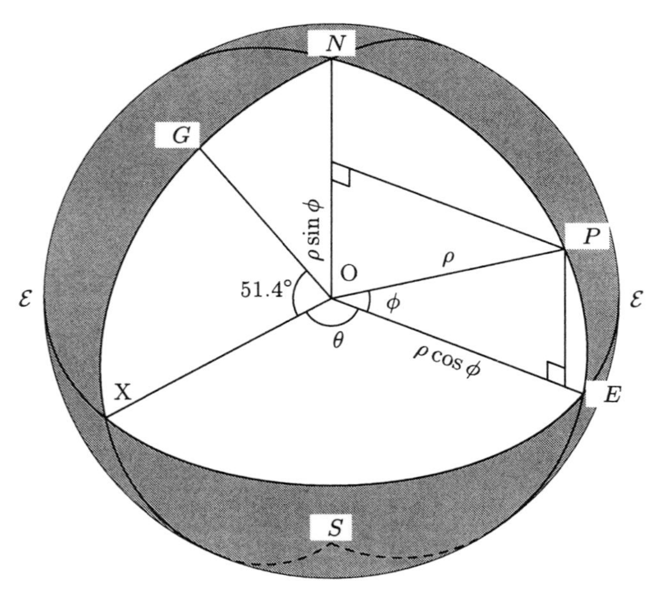

Fig. 8.6

the *longitude* of P and is denoted by θ. The values of longitude lie between $\pm\pi$ or $\pm 180°$. Points east of Greenwich have positive longitude and points west have negative longitude.

In contrast to latitude, the longitude of a point is rather difficult to calculate. It depends on having an accurate clock set to Greenwich time which is compared with local time.

A useful unit of measurement on the Earth is a *nautical mile* which is the great circle distance between two points subtending an angle of one minute of arc at the centre. Accordingly one nautical mile is

$$\frac{2\pi \times 3963}{360 \times 60} = 1.152\,79 \text{ miles.}$$

The *knot* is the unit of speed in use at sea; it is 1 nautical mile per hour.

It is convenient to take cartesian axes as follows. The origin is at the centre of the Earth, the x-axis passes through the point X where the Greenwich meridian meets the Equator, that is the point with latitude and longitude zero, the y-axis passes through the point on the Equator with longitude $+90°$ and the z-axis passes through the north pole N.

If ρ is the radius of the sphere then the triple (ρ, θ, ϕ) defines the *spherical polar coordinates* of the point P. If we project vertically onto the (x, y)-plane we obtain $(\rho\cos\phi, \theta)$ as the (plane) polar coordinates of the projected point. Similarly a horizontal projection gives $\rho\sin\phi$ as the z-coordinate. Accordingly the cartesian and spherical polar coordinates are related by the fundamental equation

$$(x, y, z) = (\rho\cos\phi\cos\theta, \rho\cos\phi\sin\theta, \rho\sin\phi).$$

As mentioned earlier when dealing with points on a sphere we shall take $\rho = 1$ for convenience and multiply by the scaling factor $\rho = 3963$ miles or $\rho = 6377.5$ kilometres for calculations.

■ The great circle distance between two points P, Q with latitude and longitude ϕ_1, θ_1 and ϕ_2, θ_2 is $\rho\psi$ where ψ is the angle in radians subtended by the two points at the centre of the Earth and is given by the formula

$$\cos\psi = \cos\phi_1\cos\phi_2\cos(\theta_1 - \theta_2) + \sin\phi_1\sin\phi_2$$

Proof The cartesian coordinates, taking $\rho = 1$, are

$$(\cos\phi_i\cos\theta_i, \cos\phi_i\sin\theta_i, \sin\phi_i)$$

for $i = 1, 2$. Then

$$\begin{aligned}\cos\psi =& P \cdot Q\\ =& \cos\phi_1\cos\theta_1\cos\phi_2\cos\theta_2 + \cos\phi_1\sin\theta_1\cos\phi_2\sin\theta_2 + \sin\phi_1\sin\phi_2\\ =& \cos\phi_1\cos\phi_2(\cos\theta_1\cos\theta_2 + \sin\theta_1\sin\theta_2) + \sin\phi_1\sin\phi_2\\ =& \cos\phi_1\cos\phi_2\cos(\theta_1 - \theta_2) + \sin\phi_1\sin\phi_2.\end{aligned}$$

□

Let us use this formula for the following.

Example 8.1

Calculate the shortest distance between Prague, 50°05′N, 14°25′E and Winnipeg, 49°55′N, 97°06′W. The formula gives

$$\begin{aligned}\cos\psi &= \cos(50°05')\cos(49°55')\cos(14°25' + 97°06') + \sin(50°05')\sin(49°55')\\ &= 0.435\,28\end{aligned}$$

where ψ is the angle they subtend at O, the centre of the Earth. (Because Winnipeg is west of Greenwich its longitude is *negative*.) So

$$\psi = 64.196\,795° = 1.120\,445\,4 \text{ radians}.$$

Which gives a distance of

$$1.120\,445\,4 \times 3963 = 4440.3253 \text{ miles}$$

or

$$1.1204454 \times 6377.5 = 7145.6409 \text{ kilometres.}$$

The distance in nautical miles is

$$64.196\,795 \times 60 = 3851.8077.$$

Exercise 8.8

Show that the cartesian coordinates ($\rho = 1$) of Prague and Winnipeg are

$$(0.621\,47, 0.159\,76, 0.766\,98) \text{ and } (-0.078\,58, -0.639\,09, 0.7651)$$

respectively. Find the length of a straight tunnel joining them.

Exercise 8.9

Find the shortest distance between Chicago, 41.5° N, 87.45° W and Washington, 38.55° N, 77°W.

8.4 Compass Bearings

But, soft! what light through yonder window breaks?
It is the east, and Juliet is the sun.

Romeo and Juliet

Consider a point P on the sphere. The tangent plane at P is the plane through P at right angles to the vector from O, the centre of the sphere. If P is not one of the poles then we can find, in a continuous way, an identification of this tangent plane with the standard plane $\mathbb{R}^2$ as follows. Consider firstly the parallel through P. We may follow this in an eastwards direction which is defined by a right-handed screw along the axis between the two poles from south to north. Identify the line of the corresponding tangent vector and its direction with the x-axis and its increasing direction in the plane. Similarly a northward direction along the meridian towards the north pole defines the correspondence with the y-axis and its increasing direction. West and south are the respective opposite directions.

Suppose the point P on the sphere has latitude ϕ and longitude θ. Let P_N, P_E be unit tangent vectors at P pointing north and east respectively. Then P_N, P_E, considered as points on the sphere, have latitude and longitude $\pi/2 - \phi, \theta + \pi$ and $0, \theta + \pi/2$ respectively. This can be seen by looking at Fig. 8.7 which is a cross-section of the Earth at right angles to P_E. The point on the sphere corresponding to P_N lies on the other side of the world and is a positive rotation of P in the plane PON through $\pi/2$. The point P_E has zero latitude and is perpendicular to the plane PON. The vectors P, P_E, P_N form a right-handed orthonormal triple.

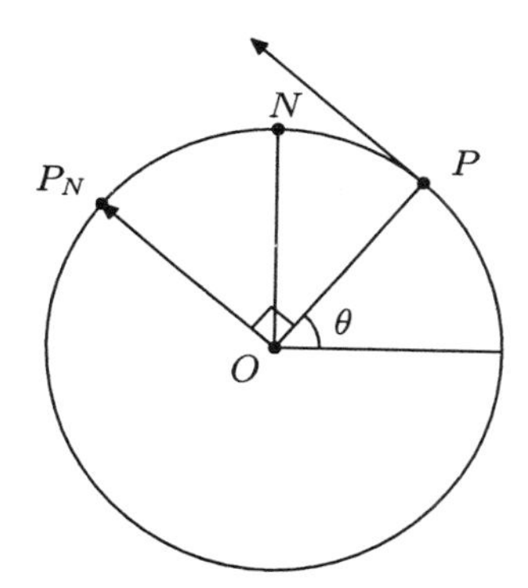

Fig. 8.7 The north pointing tangent vector P_N

Their cartesian coordinates are

$$\boldsymbol{P_N} = (-\cos\boldsymbol{\theta}\sin\boldsymbol{\phi}, -\sin\boldsymbol{\theta}\sin\boldsymbol{\phi}, \cos\boldsymbol{\phi}), \quad \boldsymbol{P_E} = (-\sin\boldsymbol{\theta}, \cos\boldsymbol{\theta}, \boldsymbol{0}).$$

Exercise 8.10

If the cartesian coordinates of P are (x, y, z), show that the cartesian coordinates of P_N, P_E are

$$P_N = \frac{1}{\sqrt{1-z^2}}(-xz, -yz, 1-z^2), \quad P_E = \frac{1}{\sqrt{1-z^2}}(-y, x, 0).$$

Exercise 8.11

Find northward pointing and eastwards pointing vectors at Chicago, 41.5°N, 87.45°W.

Exercise 8.12

I walk 1 mile south, 1 mile east, 1 mile north and am back where I started. Where am I?

At the north pole the only direction is south and at the south pole the only direction is north. East and west cannot be defined at the poles. At all other points the tangent plane can be identified with $\mathbb{R}^2$ by taking east as the x-axis and north as the y-axis. The fact that there are two points on the sphere where the tangent plane cannot be continuously identified with the standard plane $\mathbb{R}^2$ is no accident but is a consequence of the topology of the sphere. If our planet were toroidal then we could define an east and north continuously at every point. Surfaces which have this property, such as a torus or a sphere with two points removed, are called *parallelisable*. Oddly enough, although the sphere S^2 is not parallelisable, the three-dimensional hypersphere S^3 is. At any point P in S^3 we can define east, north and up. In particular if $P = (a, b, c, d)$ is a general point in S^3 then the three points

$$(-b, a, d, -c), \ (-c, -d, a, b), \ (-d, c, -b, a)$$

are at right angles to (a, b, c, d) and to themselves. So they define the three directions. This is most easily seen using quaternions.

In general, direction, other than east or west, is measured by its deviation from a north or south bearing. So N45°E corresponds by the above analysis to a direction parallel to the vector with coordinates $(1, 1)$ in the plane. Similarly S60°W corresponds to $(-\sqrt{3}, -1)$.

Since directions correspond to points on a circle these are often indicated by a binary subdivision of the circle into 32 points as shown in Fig. 8.8.

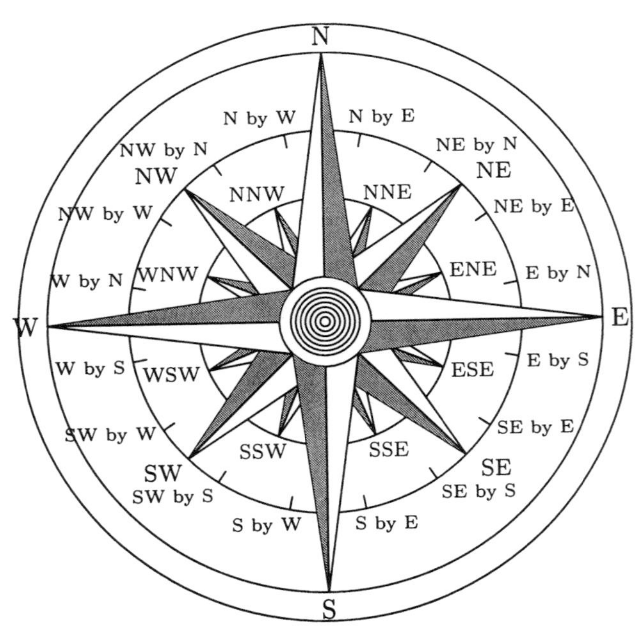

Fig. 8.8 Points of the compass

The Vikings navigated their voyages across the North Sea with the aid of a *sun compass*. The plan and side view are illustrated in Fig. 8.9. The face of the sun compass is a disc of wood held horizontally by means of a handle (h). A central pin (p) casts a shadow (s) whose tip is made to lie on a grooved track (t) by a rotation of the disc. Then north is indicated by (n). The track is a hyperbola and can be calibrated before the journey; it will of course depend on the latitude and time of year. The position of the shadow in Fig. 8.9 is shown at about 9 o'clock in the morning.

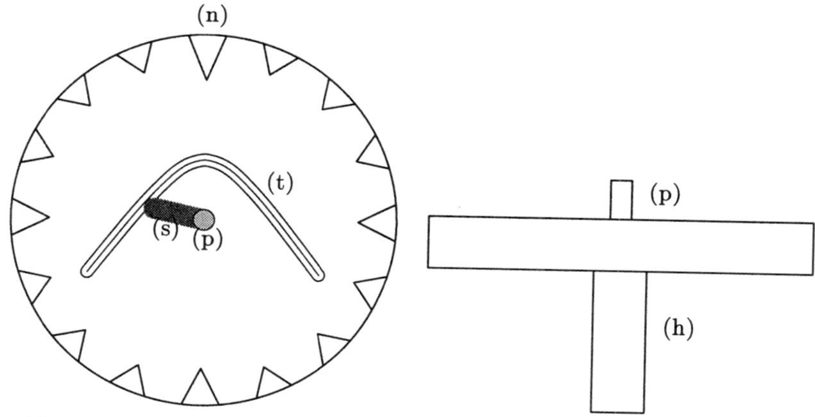

Fig. 8.9. Plan of sun compass

Fig. 8.10. Side view

Constant bearing

If two points are on the same parallel of latitude then you can get from one to the other by travelling consistently west or east. If the constant latitude is ϕ and the two longitudes are θ_1 and θ_2 then the angular distance travelled is

$$(\theta_2 - \theta_1)\cos\phi.$$

This is never less and may be considerably more than the great circle angular distance ψ where

$$\cos\psi = \cos(\theta_2 - \theta_1)\cos^2\phi + \sin^2\phi.$$

Example 8.2

Let us reconsider our journey from Prague, 50°05′N, 14°25′E to Winnipeg, 49°55′N, 97°06′W.

We calculated earlier that the shortest distance was 3851.8077 nautical miles. Suppose a pilot decided, instead of taking the great circle route, to fly with a constant westward bearing. After all we can take the latitude of both

Prague and Winnipeg to be 50° without much error. Then the flight would be along a small circle arc with distance in nautical miles given by

$$111\tfrac{31}{60} \times 60 \times \cos 50^\circ = 4300.8918.$$

This is more than 11% greater than the great circle distance. Considering the cost of aviation fuel it is small wonder that airlines take the latter if at all possible.

However there are problems with taking the shortest route. For one thing your bearing will always be changing unless you are travelling along the Equator. For example if we take the great circle journey from Prague to Winnipeg our initial bearing will be somewhat north of west. The amount of that somewhat can be calculated from the following analysis. Let

$$X = (\cos\theta_1 \cos\phi_1, \sin\theta_1 \cos\phi_1, \sin\phi_1), \ \ Y = (\cos\theta_2 \cos\phi_2, \sin\theta_2 \cos\phi_2, \sin\phi_2)$$

be points on the unit sphere subtending an angle ψ at the centre. The tangent at X which points in the direction of Y is given by $T = Y \operatorname{cosec}\psi - X \cot\psi$ as we saw earlier. If we take the dot product of this vector with the eastward pointing vector $X_E = (-\sin\theta_1, \cos\theta_1, 0)$ and simplify, we obtain the cosine of the angle that this tangent makes with east, leading to the following result.

■ The angle α which the direction of the shortest route from X to Y makes with the east is given by the formula

$$\boldsymbol{\cos\alpha = \sin(\theta_2 - \theta_1)\cos\phi_2 \operatorname{cosec}\psi.} \qquad \square$$

So in particular for the journey from Prague to Winnipeg $\cos\alpha$ is

$$\sin(-111^\circ 31') \times \cos(49^\circ 55') \times \operatorname{cosec}(64.196\,795^\circ) = -0.6654\ldots.$$

So α is about 131.711° which gives a bearing of N41°42.66′W.

There are also geographical difficulties with the shortest route. Icebergs are one hazard. During a journey along the shortest route from X to Y a northernmost (or southernmost) point will be reached. This will be the point where the tangent $X(-\sin\theta - \cot\psi\cos\theta) + Y\operatorname{cosec}\psi\cos\theta$ has zero third coordinate, where θ is the angle between a general point on the geodesic and X. That is

$$\sin\phi_1(-\sin\theta - \cot\psi\cos\theta) + \sin\phi_2 \operatorname{cosec}\psi\cos\theta = 0.$$

Suppose that the solution to this is $\theta = \beta$, then we have the following.

■ The most northerly (or southernmost) point on the shortest route from X to Y is

$$\boldsymbol{X(\cos\beta - \cot\psi\sin\beta) + Y\operatorname{cosec}\psi\sin\beta}$$

where β is given by

$$\tan\beta = \frac{\sin\phi_2}{\sin\phi_1}\operatorname{cosec}\psi - \cot\psi.$$

The extreme latitude is λ where

$$\sin\lambda = \sin\phi_1(\cos\beta - \cot\psi\sin\beta) + \sin\phi_2\operatorname{cosec}\psi\sin\beta. \quad \square$$

If $\phi_1 = \phi_2 = \phi$ these formulæ simplify to

$$\tan\beta = \operatorname{cosec}\psi - \cot\psi$$

and

$$\sin\lambda = \sin\phi\sec\beta.$$

For the journey from Prague to Winnipeg, let $\phi_1 = \phi_2 = 50°$. Then $\tan\beta = 0.6273$ and $\beta = 32.1°$. So $\sin\lambda = 0.9043$ and $64.724°$ is the most northerly latitude reached.

Exercise 8.13

Three planes take off from O'Hare airport in Chicago, 41.5°N, 87.45°W and take the shortest routes to Delhi, 28.39°N, 77.13°E, Tokyo 35.4°N, 139.45°E and Istanbul, 41.02°N, 28.57°E. Which plane flies nearest the north pole? What will be the initial direction of each plane?

Exercise 8.14

A Portuguese navigator sets sail from Lisbon, 38.44°N, 9.08°W to Washington, 38.55°N, 77°W. What saving in distance does he make by sailing the great circle route instead of a constant west bearing? If he takes the great circle route what is his initial bearing and how far north will he go? Repeat for an Irish monk who sails from Galway Bay, 53.17°N, 8°W to Goose Bay, 53.5°N, 60.2°W.

8.5 The Celestial Sphere

But let's not talk about fare-thee-wells now the night is a starry dome and they're playin' that scratchy rock'n'roll beneath the Matalla moon.

Joni Mitchell

When we look at the stars on a clear dark night we get the impression that they are points of light fixed to a hemispherical dome surrounding us. The eye fails to give us any idea of their distance. In fact the distances involved are so colossal that we cannot relate to them in any personal way. The largest distance to the Sun is an astronomical unit (au) about 149.6 million kilometres, the nearest star is about 1.275 parsec distant where one parsec (pc) is about 206,265 au. On a very clear dark night the naked eye can see the Andromeda galaxy which is a truly staggering 670,000 pc distant. Powerful telescopes can see objects whose distances away when contemplated turn the soul to ice. As a consequence the far stars appear to be fixed apart from their apparent movement due to the rotation and motion of the Earth.

Exercise 8.15

The official definition of a parsec is the distance at which 1 au subtends an angle of 1 arcsecond. Deduce the length of a parsec given above.

Exercise 8.16

Imagine that the sun is one foot (30cm) away. How far away would (a) the nearest star be, (b) the Andromeda galaxy be?

Exercise 8.17

In March 1998, an advertisement in the Sun newspaper offered binoculars with which you could see up to 35 miles. Discuss the efficiency of these instruments in comparison with the human eye.

The sphere to which the stars appear to be attached is called the *celestial sphere.* The position of a star in the sky is determined by *right ascension*, RA, and *declination*, DEC, corresponding to longitude and latitude respectively.

Declination is the angle (in degrees) of the star above the equatorial plane. Right ascension is the angle of the star around the celestial Equator. Right ascension is measured eastwards in hours and minutes. Since 24 hours corresponds to one complete revolution or 360° this can be converted to degrees, minutes and seconds of arc by the rule

$$1 \text{ hour } = 15^\circ, \; 1 \text{ minute } = 15', \; 1 \text{ second } = 15''.$$

In contrast to distance away, we can calculate the angles between stars and their relative positions on the celestial sphere with great accuracy. The apparent distance apart of two stars, called their *angular separation* is determined by the angle they subtend to an Earthly observer. This is given by the same formula as the great circle distance above and can be used for the following sort of example.

Example 8.3

Let us find the angular separation of Algol, 3h 8.2m, 40.96° in Perseus from Capella, 5h 16.7m, 46° in Auriga. The first job is to change the difference of the right ascensions to degrees: that is 2h 8.5m $= 30 + 8.5/4 = 32.125°$. From the formula given earlier for distances on the Earth and converted to celestial coordinates the angular separation is ψ where

$$\cos\psi = \cos \mathrm{DEC}_1 \cos \mathrm{DEC}_2 \cos(\mathrm{RA}_1 - \mathrm{RA}_2) + \sin \mathrm{DEC}_1 \sin \mathrm{DEC}_2.$$

So $\cos\psi = 0.76 \times 0.69 \times 0.85 + 0.66 \times 0.72 = 0.92$ and $\psi = 23.7°$.

Exercise 8.18

Find the angular separation of the Gemini twins, Castor, 7h 34.6m, 31.89° and Pollux, 7h 45.3m, 28.03°.

The next time you see the Sun and Moon in the sky together try to trace the rays of light from the Sun to the Moon. They will appear to follow a curved path (Fig. 8.11).

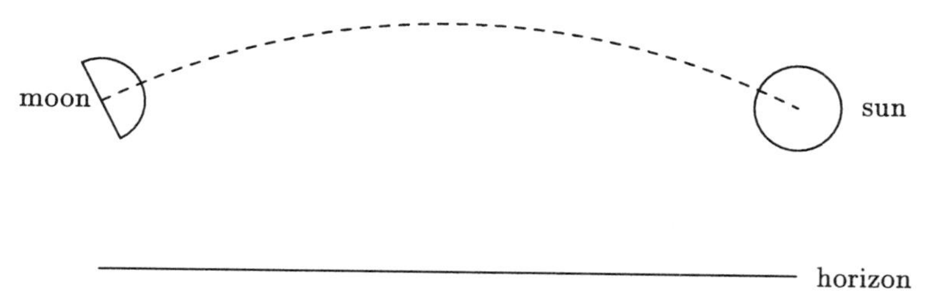

Fig. 8.11 The Sun illuminating the Moon

This is because the sky is a sphere and rays of light follow a geodesic. It is only an illusion that the horizon is flat.

Aristarchus of Samos in 260 BC noted that at half moon the triangle formed by the Earth, Moon and Sun was a right-angled triangle with hypotenuse the Earth-Sun line. By measuring the angular separation of the Sun and Moon the relative distances of the Sun and Moon from the Earth could be calculated. It is not to his discredit that his answer was one tenth of the actual answer, considering the crude methods available at the time.

8.6 Observer's Coordinates

Right ascension and declination are examples of absolute coordinates, that is they are unchanged year in year out. (Actually this is not quite true since not only is there a general change due to precession but some of the nearer stars have a slow movement at right angles to the line of sight called *proper motion.*)

Easier to understand and use are the coordinates of a star which an observer would see at any given time but which vary continually.

Consider an observer O at the centre of the celestial sphere which we will take as the unit sphere, S^2 (Fig. 8.12). The horizontal plane will be taken as $z = 0$ and the vertical line through the observer as the z-axis. The point vertically above the observer is called the *zenith* and is the point $Z = (0, 0, 1)$. The horizontal plane meets the celestial sphere in the *horizon*. So Z is a pole of the horizon. The other pole is the *nadir* $Z' = (0, 0, -1)$ which is directly below the observer.

On the horizon are the compass points east $E = (1, 0, 0)$ and north $N = (0, 1, 0)$. So that $\{E, N, Z\}$ form a right-handed unit basis for coordinates. Other compass points are south $S = (0, -1, 0)$ and west $W = (-1, 0, 0)$.

The *north celestial pole* or *pole* for short is the point P so that OP is the axis of rotation of the Earth. The Earth rotates so that a right-handed screw points from O to P. It follows contrarily that the celestial sphere appears to rotate in a left-handed manner about OP. The coordinates of P are $(0, \cos\phi, \sin\phi)$ where ϕ is the latitude of the observer. So $\angle NOP = \phi$. The points N, P, Z, S lie on a great circle called the observer's *meridian*. The poles of the meridian are E, W.

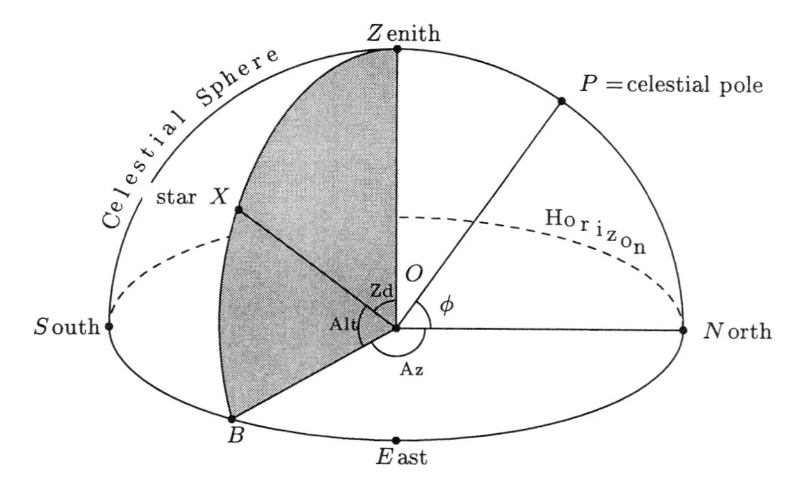

Fig. 8.12

Let $X = (x, y, z)$ be a star. Extend the geodesic ZX to meet the horizon at B. The angle $\angle BOX = \textsc{Alt}$ is called the *altitude* of the star and $\angle XOZ = \textsc{Zd}$ is called the *zenith distance.* The angle $\angle NOB = \textsc{Az}$ is called the *azimuth* of the star.

The angle $\angle XOP = d$ is called the *north polar distance* and represents the angle of separation of the star from the pole. The complementary angle $\textsc{Dec} = 90 - d$ is called the *declination* of X. The equatorial plane of the Earth which is at right angles to OP meets the celestial sphere in the *celestial equator.* So declination is the distance of X from the Equator.

Let $\theta = 90 - \textsc{Az}$. Then $\theta, \textsc{Alt}$ are spherical polar coordinates for X. It follows that

$$\begin{aligned} X &= (\cos \textsc{Alt} \cos\theta, \cos \textsc{Alt} \sin\theta, \sin \textsc{Alt}) \\ &= (\cos \textsc{Alt} \sin \textsc{Az}, \cos \textsc{Alt} \cos \textsc{Az}, \sin \textsc{Alt}). \end{aligned}$$

By taking the scalar product $X \cdot P$ we obtain a formula for the declination

$$\mathbf{\sin \textsc{Dec} = \cos \textsc{Alt} \cos \textsc{Az} \cos\phi + \sin \textsc{Alt} \sin\phi.}$$

Exercise 8.19

For a star $\textsc{Alt} = 60°$ and $\textsc{Az} = 230°$ at latitude $52°$, find its declination.

Exercise 8.20

Find the azimuth of a star when rising and setting.

Exercise 8.21

* A star has zenith distance $\textsc{Zd}_1$ on the meridian and $\textsc{Zd}_2$ on the prime vertical (the great circle running east to west through the zenith). Show that

$$\begin{aligned} \cot\phi &= \cot \textsc{Zd}_1 - \operatorname{cosec} \textsc{Zd}_1 \cos \textsc{Zd}_2 \\ \cot \textsc{Dec} &= \operatorname{cosec} \textsc{Zd}_1 \sec \textsc{Zd}_2 - \cot \textsc{Zd}_1. \end{aligned}$$

Exercise 8.22

Show that a star never sets, that is, it is *circumpolar*, if $\textsc{Dec} \geq 90 - \phi$.

8.7 Time and Right Ascension

This is the world service of the BBC. The time is ten Greenwich. Meantime here is the news.

(Alleged) mispunctuation

Long before clocks were invented, the passage of time was marked by the apparent motion of the Sun. The Earth rotates on its axis once a day and the Sun appears to cross the sky and return in a day. However the Earth moves fastest in its orbit when nearest the Sun. Since this happens in the northern winter the Sun then appears to cross the sky faster. To get round this, a fictitious or *mean Sun* is postulated which moves across the sky at the same rate, winter and summer. The mean Sun is either ahead or behind the real Sun and they coincide twice a year. Civil or mean time is marked by the mean Sun.

The Earth rotates around the Sun in one year and so the Sun moves from its highest altitude at midday in high summer to its lowest altitude in the depths of winter and back in one year. Refinements are added by other celestial bodies. The Moon orbits the Earth in one month. The reappearance of Sirius in Canis Major signals the dog days of summer and the start of the march towards winter. Once Arcturus reappears in the northern evening sky then spring cannot be far away.

The Earth completes its orbit around the Sun in 365.2422 days. Because this is slightly more than a year of 365 days the calendar has to be changed every leap year by the addition of an extra day, 29 February. A leap year occurs if the corresponding year number is divisible by 4. Century years are excluded unless the year number is divisible by 400. So 1900 was not a leap year but 2000 is. Between 1600 and 2000 there were $100-3 = 97$ leap years. As a fraction, the extra amounts to $97/400 = 0.2425$ days. This is sufficiently near to the exact number for the calender be unchanged by a day for over 3000 years.

The Babylonians, who were great astronomers, had a number system based on 60, there are $360 = 60 \times 6$ degrees in a complete cycle and there are nearly 360 days in a year: surely some connection? The historians are not sure.

The plane containing the orbit of the Earth is called the *ecliptic*. It is marked in the celestial sphere by a great circle, along which the Sun appears to move. The ecliptic makes an angle of 23.5° with the celestial equator because the axis of rotation of the Earth makes an angle of 23.5° with a normal to the ecliptic plane.

This inclination of the Earth's axis, called the *obliquity*, accounts for the seasons. Fig. 8.13 shows the Earth at the northern summer when the tilt of the axis towards the Sun is greatest.

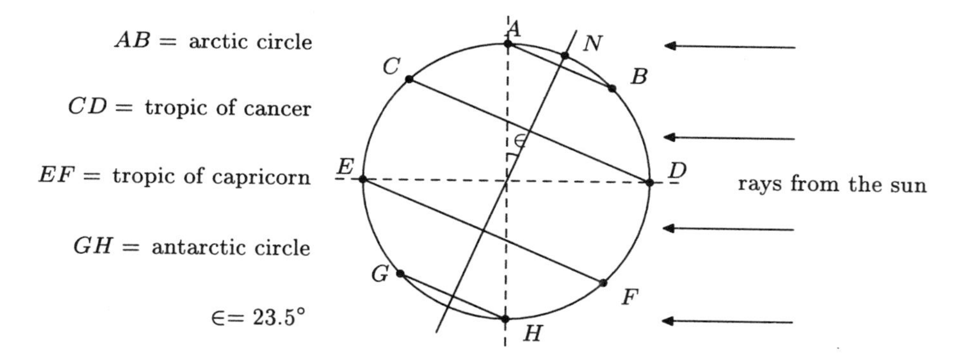

Fig. 8.13 The northern midsummer and southern midwinter

The line ED lies in the ecliptic and the line AH is perpendicular to the ecliptic plane. The obliquity is denoted by $\in$. The point D on the *tropic of Cancer* has the Sun directly overhead. The point E on the *tropic of Capricorn* has the Sun directly overhead six months later. Places inside the Arctic circle have constant Sunlight during the day. Places inside the Antarctic circle have constant dark. Six months later is the northern winter and the situation is reversed. To see what happens, imagine the rays of the Sun coming from the left.

Exercise 8.23

On midsummer's day the Sun is directly overhead in Syrene. At the same time in Alexandria, 500 miles north, the Sun has a zenith distance of 7.5°. Use this to estimate the radius of the Earth.

Exercise 8.24

Is the Moon higher or lower in the summer than in the winter?

Between winter and summer lie the *equinoxes*. As the name suggests these are the days when night and day have the same length, 12 hours, everywhere on Earth. More precisely, the position of the Earth will be the nearest point to the Sun of the plane containing the axis of rotation and a normal to the ecliptic. So both poles will have the Sun on the horizon at the same time. This means that the Sun lies in the equatorial plane. The *spring equinox* occurs around 21 March when the Sun passes from south of the Equator to the north. The *autumnal equinox* occurs around 21 September when the Sun drops below the Equator.

On the celestial sphere the intersection of the ecliptic with the equator consists of two antipodal points. The one associated with the spring equinox is called the *first point of Aries* and is denoted Υ. In the same way that Greenwich marks the zero meridian on Earth, the first point of Aries serves as the zero point of RA(Fig. 8.14). From this zero point RA is measured eastwards in hours and minutes. Since 24 hours corresponds to one complete revolution or 360° this can be converted to degrees, minutes and seconds of arc by the usual rule

$$1 \text{ hour } = 15^\circ, \ 1 \text{ minute } = 15', \ 1 \text{ second } = 15''.$$

Or conversely, one degree is 4 minutes.

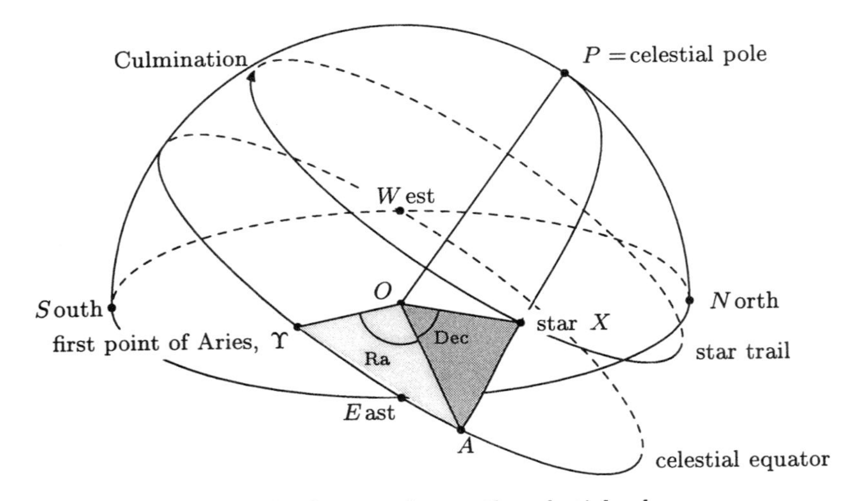

Fig. 8.14 Right ascension on the celestial sphere

The *tropical year* is defined to be the interval between the two spring equinoxes. The direction of the Earth's axis is not quite constant but changes due to *precession* by 50.3″ per year. So the Earth's axis returns to its original position after $360^\circ/50.3'' = 25765.408$ years. The *sidereal year* is defined to be the time for the Sun to return to the same position amongst the stars. This is 365.256 36 days. The difference is due to precession.

Exercise 8.25

Show that sidereal year/tropical year agrees with $1 + 50.3''/360^\circ$ up to six places of decimal.

Exercise 8.26

The first point of Aries actually lies in Pisces. The ecliptic passes from Pisces into Aries at the point with RA 1 hr 45 m and DEC 12°. Assuming that the borders of the constellations are roughly the same as in ancient times, show that the first point of Aries was named over 2000 years ago.

The positions of the stars known in ancient times can be used to give an estimate of when they were named in the Western/Mediterranean tradition. There is an unnamed region in the south of the celestial sphere where the sky viewed from the Mediterranean was constantly below the horizon. If the centre of this region is taken as the position of the south celestial pole at the time then precession gives a date of 2800 BC for their naming.

The position of stars can be used to give a definition of astronomical time. If we take a long exposure photograph at night of the area around the north celestial pole, one star, Polaris or the Pole Star, which is about 0.5° from the pole, will appear almost stationary whilst the others will leave anticlockwise circular trails with centre the pole and radius the north polar distance. This movement, remember, is only apparent and is due to the daily rotation of the Earth on its axis. So if we know the position of a star tonight, in 24 hours it will have returned to the same position. Well actually no: because in that 24 hours the Earth will have completed 1/365.2422 of its orbit and so the star will be in the same place $\frac{24\times 60}{365.2422} = 3.942\,589\,3$ minutes earlier. This difference is a day in sidereal time which is faster than ordinary or civil time.

It follows that the position of a star in the sky is repeated every 23h 56.057 411m. The zero of *local sidereal time* or LST is defined to be the moment when the first point of Aries Υ culminates, that is crosses the north/south meridian from east to west. A star will culminate RA later, where RA is its *right ascension*.

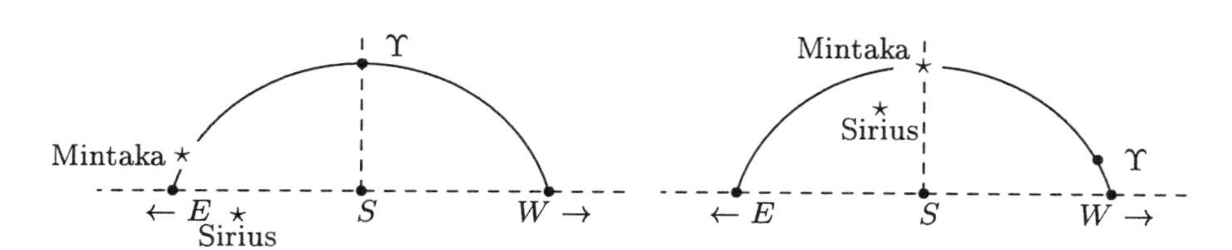

Fig. 8.15 Position of Υ at 0 LST and at 5h 32m LST

Fig. 8.15 illustrates the motion of Mintaka and Sirius. Mintaka is at the end of Orion's belt, has right ascension 5h 32m and virtually lies on the celestial equator. Sirius has right ascension 6h 45.1m.

Notice that at 0 LST Mintaka is just rising and Sirius is still below the horizon. At 5h 32m LST Sirius has risen and Mintaka is culminating.

Exercise 8.27

If ψ is the angle which a star's path at rising makes with the horizon, prove that

$$\cos\psi = \sin\phi \sec \text{DEC}.$$

If we know the LST at one place then we can find it elsewhere provided we know the change in longitude between the places. For example if it is 5h 12m at Manchester which has longitude $2°18.4'$W then to find the LST at Greenwich which of course has longitude zero we must add the longitude of Manchester converted into time. This is 9m 13.6s. So the LST at Greenwich is 5h 21m 13.6s.

In order to fix the position of a star anywhere in the world we need a time at a standard place. This is chosen to be Greenwich. *Greenwich mean time* or GMT is defined to be the civil time in Britain ignoring summer time. Another name for GMT is *universal time* UT. For anywhere else in the world *zone time* is GMT incremented by the number of time zones of 15° crossed. For example a place with longitude 163°E has zone time GMT + 11h since $163 = 10 \times 15 + 13$. For longitude 64°W the zone time is GMT − 4h.

The local sidereal time at Greenwich is called *Greenwich sidereal time* or GST. Its relationship with GMT can be found from an almanac. However only one reading is necessary for the rest can then be calculated. We note that at noon GMT in Greenwich on 1 January 1998 the sidereal time GST is 18h 42.5m. Every day after (before), GST is incremented by +3.942 589 3m (−3.942 589 3m). For example on February 14 at noon GMT, GST is 21h 35.973 93m.

Exercise 8.28

What is the GMT on 22 March 1998 when the first point of Aries crosses the Greenwich meridian from east to west?

The *hour angle*, HA, of a star X is the angle XSZ measured in sidereal time. So

$$\text{HA} = \mathsf{LST} - \text{RA}.$$

We are now in a position to calculate the hour angle of any star at any time anywhere in the world.

Consider an observer with longitude $64°28'$W who wishes to calculate the hour angle of Betelgeuse on 14 February 2000 at 22h zone time. For simplicity we will round off the minutes to 2 decimal places. The first thing to do is to find GMT. This is $22 + 4 = 26$ and is 14h after noon on 14 February. The next thing to do is to convert these hours to sidereal time by adding $14 \times 3.94/24 = 2.3$m. Now the GST at noon on 14 February is 21h 34.05m. This can be looked up in an almanac or deduced from our earlier example on the same day by subtracting $2 \times 0.2422 \times 3.942\,589\,3$m since 2 years have passed without a leap day. The explanation for the factors is as follows: 0.2422 is the extra fraction of a day which is the length of one year and 3.942 589 3m is the extra time of a sidereal day. It follows that the GST is 35h 36.35m. Subtract the longitude converted to time, 4h 17.87m and the RA of Betelgeuse which is 5h 53.82m. It follows that the HA of Betelgeuse is 25h 24.66m or 1h 24.66m. The calculations are set out below.

	h m
Zone time	22 00.00
plus zone	04 00.00
GMT	26 00.00
less 12h	12 00.00
hours after noon	14 00.00
add ST conversion	00 02.30
sidereal hours after noon	14 02.30
GST at noon	21 34.05
GST	35 36.35
less longitude	04 17.87
LST	31 18.48
less RA of Betelgeuse	05 53.82
HA of Betelgeuse	25 24.66
or	01 24.66

Exercise 8.29

Find the hour angle of Aldebaran, RA =4h 35.9, at 21h on 21 March 2000 at longitude $36°16'$E.

Exercise 8.30

Rigel, RA = 5h 14.53m, has just culminated. Your chronometer says that GMT is 16h 20.2m on 4 February 1999. What is your longitude? What is your zone time?

Finally in this chapter we show how the hour angle and declination are related to the altitude and azimuth of a star.

We use the coordinates given earlier for the zenith $Z = (0,0,1)$, the pole $P = (0, \cos\phi, \sin\phi)$, etc. Let S' be the highest point on the Equator. Then $S' = (0, -\sin\phi, \cos\phi)$. Let X' be the projection of the star X from P onto the

Equator. Then $S'X' = \text{HA}$ and $X'X = \text{DEC}$. So

$$\begin{aligned} X' &= \cos \text{HA} S' + \sin \text{HA} W \\ &= (-\sin \text{HA}, -\cos \text{HA} \sin \phi, \cos \text{HA} \cos \phi) \end{aligned}$$

and

$$\begin{aligned} X &= \cos \text{DEC} X' + \sin \text{DEC} P \\ &= (-\cos \text{DEC} \sin \text{HA}, -\cos \text{DEC} \cos \text{HA} \sin \phi + \sin \text{DEC} \cos \phi, \\ &\quad \cos \text{DEC} \cos \text{HA} \cos \phi + \sin \text{DEC} \sin \phi). \end{aligned}$$

But in terms of the observer's coordinates

$$X = (\cos \text{ALT} \sin \text{AZ}, \cos \text{ALT} \cos \text{AZ}, \sin \text{ALT}).$$

Equating the coordinates we get

$$\cos \text{ALT} \sin \text{AZ} = -\cos \text{DEC} \sin \text{HA} \tag{1}$$

$$\cos \text{ALT} \cos \text{AZ} = -\cos \text{DEC} \cos \text{HA} \sin \phi + \sin \text{DEC} \cos \phi \tag{2}$$

$$\sin \text{ALT} = \cos \text{DEC} \cos \text{HA} \cos \phi + \sin \text{DEC} \sin \phi \tag{3}$$

from which $\sin \text{HA}$ can be calculated from ALT and AZ, DEC having been calculated from the earlier formula

$$\sin \text{DEC} = \cos \text{ALT} \cos \text{AZ} \cos \phi + \sin \text{ALT} \sin \phi. \tag{4}$$

Conversely ALT and $\sin \text{AZ}$ can be calculated from HA and DEC.

In order to find AZ the sign of $\cos \text{AZ}$ must be found from equation (2). The left-hand side will be positive if

$$\sin \text{DEC} \cos \phi > \cos \text{DEC} \cos \text{HA} \sin \phi.$$

Since $\cos \phi$ and $\cos \text{DEC}$ are both positive this condition can be written

$$\tan \text{DEC} > \cos \text{HA} \tan \phi.$$

Let ψ be the solution to

$$\sin \psi = \sin \text{AZ} = -\cos \text{DEC} \sin \text{HA} \sec \text{ALT}$$

which lies between $-\pi/2$ and $+\pi/2$. Then AZ is given by the following table.

$\tan \text{DEC} > \cos \text{HA} \tan \phi$	$\tan \text{DEC} < \cos \text{HA} \tan \phi$
$\text{AZ} = \psi$	$\text{AZ} = \pi - \psi$

Something similar can be given for finding HA. Let ξ be the solution to

$$\sin \xi = \sin \text{HA} = -\cos \text{ALT} \sin \text{AZ} \sec \text{DEC}$$

which lies between $-\pi/2$ and $+\pi/2$. Then HA is given by the following table.

$\sin \text{DEC} \cos\phi > \cos \text{ALT} \cos \text{AZ}$	$\sin \text{DEC} \cos\phi < \cos \text{ALT} \cos \text{AZ}$
$\text{HA} = \xi$	$\text{HA} = \pi - \xi$

Consider the following example. Suppose at latitude 55°N the hour angle of Betelgeuse is 20h 46m. From a star catalogue we see that its declination is 7.41°. Converting 20h 46m to 311.5° and substituting in the third equation above gives $\sin \text{ALT} = 0.482\,534\,1$. Since ALT lies in the range ± 90, $\text{ALT} = 28.85°$.

From the first equation $\sin \text{AZ} = 0.847\,942\,5$ so $\text{AZ} = 57.99°$ or $122.01°$. Since $\tan \text{DEC} < \cos \text{HA} \tan\phi$ it follows that $\cos \text{AZ}$ is negative and so $\text{AZ} = 122.01°$.

Exercise 8.31

Your observatory, 39.22N, 20.6E reports a strange object with altitude 26.61° and azimuth 215.66° at 23h local time on 10 January 2000. What are the celestial coordinates of the object? Should you get excited?

Exercise 8.32

Deduce equation (1) from the sine formula for the spherical triangle PZX. Be careful of signs!

Exercise 8.33

If HA is the hour angle of a star at rising, show that

$$\cos \text{HA} = -\tan \text{DEC} \tan\phi.$$

Answers to Selected Questions in Chapter 8

8.1 Since $(1,3,-2) = (1,1,0) + (0,2,-2)$ it follows that the unit vectors in those directions all lie in a plane through the origin. Since

$$T \cdot X = (1/\sqrt{2}, 1/\sqrt{2}, 0) \cdot (-1/\sqrt{2}, 1/\sqrt{2}, 0) = 0,$$

T is a unit tangent vector at X. (And X is a unit tangent vector at T.)

8.2 The substitution $z = 2$ has just one solution $(-1, 2, 2)$ which is the point of contact.

8.3 Since $X \cdot Y = 0$ the angle ψ is $\pi/2$. So the general point is

$$\left(\frac{\cos\theta - \sin\theta}{2}, \frac{\cos\theta - \sin\theta}{2}, \frac{\cos\theta + \sin\theta}{\sqrt{2}}\right).$$

Since $(\cos\theta - \sin\theta)/2 = \sin(\pi/4 - \theta)/\sqrt{2}$ this is largest when $\theta = -\pi/4$. So $P = (1/\sqrt{2}, 1/\sqrt{2}, 0)$ has this property. Since $P \cdot Y = -1/\sqrt{2}$ the angle between them is $3\pi/4$. The required direction is $Y\sqrt{2} - P(-1) = (0,0,1)$.

8.4 Using the formulæ above, $\operatorname{cosec}\psi = 2/\sqrt{3}$ and $\cot\psi = -1/\sqrt{3}$. The general point is

$$\left(\frac{\cos\theta}{2} - \frac{\sin\theta}{\sqrt{3}}, \frac{\cos\theta}{2} - \frac{\sin\theta}{\sqrt{3}}, \frac{\cos\theta}{2} - \frac{\sin\theta}{\sqrt{3}}, \frac{\cos\theta}{2} + \sqrt{3}\sin\theta\right).$$

The last coordinate is largest when $\tan\theta = 2\sqrt{3}$ giving the point

$$(-3/2\sqrt{13}, -3/2\sqrt{13}, -3/2\sqrt{13}, \sqrt{13}/2).$$

The required direction is $(-1,-1,-1,3)/2\sqrt{3}$.

8.5 $A = (1,0,0),\ B = (0,1,0),\ C = (0,0,1),\ \pi/2$.

8.6 $c = 83°46'32''$, $\alpha = 57°40'45''$ and $\beta = 72°49'50''$.

8.8 4209.5 miles.

8.9 589 miles.

8.11 $P_N = (-0.029, 0.662, 0.749)$, $P_E = (0.999, 0.045, 0)$.

8.12 The Christmas cracker answer is the north pole. However there are infinitely more possibilities. Consider a parallel south of the equator with circumference one mile. Take any point due north and one mile from this. Repeat for parallels of length a half mile, a third of a mile and so on.

8.13 For the plane to Delhi the extreme latitude is 79.34° and the bearing is N14.29E. For the other two planes the figures are 63.51°, N36.55W and 58.87°, N43.64E.

8.14 For the Portuguese navigator the two distances are 3874.58 and 4042.33 nautical miles. The initial bearing is N59.76°W and the most northern latitude is 47.42°. For the Irish monk the figures are 2347.2, 2443.49, N61.08W and 58.35°. In both cases it hardly seems worth the bother.

8.15 One parsec$= 1/\tan 1''$ etc.

8.16 (a) About 50 miles or 80 kilometres: the distance from Brighton to London, (b) 42 million kilometres: a quarter of the way to the Sun.

8.18 4.5°.

8.19 DEC $= 28.98°$

8.20 Put $\text{ALT} = 0$ in the above equation and the azimuth at rising is the angle ψ between 0 and π which satisfies $\cos\psi = \sin \text{DEC} \sec\phi$. The azimuth at setting is $2\pi - \psi$.

8.21 Let Z_1 be the point where the star crosses the meridian. Then $Z_1 = \cos \text{ZD}_1 Z + \sin \text{ZD}_1 S = (0, -\sin \text{ZD}_1, \cos \text{ZD}_1)$. Let Z_2 be the point where the star crosses the prime vertical. Then

$$Z_2 = \cos \text{ZD}_2 Z + \sin \text{ZD}_2 E = (\sin \text{ZD}_2, 0, \cos \text{ZD}_2).$$

Because $Z_2 - Z_1$ is perpendicular to OP we have $Z_1 \cdot P = Z_2 \cdot P$ which when expanded reduces to the first equation.

Let $S' = (0, -\sin\phi, \cos\phi)$ be the point on the meridian at right angles to P. Then Z_1 can also be written as $Z_1 = \cos \text{DEC} S' + \sin \text{DEC} P$. If we compare the coordinates of the two versions for Z_1 we obtain

$$\begin{aligned} \cos \text{DEC} &= \cos \text{ZD}_1 \cos\phi + \sin \text{ZD}_1 \sin\phi \\ \sin \text{DEC} &= \cos \text{ZD}_1 \sin\phi - \sin \text{ZD}_1 \cos\phi. \end{aligned}$$

So

$$\begin{aligned} \cot \text{DEC} &= \frac{\cos \text{ZD}_1 \cos\phi + \sin \text{ZD}_1 \sin\phi}{\cos \text{ZD}_1 \sin\phi - \sin \text{ZD}_1 \cos\phi} \\ &= \frac{\cos \text{ZD}_1 \cot\phi + \sin \text{ZD}_1}{\cos \text{ZD}_1 - \sin \text{ZD}_1 \cot\phi}. \end{aligned}$$

If we now substitute the value of $\cot\phi$ from the first equation and simplify we get the second equation.

8.22 On the great circle $NPZS$ let X be the point where the star crosses. The angle $\angle XOP$ is $90 - \text{DEC}$. Then the star will never set if $90 - \text{DEC} < \angle NOP = \phi$.

8.23 The circumference of the Earth is $500 \times (360/7.5) = 24000$ miles and so the radius is $24000/2\pi$, which is about 3820 miles. This was an actual method used by Eratosthenes in 270 BC. His estimate is remarkably close to the actual value of 3963.

8.24 The full Moon is on the opposite side of the Earth to the Sun so that in the diagram for summer it is in the Sun's winter position. So the Moon is lower in the summer than in the winter.

8.26 Let ψ be the angular separation of this point in Aries from the present first point of Aries (RA zero, DEC zero). The formula for angular separation gives $\cos\psi = \cos(12)\cos(26.25) + 0 = 0.87727\ldots$. So ψ is $28.6847\ldots$ degrees. The number of years past is $(\psi \times 60 \times 60)/50.3 = 2052.9844\ldots$

8.27 Let R be the point on the horizon where the star rises. The coordinates of R are $R = (\sin \text{Az}, \cos \text{Az}, 0)$ and we have already seen that $\cos \text{Az} = \sin \text{Dec} \sec \phi$. The direction of rising is perpendicular to R and P and so is in the direction $U = R \times P = (\cos \text{Az} \sin \phi, -\sin \text{Az} \sin \phi, \sin \text{Az}, \cos \phi)$. The length of U is easily calculated to be $\cos \text{Dec}$. Since $T = (\cos \text{Az}, -\sin \text{Az}, 0)$ is a unit tangent vector to the horizon at R it follows that $\cos \psi = U \cdot T / \cos \text{Dec} = \sin \phi \sec \text{Dec}$.

8.28 80 days have passed since 1 January so GST then is incremented by $80 \times 3.94 =$ 5h 15.2m giving 23h 57.7m. So 2.3 sidereal minutes after noon Υ will cross the meridian. This is $2.3 \times 1.0027454 = 2.3063144$ civil minutes later. On or around this date each year the Sun will rise at 6 o'clock almost at W. That is the position of the Sun will coincide with Υ. Hence approximately 6h later at noon Υ the Sun will cross the meridian. The fact that this is not quite true is due to the vaguaries of the calendar.

8.29 The setting out and calculations take the same form as in the worked example

	h m
Zone time	21 00.00
less zone	03 00.00
GMT	18 00.00
less 12h	12 00.00
hours after noon	6 00.00
add ST conversion	00 00.99
sidereal hours after noon	6 00.99
GST at noon	23 49.78
GST	29 50.77
less longitude	02 25.07
LST	27 25.70
less RA of Aldebaran	04 35.90
HA of Aldebaran	22 49.80

8.30 The setting out is shown below. There are 34 days after 1 January so GST is incremented by 34×3.94m= 2h 13.96m

	h m
GMT	16 20.20
less 12h	12 00.00
hours after noon	04 20.20
add ST conversion	00 00.71
sidereal hours after noon	04 20.91
GST at noon	20 56.55
GST	25 17.46

less LST	05 14.53
longitude W	20 2.93
longitude E	03 57.07

Converting the longitude to degrees we get 59.07°. The zone time is 16h 20.2m + 4h which is 20h 20.2m.

8.31 Using formula (4) we find DEC = −5.45°. Using the other formulæ gives HA = 31.77° or 2h 7.08m. GST is 6h 20.08m and LST is 7h 42.48m giving RA = 5h 35.4m. No need to get excited, these are the coordinates of the great Orion nebula.

8.32 The angle at P is HA, the angle at Z is −AZ. The distances ZX and XP are 90 − ALT and 90 − DEC respectively.

8.33 Putting ALT = 0 in (2) yields

$$\cos \mathrm{AZ} = -\cos \mathrm{DEC} \cos \mathrm{HA} \sin \phi + \sin \mathrm{DEC} \cos \phi.$$

Exercise 8.20 found that on rising

$$\cos \mathrm{AZ} = \sin \mathrm{DEC} \sec \phi.$$

Subsituting this in the above gives the answer.

9
Quaternions and Octonions

I pulled out on the spot a pocket book, which still exists, and made an entry there and then. Nor could I resist the impulse – unphilosophical as it may have been – to cut with a knife on a stone of Brougham Bridge, as we passed it, the fundamental formula with the symbols i, j, k:

$$i^2 = j^2 = k^2 = ijk = -1,$$

which contains the solution of the Problem, but of course, as an inscription has long since mouldered away.

W.R. Hamilton

9.1 Extended Complex Numbers

We have all seen how useful complex numbers are for describing isometries of the plane. We will now look at the quaternionic numbers and how useful they are for describing rotations of $\mathbb{R}^3$. Indeed computer-game programmers have taken to using quaternions as a quick way to turn the picture round when heroes need to defend themselves.

If we think of complex number multiplication as multiplying pairs of real numbers (a, b) where $Z = a + ib$ then quaternionic multiplication is about multiplying quadruples of real numbers.

The discovery of quaternions is attributed to William Rowan Hamilton (1805–1865). However, the law of multiplication of quaternions had already been discovered by Gauss in 1820 who did not bother to publish his results.

We know that Hamilton made his discovery on 16 October, 1843 because of the entries in his notebooks and correspondence with his son and his friend John Graves (1806–1870).

Hamilton knew that multiplication of complex numbers defined a multiplication of pairs of real numbers respecting the norm or length function. Hamilton called this the *law of the moduli* and it can be written algebraically as

$$|\boldsymbol{XY}| = |\boldsymbol{X}|\,|\boldsymbol{Y}|.$$

He therefore endevoured to find a multiplication for triples. In a letter to his son Archibald he wrote:

> *Every morning in the early part of October 1843, on my coming down to breakfast, your brother William Edward and yourself used to ask me: "Well, Papa, can you multiply triples?" Whereto I was always obliged to reply, with a sad shake of the head, "No, I can only add and subtract them".*

In analogy to the complex numbers $a + ib$, Hamilton wrote his triplets as

$$a + ib + jc$$

where $i^2 = j^2 = -1$.

In multiplying the triples the difficulty is deciding on the value of the product ij. Hamilton tried various values for ij but the breakthrough came when he put $ij = k$ where k is some unknowm element whose properties can be deduced from the algebra. It quickly became apparent that k was a third root of -1 with an existence just as valid as that of i, j. Moreover in order not to reinvent the complex numbers it is necessary that $ij \neq ji$, in fact $ji = -ij$.

9.2 Multiplying Quaternions

Consider an element $Q = (t, x, y, z)$ of $\mathbb{R}^4$ which we will write in quaternion form as $t + ix + jy + kz$ where i, j, k are generalised square roots of -1. Just as there is a dichotomy between the sets $\mathbb{R}^2$ and $\mathbb{C}$ so we will be ambiguous over the rôle of $\mathbb{R}^4$ and $\mathbb{H}$ the set of quaternions. If we want to consider a quadruple of real numbers as a point of $\mathbb{R}^4$ then we write it as (t, x, y, z). If we want to consider it as a quaternion then we write it as $t + ix + jy + kz$ although this distinction will often be blurred. To decide how to multiply quaternions we define the multiplication of i, j, k by using the following table. This means that any two elements in clockwise order in $\overset{i \quad j}{k}$ multiply to give the third. Whereas in anticlockwise order they multiply to give minus the third. General quaternions are multiplied using distributivity.

$$\begin{array}{c} \\ i \\ j \\ k \end{array}\begin{array}{c} \begin{array}{ccc} i & j & k \end{array} \\ \begin{pmatrix} -1 & k & -j \\ -k & -1 & i \\ j & -i & -1 \end{pmatrix} \end{array}$$

Example 9.1

$$\begin{aligned}(4+2j)(i-k) &= 4i - 4k + 2ji - 2jk \\ &= 4i - 4k - 2k - 2i \\ &= 2i - 6k.\end{aligned}$$

Exercise 9.1

Evaluate $(2+i)(3-j+k)$.

Exercise 9.2

* Show that the elements $\{\pm 1, \pm i, \pm j, \pm k\}$ form a group of order 8, usually called the *quaternion group*. Show that quaternion multiplication is associative and that with addition the quaternions form a (non-commutative, associative) ring.

Exercise 9.3

* Let Q be any non-zero quaternion. Show that $\{Q, iQ, jQ, kQ\}$ form a basis for $\mathbb{R}^4$.

9.3 Inverses of Quaternions

The inverse of a quaternion is calculated by much the same trick as is used to find the inverse of a complex number. Define the *quaternionic conjugate* $\overline{Q}$ of a quaternion $Q = t + ix + jy + kz$ to be $\overline{Q} = t - ix - jy - kz$. The map $Q \to \overline{Q}$ can be thought of as reflection in the real line $\mathbb{R}$. The equality $\overline{Q} = Q$ holds if and only if Q is real. It can be easily verified that conjugation respects addition and anti-respects multiplication.

■ For any quaternions Q, Q' quaternionic conjugation satisfies

$$\overline{Q + Q'} = \overline{Q} + \overline{Q'}, \quad \overline{QQ'} = \overline{Q'}\,\overline{Q}.$$ □

As usual let $|Q|^2 = t^2 + x^2 + y^2 + z^2$. Then because cross terms cancel we have

$$Q\overline{Q} = \overline{Q}Q = |Q|^2.$$

This allows us to prove Hamilton's law of the moduli,

$$\begin{aligned} |QQ'|^2 &= (QQ')(\overline{QQ'}) \\ &= (QQ')(\overline{Q'}\,\overline{Q}) \\ &= Q(Q'\overline{Q'})\overline{Q} \\ &= |Q'|^2 Q\overline{Q} \\ &= |Q|^2|Q'|^2. \end{aligned}$$

Notice we have used the fact that quaternionic multiplication is associative and that real numbers commute with quaternions.

■ If $Q \neq 0$ then Q has an inverse $Q^{-1} = \overline{Q}|Q|^{-2}$. □

Exercise 9.4

Find $(2 + 3i - k)^{-1}$.

Exercise 9.5

Use the law of moduli to show that if two integers are a sum of four squares then so is their product.

Call a quaternion Q a *unit quaternion* if $|Q| = 1$. The unit quaternions lie on the unit 3-sphere, $S^3 \subset \mathbb{R}^4$.

Exercise 9.6

* Show that the set of non-zero quaternions $\mathbb{H} - \{\mathbb{O}\}$ and the set of unit quaternions S^3 form groups under quaternion multiplication.

Just as we think of a complex number as a pair of real numbers, we can think of a quaternion as a pair of complex numbers. So

$$Q = t + ix + jy + kz = Z_1 + Z_2 j \text{ where } Z_1 = t + ix,\ Z_2 = y + iz.$$

Multiplication is given by

$$(Z_1 + Z_2 j)(Z_1' + Z_2' j) = Z_1 Z_1' - \overline{Z_2'} Z_2 + (Z_2' Z_1 + Z_2 \overline{Z_1'}) j$$

or in terms of pairs

$$\boldsymbol{(Z_1, Z_2)(Z_1', Z_2') = (Z_1 Z_1' - \overline{Z_2'} Z_2, Z_2' Z_1 + Z_2 \overline{Z_1'}.)}$$

The curious order in the products above is to match that of octonion multiplication considered later.

Exercise 9.7

Show that any quaternion may be identified with the 2×2 complex matrix

$$\begin{pmatrix} Z_1 & Z_2 \\ -\overline{Z_2} & \overline{Z_1} \end{pmatrix}$$

with quaternion multiplication corresponding to matrix multiplication.

9.4 Real and Pure Parts of Quaternions

If $Q = t + ix + jy + kz$ then t is called the *real* or *temporal* part of Q and $ix + jy + kz$ is called the *pure* or *spatial* part. The decomposition $Q = (Q + \overline{Q})/2 + (Q - \overline{Q})/2$ writes each quaternion as the sum of its two parts.

A pure quaternion can be thought of as an element of space $\mathbb{R}^3$ and has the usual scalar and vector products. Two pure quaternions P, P' have an interesting formula for their quaternion product in terms of the other products namely

$$\boldsymbol{PP' = -P \cdot P' + P \times P'.}$$

So $-P \cdot P'$ is the real part of PP' and $P \times P'$ is the pure part.

If $Q = t + P$ and $Q' = t' + P'$ then the general formula is

$$\boldsymbol{QQ' = tt' - P \cdot P' + tP' + t'P + P \times P'.}$$

The pure unit quaternions lie in the sphere $S^2 \subset \mathbb{R}^3$. If $H \in S^2$ then $H \cdot H = 1$ and so $H^2 = -H \cdot H + H \times H = -1$. Conversely if Q is a quaternion which satisfies $Q^2 = -1$ then clearly Q is of unit length and $Q^{-1} = -Q = \overline{Q}$ so Q is also pure. We may sum this up as follows.

■ For quaternions $\sqrt{-1} = S^2$. □

Just as for complex numbers a quaternion Q can be written in *polar form:*

$$\boldsymbol{Q = r(\cos\theta + H\sin\theta)} \textbf{ where } \boldsymbol{r = |Q|} \textbf{ and } \boldsymbol{H^2 = -1}$$

The angle θ is a generalisation of the argument of a complex number. A quaternion is real if and only if its argument is 0 or π.

If Q is a non-real quaternion let $\mathbb{C}_Q$ denote the two-dimensional subspace of R^4 spanned by $1, H$ where H is as above.

■ Every non-real quaternion determines a copy of the complex numbers $\mathbb{C}_Q$. If $Q = r(\cos\theta + H\sin\theta)$ then $\mathbb{C}_Q$ is isomorphic to $\mathbb{C}$ by the correspondence

$$x + iy \leftrightarrow x + Hy.$$ □

Exercise 9.8

Define the *commutator* of two quaternions $[Q, Q']$ by the rule $[Q, Q'] = QQ' - Q'Q$. So $[Q, Q'] = 0$ if and only if Q, Q' commute. Show that $\overline{[Q, Q']} = -[Q, Q']$ and that $[Q, Q']$ is a pure quaternion.

Deduce that Q, Q' commute if and only if either one is real or if $Q' \in \mathbb{C}_Q$.

9.5 Multiplying Quaternions and Linear Transformations of $\mathbb{R}^4$

Let $Q = t + ix + jy + kz = r(\cos\theta + H\sin\theta)$ be a fixed non-zero quaternion. Then there are two linear transformations of $\mathbb{R}^4$ determined by right and left multiplication by Q, namely $R_Q(X) = XQ$ and $L_Q(X) = QX$. The matrix of R_Q has rows determined by its action on $1, i, j, k$ respectively; that is

$$\begin{pmatrix} t & x & y & z \\ -x & t & -z & y \\ -y & z & t & -x \\ -z & -y & x & t \end{pmatrix}.$$

Since multiplication by a real number corresponds to a scale change we may assume that Q is non-real and for similar reasons that Q has unit length, so $r = 1$. In this case it turns out that R_Q and L_Q have two mutually orthogonal invariant subspaces on which they act by rotations.

We shall now describe these subspaces explicitly. One is $\mathbb{C}_Q$ defined earlier and the other is its orthogonal complement $\mathbb{C}_Q^{\perp}$. This latter subspace is the set of all combinations

$$\mathbb{C}_Q^{\perp} = xH' + yH''$$

where x, y are arbitary real numbers and $\{H', H''\}$ is a basis of the two dimensional subspace of $\mathbb{R}^3$ orthogonal to H. It will be convenient to assume that $\{H, H', H''\}$ is a right-handed orthonormal basis and so $H'' = H \times H'$.

Let $X = x + yH \in \mathbb{C}_Q$, then a simple calculation gives

$$XQ = x\cos\theta - y\sin\theta + (x\sin\theta + y\cos\theta)H$$

which also lies in $\mathbb{C}_Q$ and represents a rotation through θ about the origin.

Let $X = xH' + yH''$ then a calculation, using the vector triple product expansion, shows that

$$XQ = (x\cos\theta + y\sin\theta)H' + (-x\sin\theta + y\cos\theta)H''$$

which lies in $\mathbb{C}_Q^{\perp}$ and represents a rotation through $-\theta$ about the origin.

A similar calculation can be done for left multiplication. Putting the two results together we have the following.

■ Right multiplication by a non-real unit quaternion Q rotates $\mathbb{C}_Q$ through $+\theta$ and $\mathbb{C}_Q^{\perp}$ through $-\theta$.

Left multiplication consists of rotations of $\mathbb{C}_Q$ and $\mathbb{C}_Q^{\perp}$ both through θ. □

Exercise 9.9

Show that the matrix of L_Q is

$$\begin{pmatrix} t & x & y & z \\ -x & t & z & -y \\ -y & -z & t & x \\ -z & y & -x & t \end{pmatrix}.$$

Combine the two multiplications and define the operation M_Q by

$$\boldsymbol{M_Q(X) = QX\overline{Q}}$$

where $|Q| = 1$. Then M_Q is a combination of a left and right multiplication. The order is immaterial since quaternion multiplication is associative.

The above result means that M_Q can be described as a rotation of space.

▮ The linear map M_Q rotates $\mathbb{C}_Q^{\perp}$ through 2θ and is fixed on $\mathbb{C}_Q$. In particular the reals $\mathbb{R}$ are kept fixed and the pure quaternions $\mathbb{R}^3$ are rotated about H through 2θ.

Conversely any rotation of $\mathbb{R}^3$ is the restriction of some M_Q where Q is unique up to sign. □

Exercise 9.10

Find the quaternion Q such that M_Q is the rotation of $\mathbb{R}^3$ given in coordinate form by $X \to X\mathbf{M}$ where M is the matrix

$$\mathbf{M} = \begin{pmatrix} 0 & 1 & 0 \\ 0 & 0 & 1 \\ 1 & 0 & 0 \end{pmatrix}.$$

The linear map M_Q defines a representation, r, of the unit quaternions S^3 onto the group of orthogonal rotations of $\mathbb{R}^3$. Denoting this group of rotations by $SO(3)$ the homomorphism $r : S^3 \to SO(3)$ is a topological double cover.

The group $SO(3)$ is called a *Lie group* after Sophus Lie (1842–1899). This is because it is a manifold (of dimension 3) and the group operations are continuous (in fact C^∞). A description of $SO(3)$ as the 3-manifold RP^3, the real projective space, can be given as follows. Let B_π denote the 3-ball of radius π, so $B_\pi = \{X \in \mathbb{R}^3 | \, |X| \leq \pi\}$. Associate to $X \in B_\pi$ the identity element of $SO(3)$ if $X = O$. Otherwise associate to X the rotation about OX through an angle $|X|$ in a positive screw sense. If $|X| = \pi$ then the two rotations corresponding to X and $-X$ are equal. So we have

▮ The space $SO(3)$ is obtained from a 3-ball by identifying antipodal points on the boundary. □

Exercise 9.11

Can every orthogonal linear map of $\mathbb{R}^4$ with determinant 1, be realised by $X \to Q'XQ$ for some quaternions Q, Q'?

9.6 Octonions

Octonion multiplication was discovered by Graves, Hamilton's friend, in 1843. Octonions were independently discovered in 1845 by Arthur Cayley (1821–1895) and are sometimes called Cayley numbers in consequence.

In analogy with complex numbers and quaternions we can write any element $(t, x, y, z, s, u, v, w) \in \mathbb{R}^8$ as an octonion (or Cayley number)

$$A = t + ix + jy + kz + ls + mu + nv + ow$$

where multiplication is defined by the following table.

$$\begin{array}{c} \\ i \\ j \\ k \\ l \\ m \\ n \\ o \end{array}\begin{array}{c} \begin{array}{ccccccc} i & j & k & l & m & n & o \end{array} \\ \begin{pmatrix} -1 & k & -j & m & -l & -o & n \\ -k & -1 & i & n & o & -l & -m \\ j & -i & -1 & o & -n & m & -l \\ -m & -n & -o & -1 & i & j & k \\ l & -o & n & -i & -1 & -k & j \\ o & l & -m & -j & k & -1 & -i \\ -n & m & l & -k & -j & i & -1 \end{pmatrix}\end{array}$$

Alternatively the points of the Fano plane can be labelled by the seven generators and the lines of the Fano plane can be oriented so that the product of any two is the third on the line with sign determined by the orientation of the line. For example the line containing n, i and o is oriented from right to left so that $ni = o, no = -i$, etc.

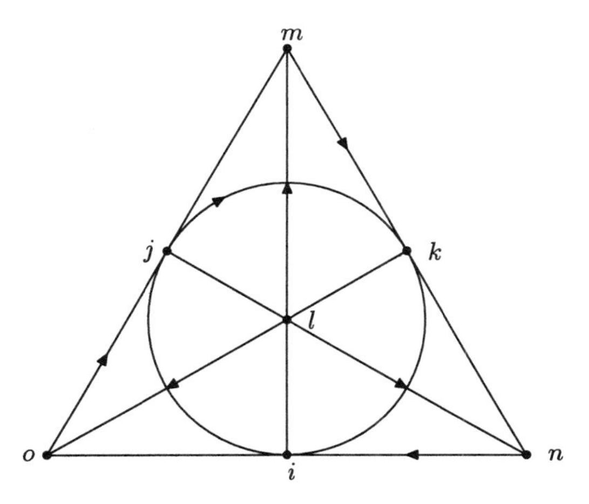

Fig. 9.1 Multiplication of octonions using the Fano plane

Things to note about the set $\{\pm 1, \pm i, \pm j, \pm k, \pm l, \pm m, \pm n, \pm o\}$ are

- it is closed under multiplication
- every element has a multiplicative inverse
- multiplication is anti-symmetric
- multiplication is not always associative, e.g. $(jk)l = -j(kl)$.

We can work out products without resorting to the above if we use the following rules. Split the set $\{i, j, k, l, m, n, o\}$ into $\{i, j, k\}$ and $\{l, m, n, o\} = \{1, i, j, k\}l$. The first set multiplies as for quaternions and in general $ab = -ba$ if a, b are distinct. For products of three elements $(ab)c = \pm a(bc)$. The plus sign is taken if a, b, c are quaternions or have a repetition. Otherwise the minus sign is taken. For example $kn = k(jl) = -(kj)l = -(-i)l = il = m$ or $ko = k(kl) = k^2l = -l$.

The above means we can write any octonion as

$$\boldsymbol{A} = \boldsymbol{Q} + \boldsymbol{R}l = \boldsymbol{Q} + l\overline{\boldsymbol{R}}$$

where $Q = t + ix + jy + kz$ and $R = s + iu + jv + kw$ are quaternions. Then multiplication is given by

$$(\boldsymbol{Q} + \boldsymbol{R}l)(\boldsymbol{Q}' + \boldsymbol{R}'l) = \boldsymbol{Q}\boldsymbol{Q}' - \overline{\boldsymbol{R}'}\boldsymbol{R} + (\boldsymbol{R}'\boldsymbol{Q} + \boldsymbol{R}\overline{\boldsymbol{Q}'})l.$$

We must be careful when using distributivity in the above example to remember that multiplication is non-associative. The following rules will help,

$$Q(R'l) = (R'Q)l, \qquad (Rl)Q' = (R\overline{Q'})l, \qquad (Rl)(R'l) = -\overline{R'}R.$$

As with quaternions, define the conjugate of an octonion by

$$\overline{t + ix + jy + kz + ls + mu + nv + ow} = t - ix - jy - kz - ls - mu - nv - ow$$

or

$$\overline{Q + Rl} = \overline{Q} - Rl.$$

We have the following.

■ For octonions, modulus and conjugation are related by the following laws.

$$\boldsymbol{A}\overline{\boldsymbol{A}} = |\boldsymbol{A}|^2, \qquad \overline{\boldsymbol{A}\boldsymbol{B}} = \overline{\boldsymbol{B}}\,\overline{\boldsymbol{A}}, \qquad |\boldsymbol{A}\boldsymbol{B}| = |\boldsymbol{A}||\boldsymbol{B}|$$

Proof The first identity is easy to verify. The others will follow from general results in Section 9.9. □

■ Every non-zero octonion A has a multiplicative inverse $\overline{A}/|A|^2$. □

Exercise 9.12

Find the inverse to $2 + i - m + 3o$.

As with complex numbers and quaternions, any octonion can be written in polar form as $A = r(\cos\theta + H\sin\theta)$ where $r = |A|$ and $H^2 = -1$. So if A is not real then it defines a copy of the complex numbers $\mathbb{C}_A$ which is the two-dimensional space spanned by A and 1.

Exercise 9.13

Show that for octonions $\sqrt{-1} = S^6$.

9.7 Vector Products in $\mathbb{R}^7$

Seven-dimensional euclidean space and three-dimensional euclidean space are the only euclidean spaces to have a vector product. We can write an octonion as the sum of a real part, t and a pure part, P in $\mathbb{R}^7$. So $A = t + P$. This leads to a vector product on $\mathbb{R}^7$ defined by $P \times P' = PP' + P \cdot P'$. In coordinate form this is given by

$$(x_1, x_2, x_3, x_4, x_5, x_6, x_7) \times (y_1, y_2, y_3, y_4, y_5, y_6, y_7) =$$

$$\begin{array}{cccccccc}
(& & +x_2y_3 & -x_3y_2 & +x_4y_5 & -x_5y_4 & -x_6y_7 & +x_7y_6, \\
 & -x_1y_3 & & +x_3y_1 & +x_4y_6 & +x_5y_7 & -x_6y_4 & -x_7y_5, \\
 & +x_1y_2 & -x_2y_1 & & +x_4y_7 & -x_5y_6 & +x_6y_5 & -x_7y_4, \\
 & -x_1y_5 & -x_2y_6 & -x_3y_7 & & +x_5y_1 & +x_6y_2 & +x_7y_3, \\
 & +x_1y_4 & -x_2y_7 & +x_3y_6 & -x_4y_1 & & -x_6y_3 & +x_7y_2, \\
 & +x_1y_7 & +x_2y_4 & -x_3y_5 & -x_4y_2 & +x_5y_3 & & -x_7y_1, \\
 & -x_1y_6 & +x_2y_5 & +x_3y_4 & -x_4y_3 & -x_5y_2 & +x_6y_1 &).
\end{array}$$

The vector product in $\mathbb{R}^7$ has all the properties enjoyed by its three-dimensional cousin except one, namely the Jacobi identity. So in general for pure octonions

$$P \times (P' \times P'') + P' \times (P'' \times P) + P'' \times (P \times P') \neq O$$

or equivalently the triple vector product identity fails

$$P \times (P' \times P'') \neq (P \cdot P'')P' - (P \cdot P')P''.$$

The identities which it does satisfy are as follows.

■ Let P, P', P'' be pure octonions and let θ be the angle between P and P'. If x is an arbitary real number then the following identities hold for vector products in $\mathbb{R}^7$.

$$P \times (P' + P'') = P \times P' + P \times P'' \tag{1}$$
$$P \times P = O \tag{2}$$
$$P \times P' = -P' \times P \tag{3}$$
$$x(P \times P') = (xP) \times P'P \cdot (P \times P') = 0 \tag{4}$$
$$P \cdot (P \times P') = 0 \tag{5}$$
$$|P \times P'| = |P|\,|P'| \sin\theta \tag{6}$$
$$(P \times P') \cdot P'' = (P' \times P'') \cdot P = (P'' \times P) \cdot P'' \tag{7}$$
$$P \times (P \times P') = (P \cdot P')P - (P \cdot P)P'. \tag{8}$$

Proof Identity (1) is straightforward. For (2) we have $P \times P = P \cdot P + P^2 = P \cdot P - P \cdot P = O$. Expanding the zero vector $(P + P') \times (P + P')$ gives $P \times P' + P' \times P$ and proves (3). Identity (4) is straightforward. For (5) the expression $P \cdot (P \times P')$ can be verified to be zero by using the explicit formula above for the cross product and checking that each term such as $+x_2y_3x_1$ occurs also as $-x_1y_3x_2$. Consider now identity (6). We may assume that P and P' have unit length. If they are orthogonal then $\sin\theta = 1$ and

$$|P \times P'|^2 = (P \cdot P' + PP')\overline{(P \cdot P' + PP')} = (PP')\overline{(PP')} = |P|\,|P'|.$$

Otherwise $P'' = P' - \cos\theta P$ is orthogonal to P and has length $\sqrt{P'' \cdot P''} = \sin\theta$. Then $|P \times P'| = |P \times P''| = \sin\theta$. For the general case let $P = |P|\hat{P}$ and $P' = |P'|\hat{P'}$ where $\hat{P}$ and $\hat{P'}$ have unit length. So

$$|P \times P'| = |P||P'||\hat{P} \times \hat{P'}| = |P|\,|P'| \sin\theta.$$

Identity (7) is obtained from (4) by replacing P by $P + P''$ and expanding. Finally consider (8). Identity (6) may be written as

$$(P \times P') \cdot (P \times P') = (P \cdot P)(P' \cdot P') - (P \cdot P')^2.$$

If we replace P' by $P' + P''$ in the above and expand out the terms we deduce that

$$(P \times P') \cdot (P \times P'') = (P \cdot P)(P' \cdot P'') - (P \cdot P')(P \cdot P'').$$

Using (7) this can be written as

$$(P \times (P \times P')) \cdot P'' = ((P \cdot P')P - (P \cdot P)P') \cdot P''.$$

Since P'' is arbitrary, (8) follows. □

The last identity (8) will be important to us. It implies that any two independant pure octonians P, P' span a three-dimensional space with $P \times P'$ which has all the vector product properties of $\mathbb{R}^3$.

9.8 Octonions and Associativity

Although octonians are not associative in general, for example $(jk)l = -j(kl)$, nevertheless there are special useful circumstances where associativity holds. For example $(AA)B = A(AB)$ always holds. So powers such as A^n make sense. This is a special case of the following result.

■ Any two octonians define an associative subalgebra.

Proof We have seen that two independant vectors in $\mathbb{R}^7$ define with their vector product a copy of $\mathbb{R}^3$. It follows that any two octonians and all their products lie in a copy of the quaternions which have associative products. □

Define the *associator* of three octonions $[A, B, C]$ by the rule

$$[\boldsymbol{A}, \boldsymbol{B}, \boldsymbol{C}] = (\boldsymbol{AB})\boldsymbol{C} - \boldsymbol{A}(\boldsymbol{BC}).$$

So $[A, B, C] = 0$ if and only if A, B, C associate. As we have seen above, $[A, A, C] = 0$. Expanding the associator identity $[A + B, A + B, C] = 0$ we have

$$[A, B, C] + [B, A, C] = 0.$$

Similarly

$$[A, B, C] = -[A, C, B] = -[C, B, A].$$

That is, the associator is *alternating*. Another useful property is

$$\begin{aligned} \overline{[A, B, C]} &= \overline{(AB)C - A(BC)} \\ &= \overline{C}(\overline{B}\overline{A}) - (\overline{C}\overline{B})\overline{A} \\ &= -[\overline{C}, \overline{B}, \overline{A}] \\ &= [\overline{A}, \overline{B}, \overline{C}]. \end{aligned}$$

Using the fact that $[A, B, C] = 0$ if any factor is real we have $[A + \overline{A}, B, C] = 0$ or

$$[\overline{A}, B, C] = -[A, B, C].$$

So $\overline{[A, B, C]} = [\overline{A}, \overline{B}, \overline{C}] = -[A, B, C]$. That is, $[A, B, C]$ is always pure.

Exercise 9.14

Show that

$$\begin{aligned}[AB, C] - A[B, C] - [A, C]B &= [A, B, C] - [A, C, B] + [C, A, B] \\ &= 3[A, B, C].\end{aligned}$$

Exercise 9.15

Show that the super associator defined as

$$[A, B, C, D] = [AB, C, D] - B[C, D, A] - [B, C, D]A$$

is also alternating.

9.9 Hexadecanions?

One consequence of the law of moduli is that

$$AB = O \text{ implies that either } A = O \text{ or } B = O.$$

The technical name for this is that the reals, complex numbers, quaternions and octonions are all examples of *division algebras*. So the obvious question is: can we make $\mathbb{R}^{16}$ into a division algebra? The answer is no. The proof requires quite deep results from algebraic topology and was proved by Adams in England and Bott and Milnor in the USA.

Let us look at the problem of defining a multiplication on $\mathbb{R}^{2^n}$ (2^n-onions) using a multiplication defined on $\mathbb{R}^{2^{n-1}}$ (2^{n-1}-onions). As n increases so the properties decline. For example the complex numbers, $\mathbb{C}$, do not form an ordered field, the quaternions $\mathbb{H}$ are not commutative, the octonions are not associative and the hexadecanions do not form a division algebra. Moreover this last fact is a direct consequence of the fact that the octonions are not associative.

Let R be an algebra with elements denoted by A, B, C, etc. It has addition and multiplication which distributes over addition. We do not assume that the multiplication is commutative or associative but we allow real numbers to commute with any element of R. We also assume a "conjugation", $A \to \overline{A}$ satisfying

$$\overline{A + B} = \overline{A} + \overline{B}, \quad \overline{AB} = \overline{B}\,\overline{A}$$

,

$$\overline{\overline{A}} = A \text{ and } \overline{A} = A \text{ if and only if } A \text{ is real.}$$

The "square length" of any element is defined by $|A|^2 = A\overline{A} = \overline{A}A$ and is always required to be real and positive (unless $A = O$ when $|A| = 0$).

The commutator $[A, B] = AB - BA$ and the associator $[A, B, C] = (AB)C - A(BC)$ have all the properties such as being alternating that we proved earlier.

Let R^2 be the set of pairs of R with addition and multiplication defined by

$$(A, B) + (C, D) = (A + B, C + D), \ (A, B)(C, D) = (AC - \overline{D}B, DA + B\overline{C}).$$

The real numbers are embedded in R^2 as the set of all pairs (A, O) where A is real.

Conjugation is defined by

$$\overline{(A, B)} = (\overline{A}, -B).$$

It is now the work of a moment to check that

$$\overline{(A, B) + (C, D)} = \overline{(A, B)} + \overline{(C, D)} \text{ and } \overline{(A, B)(C, D)} = \overline{(C, D)}\ \overline{(A, B)}.$$

The square length of a pair is

$$|(A, B)|^2 = (|A|^2 + |B|^2, O)$$

and so

$$\begin{aligned} (A, B)\overline{(A, B)} &= (A, B)(\overline{A}, -B) \\ &= (A\overline{A} + \overline{B}B, -BA + BA) \\ &= (A\overline{A} + B\overline{B}, O) \\ &= |(A, B)|^2. \end{aligned}$$

In particular every non-zero element has an inverse given by

$$(A, B)^{-1} = \overline{(A, B)}/|(A, B)|^2.$$

The commutator of two pairs is

$$[(A, B), (C, D)] = ([A, C] - (\overline{D}B - \overline{B}D), DA - D\overline{A} + B\overline{C} - BC).$$

This will always be zero only if R is commutative and $\overline{A}$ is always A. For example if R is the reals and R^2 is the complex plane.

Now let us look at associativity:

$$\begin{aligned} ((A, B)(C, D))(E, F) &= (AC - \overline{D}B, DA + B\overline{C})(E, F) \\ &= ((AC)E - (\overline{D}B)E - \overline{F}(DA) - \overline{F}(B\overline{C}), \\ &\qquad F(AC) - F(\overline{D}B) + (DA)\overline{E} + (B\overline{C})\overline{E}). \end{aligned}$$

On the other hand,

$$\begin{aligned} (A, B)((C, D)(E, F)) &= (A(CE) - A(\overline{F}D) - (\overline{C}\,\overline{F})B - (E\overline{D})B, \\ &\qquad (FC)A + (D\overline{E})A + B(\overline{E}\,\overline{C}) - B(\overline{D}F)). \end{aligned}$$

So in order for R^2 to be associative clearly R must be associative and commutative.

Finally consider the law of moduli. Suppose that R is associative and a division algebra. Then

$$\begin{aligned}|(A,B)(C,D)|^2 &= (A,B)(C,D)\overline{(A,B)(C,D)} \\ &= (AC-\overline{D}B, DA+B\overline{C})(\overline{C}\,\overline{A}-\overline{B}D, -DA-B\overline{C}) \\ &= (|A|^2|C|^2+|A|^2|D|^2+|B|^2|C|^2+|B|^2|D|^2+X, O)\end{aligned}$$

where

$$\begin{aligned}X &= -AC\overline{B}D + C\overline{B}DA + \overline{A}\,\overline{D}B\overline{C} - \overline{D}B\overline{C}\,\overline{A} \\ &= -[A, C\overline{B}D] + [\overline{A}, \overline{D}B\overline{C}] \\ &= -[A, C\overline{B}D] + [A, \overline{\overline{D}B\overline{C}}] \\ &= O\end{aligned}$$

by the properties of commutators. So the law of moduli holds for R^2 which is therefore a division algebra.

On the other hand, note that

$$(i,l)(-k,n) = (O,O)$$

so hexadecanions do not form a division algebra.

Answers to Selected Questions in Chapter 9

9.1 $6+3i-3j+k$.

9.4 $(2-3i+k)/14$.

9.5 The law of moduli can be written

$$\begin{aligned}(a^2+b^2+c^2+d^2)(x^2+y^2+z^2+t^2) &= (ax-by-cz-dt)^2 \\ &\quad + (ay+bx+ct-dz)^2 \\ &\quad + (az-bt+cx+dy)^2 \\ &\quad + (at+bz-cy+dx)^2.\end{aligned}$$

Clearly if each factor is a sum of four squares then so is their product.

9.7 Having identified the quaternion $Q = t+ix+jy+kz$ with the pair of complex numbers Z_1+Z_2j where $Z_1 = t+ix$, $Z_2 = y+iz$ we can further identify this pair of complex numbers with the matrix above. The correspondence respects addition and multiplication of quaternions. The result is an isomorphism of the ring of quaternions with a subring of 2×2 complex matrices.

9.8 With the usual notation $[Q, Q'] = 2\sin\theta\sin\theta' H \times H'$ and the first part of the question follows easily.
$[Q, Q']$ is zero only if $\sin\theta = 0$ or $\sin\theta' = 0$ (Q or Q' real) or $H \times H' = O$ (H and H' are dependant). Since $\mathbb{C}_Q$ is algebraically like the complex plane and complex numbers commute it follows that anything in $\mathbb{C}_Q$ commutes with Q. Conversely suppose $Q' \notin \mathbb{C}_Q$. Then $H \times H' \neq O$ and Q, Q' cannot commute.

9.10 $Q = (1 + i + j + k)/2$.

9.12 $(2 - i + m - 3o)/15$.

Bibliography

J.F. Adams, Vector Field on Spheres, Bull. Amer. Math. Soc. 68(1962) 30–41

M. Berger, Geometry I and II (1987) Springer-Verlag. *An encyclopaedic trip through geometry.*

R. Bott and **J. Milnor**, On the parallelisability of spheres, Bull. Amer. Math. Soc. 64 (1958) 87–89

C. Caratheodory, Theory of Functions of a Complex Variable, Volume 1 (1958) Chelsea. *Contains a masterly introduction to complex numbers.*

H.S.M. Coxeter, Introduction to Geometry (1961) Wiley. *The classic book on elementary geometry.*

H.S.M. Coxeter, Regular Polytopes (1978) Dover. *All you wanted to know and more about polyhedra.*

W.L. Edge, The Theory of Ruled Surfaces (1931) Cambridge

D. Fowler, The Mathematics of Plato's Academy, A New Reconstruction, 2nd augmented edition (1999), Oxford University Press

L. Hogben, Mathematics for the Million (1942) Unwin. *An interesting and idiosyncratic book full of marvellous examples and diagrams. First published in 1936 and still in print.*

G. Ifrah, The Universal History of Numbers (1998) Harvill Press. *A fascinating history of numbers and their notation.*

C.G. Jung, Synchronicity – An Acausal Connecting Principle (1991) Arc Paperbacks. *In this book Jung argues that space and time only become fixed concepts when they are measured (surely not an original thought). In other*

words they are psychic in origin. Events in the world, causal or otherwise, can be assigned a **tao** *or meaning. Synchronicity is an acausal juxtaposition of two meaningfully related events. In this category come coincidences, predictive dreams and possibly astrology. Synchronicitic events being at a more primitive or deeper level are beyond space and time. For example if I say that someone is born under Mars and is therefore warlike my statement may have no rational basis but may be statistically true.*

D.E. Knuth, The TeXbook (1984) Addison-Wesley. *The book about the computer program which made life so much easier for mathematicians.*

S. Lipschutz, Linear Algebra (1968) Schaum. *All you need to know about linear algebra. Nicely written with lots of examples and cheap!*

H. Rademacher, Topics in Number Theory (1973) Springer–Verlag

D. Sobel, Longitude (1998) Fourth Estate. *A readable account of the search for a solution to the thorniest scientific problem of the 18th century: how to determine longitude. Many thousands of lives had been lost at sea over the centuries due to the inability to determine an east–west position. This is the story of the clockmaker, John Harrison, who solved the problem.*

J.A. Todd, Projective and Analytical Geometry (1947) Pitman

B.L. van der Waerden, A History of Algebra (1985) Springer–Verlag

Index

$a, b, c, \ldots$ real numbers 217
a major axis of ellipse or planet's orbit 232
x, y, z spatial coordinates 137
f_n Fibonacci numbers 23
a, b, c sides of a triangle 88
$b^2 - 4ac$ quadratic discriminant 40
d directrix 225
ℓ a line 54
i complex number $\sqrt{-1}$ 100
i, j, k quaternions 287
i, j, k, l, m, n, o octonions 295
l semi-latus rectum 225
m slope of a line 52
$n = 2\pi/p$ rate of a planet 235
p period of a planet 235
$\{p\}, \{\frac{p}{q}\}$ Schläfli symbols 70,71
r modulus of complex number 105
r radius of incircle 89
r_A, r_B, r_C radii of outcircles 89
s semiperimeter 88
$A, B, C, \ldots$ points in some euclidean space 30
except R radius of circumcircle 89
D, E, F feet of altitudes 84
G centroid 83
H orthocentre 84
I, J, K unit points on the axes in $\mathbb{R}^3$ 139
I incentre 86
I_A, I_B, I_C outcentres 86
N nine point circle centre 91
O origin 30 or circumcentre 83
$\hat{P}$ point with unit modulus 139
Q quaternion 99, 287
S^1 unit circle 104
S^2 unit sphere 253
S^n n dimensional unit sphere 254
S, T transformations 108
Z zenith 272
$A \times B$ vector or cross product 142, 291
$[ABC]$ scalar triple product 144
$|X|$ length of X 38
$|X - Y|$ distance between X and Y 38
$X \cdot Y$ dot product of X and Y 39
$\overrightarrow{PQ}$ vector from P to Q 32
Z complex number 100
$\bar{Z}$ complex conjugate 101
$\bar{Q}$ conjugate of a quaternion 289
$\bar{A}$ conjugate of an octonion 296
$|Z|$ modulus 103
$\mathbf{E}$ identity matrix 100, 160
$\mathbf{M}$ general 3×3 matrix 160
Möb group of Möbius transformations 120
A= $[A]$ etc, points in projective plane 184
XYZ triangle of reference 184
U= $[1, 1, 1]$ unit point 184
$\angle XOY$ angle between the lines OX and OY 53
α true anomaly 233
β eccentric anomaly 233
α, β, γ typically internal angles of a triangle 67
$\hat{\alpha}, \hat{\beta}, \hat{\gamma}$ external angles of a triangle 67
ε eccentricity 225
r, θ polar coordinates of a point in the plane 44
$re^{i\theta}$ polar form of a complex number 105

ρ, θ, ϕ spherical polar coordinates of a point in space 140
λ eigenvalue 157, 202, 224, 245
λ extreme latitude 269
ψ length of geodesic 255
ϕ latitude 260
θ argument of complex number 104
θ longitude 260
τ golden ratio 21
ϖ a plane 151
Υ first point of Aries 276
∞ infinity for $\mathbb{R}$ 16 or $\mathbb{C}$ 111
$\mathbb{C}$ set of complex numbers 100
$\mathbb{C}^+$ extended complex plane 111
$\mathbb{E}$ euclidean plane 185
$\mathbb{N}$ set of natural numbers 2
$\mathbb{Q}$ set of rational numbers 13
$\mathbb{R}$ set of real numbers 15
$\mathbb{R}^n$ euclidean space of dimension n 30
$\mathbb{Z}$ set of integers 10
$\mathbb{Z}_2P^2$ Fano plane 207, 295
$\mathcal{E}$ equator 261
$\begin{vmatrix} a_1 & a_2 & a_3 \\ b_1 & b_2 & b_3 \end{vmatrix} = [a_2b_3 - a_3b_2, a_3b_1 - a_1b_3, a_1b_2 - a_2b_1]$ 186
(AB, CD) cross ratio 120, 190
$\mathsf{A}\wedge\mathsf{B}$ projectivity 195
$\mathsf{A} \overset{\mathsf{P}}{\wedge} \mathsf{B}$ perspectivity 195
$[Q, Q']$ commutator of two quaternions 292
$[A, B, C]$ associator of three octonions 299
ALT altitude 273
AZ azimuth 273
DEC declination 273
GMT Greenwich mean time 278
GST Greenwich sidereal time 278
LST local sidereal time 277
HA hour angle 278
NPD north polar distance 273
RA right ascension 276, 277
ZD zenith distance 273
$\in$ obliquity 275

Abel 133
absolute 41
acute-angled 68
Adams 300
adjacent angles 43, 65
affine 38, 108
Alberti, L.B. 206
alternate and corresponding angles 66
altitude 84, 273
angle 40–42
angle bisector 79
angle trisector 94
angular separation 270
Antarctic circle 275
antipode 253
apex 167, 193, 212
aphelion 232
apostrophus 3
Arabs 2
Archimedes 63, 67, 170
Arctic and Antarctic circles 261
Arctic circle 275
area 48, 260
area of a circle 50
area of a lune 258
area of a triangle 48
area of the ellipse 218
area of the parallelogram 50, 144
area of the triangle 144
argument 104
Aristarchus of Samos 271
associative 101
associativity 299
associativity 5, 34
associator 299
astronomical unit 235, 270
asymptote 31, 220
asymptotic surface 242
autumnal equinox 275
auxiliary circle 217, 233
axis 193, 212, 218
azimuth 273

Babylonians 2
barycentre 84
base 167
based 32
basis 35, 139, 144
Bible 20
binary 2
Borromean rings 174
Bosse, A. 188
Bott 300
Brahe, T. 232
branches 219

calculus 3, 30
calendar 274
Cardano 99
cartesian coordinates 137
Cassegrain focus 231
casus irreducibilis 132
Cauchy–Schwarz 39

Cayley, A. 295
celestial equator 273
celestial sphere 270
centre 78
centre of gravity 84
centre of inversion 115
centroid 83
Chinese 6
chord 78
cipher 9
circle 19, 78, 107, 212, 215
circular measure 43
circumcentre 83
circumcircle 83
circumference of a circle 18
circumpolar 273
civil 274
codimension 37
collinear 43
commutative 101
commutativity 34
commutator 292
compass bearings 264
complex conjugation 101
complex number 99
complex number 99, 100
compound 5
concurrent 67
cone 167, 212
configurations 54
configurations of lines 151
conformal 114
congruences 33
congruences and similarities 73
conic sections 211
constant bearing 267
continued fraction 13
contour 238
convex hull 37
coordinate axes 138
coordinates 29
Copernicus 231
cosine 88
cosine rule 257
Coxeter group 160
Cramer 51
cross product 142
cross ratio 120, 190
cube 7, 170
cubic and biquadratic polynomials 130
cuboctahedron 170
culminate 277
cyclic 70
cyclic quadrilateral 82

Dürer, A. 204
Dantzig 9
date line 261
decagon 70
declination 270, 273
degenerate conic 213
Desargues' Theorem 188
Desargues, G. 188
Descartes, R. 29
determinant 144
diameter 78
dilation 107
dimension 35
direct 73, 107, 155
direction cosines 139
directrix 225
discriminant 40
distance 38, 139
distributive 101
distributivity 5
division 107
division algebras 300
dodecahedron 171
dot product 39, 141
drawing the ellipse 227
dual 167
duality 186

eccentric anomaly 233
eccentricity 225
ecliptic 274
edges 69
Egyptian 7
eigenvalue 202, 245
eigenvector 245
eigenvectors 202
ellipse 212, 217
ellipsoid 238
elliptic paraboloids 242
equator 261
equator 253
equatorial plane 253
equiangular spiral 47
equianharmonic set 124
equilateral 67, 166
equinox 275
equivalence class 11
equivalence relation 11, 42
Eratosthenes 63
euclidean algorithm 13
Euclidean Plane 63
Euler 8, 21, 99

Euler line 85
Euler's constant 24
extended 111

Fano Plane 207
Fano plane 295
feet 32
de Fermat, P. 30
Fermat's problem 92
Fibonacci numbers 23
field 14
fifth 64
first point of Aries 276
fixed point 202
focus 225
foot 84
footprints 74

Galois 133
Gauss 288
generalised plane 37
generating 35
generator 212
geodesic 255
geodesic triangle 256
geoid 239
Giza 7
glide reflection 73, 108
God 2, 20
god 232
golden ratio 21, 22, 173
graph 31
Graves, J. 288
great circle 253
Greenwich 261
Greenwich mean time 278
Greenwich sidereal time 278
group 34, 42
guitar 46

half turn 32, 33, 36
Hamilton, W.R. 288
harmonic 199
harmonic set 123
head 32
heptagon 70
hexadecanions 300
hexagon 70
hexahedron 166
homeomorphism 33, 36
homogeneous coordinates 184
homology theory 170
horizon 272
hour angle 278
Housman, A.E. 4
hyperbola 212, 219
hyperboloid of one sheet 240
hyperboloid of two sheets 241
hyperplane 37, 51, 55
hypersphere 39
hypotenuse 67

icosahedron 171
Ifrah, G. 1
imaginary 102
incentre 86
incircle 86
India 2
indirect 73, 107
indirect isometries 157
induction 2
infinity 31, 111
integers 2
interior and exterior angles of a triangle 67
intersection of two lines 52
inverse 34, 103
inversion 114, 115
iron pyrites 171
isometries 33, 73
isometries of space 154
isometry 108
isosceles 67
isosceles triangles 76

Jacobi identity 297
Jung, C.G. 38

Kepler 232
Kepler's equation 235
Klein 74
knot 262
Kronecker, L. 2

latitude 260
latus rectum 225
law of the moduli 288
laws 232
laws of indices 9
leap year 274
left handed 139
lemniscate 45, 46
Lie group 42, 294
Lie, S. 294
limaçon 46
limit 3
line 106
line 31, 35–38, 43, 45, 51–54, 185

line at infinity 185
linear combination 34, 139
linearly independent 34
local sidereal time 277
longitude 260
lune 258

man 2
Maple 9
Mathematica 9
mean Sun 274
mean time 274
medians 83
meridian 260, 272
metres 32
midpoint 35, 149
Milnor 300
mindless machine 10, 14
Minkowski 40
Möbius transformation 107, 118
modulo 42
modulus 103, 139
multiplication 101, 105
multiplying quaternions 288

nadir 272
naming 277
Napoleon's Theorem 92
nautical mile 262
Newton 232
nine-point circle 90
non-desarguian plane 189
non-euclidean geometries 64, 68
nonagon 70
north celestial pole 272
north polar distance 273
north pole 253
northern summer 274
northern winter 274

oblate spheroid 239
obliquity 274
oblong 70
obtuse-angled 68
octagon 70
octahedron 167, 170
octonion 99
octonions 295
one-point compactification 112
opposite angles 43, 65
order 166
origin 29
origin 30
orthocentre 84
orthogonal 41, 117, 154, 215
orthogonal matrices 33
orthogonal matrix 160
orthogonal projection 162
orthogonal transformation 160
orthonormal 142
orthonormal basis 161
outcentre 87
outcircle 87

pair of complex numbers 291
Pappus line 198
Pappus' Theorem 202
Pappus' theorem 198
parabola 212, 218
parallel 52, 260
parallelepiped 145
parallelisable 266
parallelogram 70
parameter 36
parametric 220
parametric form 149
parsec 270
pencil 193
pentagon 130
pentagonal number 8
perihelion 232
period 46
perpendicular 41, 84
perspective drawing 204
perspectivity 195
plane 29, 30, 32, 33, 36, 41, 44, 53, 54, 147
planet 232
Plato 63
Platonic Polyhedra 171
Playfair 64
Poincaré 170
points of the compass 266
polar coordinates 44
polar form 105, 292, 297
pole 253
polygon 69
polyhedra 166
powers 14
precession 276
prime 5
prism 167
Proclus 63
projective plane 184
projective transformation 200
projectivity 195
proper motion 272

Ptolemy 63, 112, 232
pure 291
pyramid 167
pyramidal numbers 7
Pythagoras 16, 67
Pythagoras' theorem 139

quadratic equation 214
quadric surface 238
quadrilateral 70, 199
quaternion 99
quaternion group 289
quaternionic conjugate 289
quaternions 287

rack 157
Rademacher 8
radian 41
radical axis 215
radicals 133
radius 78
range 193
rational 218
rational parameterisation 217
real 291
real axis 102
real cross ratio theorem 121
real line 30
rectangle 70
rectangular 138
rectangular hyperbola 45
reflection 107
reflection 73, 102
reflexivity 11
regular 70
regular polyhedra 174
retrograde motion 237
rhombic dodecahedron 168
rhombus 70
Riemann sphere 111
Riemann, B. 111
right 167
right angle 41
right ascension 274
right ascension 270, 277
right handed 139
right-angled 67
right-handed screw 139
Roller, M. 171
Roman numerals 2
root of unity 127
rosettes 46
rotary reflection 157
rotation 73, 107, 154
ruler and compass 94, 129

sailor 77
scalar 141
scalar triple product 144
scalene 67
Schläfli 175
semi-angle 212
semi-latus rectum 225
semiperimeter 88
septagon 70
sidereal time 277
sidereal year 276
similar 75
similarities 73
sine 88
sine rule 258
skew 151
slope 52
small circle 253
south pole 253
Spain 2
special value 123
spherical polar coordinates 141, 262
spherical triangle 256
spring equinox 275
square 70
square numbers 5
star 71
star convex 72
starred 8
stereographic projection 111
subtended 67
sun compass 266
suspension 168
symmetric form 149
symmetric matrix 245
symmetries of a cube 158
symmetries of a regular tetrahedron 160
symmetry 11

tail 32
tally marks 3
tangent 78
tangent line 218
tangent plane 253
tangent vector 253
tetrahedron 166, 170
time 274
transcendental 235
transitivity 11
translation 32, 73
transversal 193

trapezium 70
triangle 51, 67
triangle inequality 39, 77
triangle of reference 184
triangular numbers 6
tropic of Cancer 275
tropic of Capricorn 275
tropical year 276
tropics 261
true anomaly 233
truncation 169

unimodular 104
unit circle 104
unit point 184
unit quaternion 290
unit sphere 139, 253
universal time 278

vanishing point 185
vector 32
vector equation of line 150
vector product 142
vector space 33
vector triple product 146
Venn diagram 70
vertex 212
vertices 67, 69
Vikings 266
da Vinci, Leonardo 204
Viète 20
volume of a tetrahedron 145
volume of the parallelepiped 145

van der Waerdan 99
Wallis 21

zenith 272
zenith distance 273
zero 9
zone time 278